D0161368

Systems Analysis and Design

The Irwin Series in Information and Decision Sciences

Systems Analysis and Design

Elias M. Awad

McIntire School of Commerce
University of Virginia

1985 Second Edition

Homewood, Illinois 60430

ISBN 0-256-02824-9

Library of Congress Catalog Card No. 84-82025

Printed in the United States of America

4 5 6 7 8 9 0 K 2 1 0 9 8 7 6

Preface

A major contribution of the first edition was to conceptualize and define the scope and domain of systems analysis and design. The text was well received by the information systems academic community. Today's fast-paced technology, however, makes it difficult for most publications to stay up-to-date. This edition is virtually a new book. It is a major update of the first edition of *Systems Analysis and Design*, which focuses on the system development life cycle using conventional and structured tools. The material goes beyond the classroom theory and concepts. It is practice oriented with examples and applications that *demonstrate* systems analysis and design. The coverage meets the curriculum recommendations for the systems analysis course for the Data Processing Management Association (CIS-4) and the Association for Computing Machinery.

The text goes beyond the mechanics of systems development. It addresses the broader information systems environment of the 1980s, such as the use of data bases for the microcomputer, quality assurance, systems auditability, prototyping, disaster recovery planning, and ethics in systems development. These are important issues requiring special treatment. The coverage trains persons in the what, why, and how of systems analysis and design. The tools and techniques are current, and the illustrations and cases ending each chapter are based on real installations. A safe deposit tracking system installed for a commercial bank is presented in modules in the life-cycle chapters, 4–12.

Important features of the text are:

1. Eleven chapters dealing with topics such as the role of the analyst, the tools of structured analysis and design, data base design, and hardware/software selection improve the student's understanding of the systems development life cycle and provide a comprehensive framework for systems development.
2. "At a Glance" provides a brief preview of the material covered in each chapter.
3. A summary of the main points, key words, and review questions follow each chapter.
4. Case studies at the end of each chapter are business situations which the author has been involved in as a consultant or analyst. Working through the cases will help you acquire the principles and gain experience in making decisions that can be useful in similar situations in the future.

The new chapters cover the following:

1. Chapter 2 reviews the systems development life cycle.
2. Chapter 3 elaborates on the multifaceted role of the systems analyst and the requirements for success in the field.
3. Chapter 4 discusses the importance of planning and the major steps in launching an initial investigation.
4. Chapter 6 describes the tools of structured analysis. They are the data flow diagram, data dictionary, decision tree, and structured English.
5. Chapter 7 explains the steps in feasibility analysis and the feasibility report.
6. Chapter 8 is a completely revised section on cost/benefit analysis focusing on the procedures for cost/benefit determination and alternative evaluation methods such as net present value, payback analysis, and cash flow analysis.
7. Chapter 9 focuses on design methodologies—structured design and the structure chart, IPO charts and structured walkthrough. Audit trail and documentation control are also discussed.
8. Chapter 11 is a review of file organization methods and data base organization. A data base software package for the microcomputer is available to support the chapter.
9. Chapter 14 deals with the procedure for hardware/software selection, financial considerations in selection, and how to negotiate a computer contract.
10. Chapter 15 is about project management and the use of planning tools and project management software for system installations.
11. Chapter 16 caps the systems development life cycle by discussing the various threats to system security, how to do risk analysis, the impor-

tance of disaster recovery planning, and the role of ethics in system development.

The text is designed to be used in a semester or a quarter course in systems analysis and design. Although no specific background is required, a student should have had a course in Introduction to Computers and a general understanding of business organizations. The text is written in a manner that is logical to the student. The early chapters focus on user need determination and feasibility studies, and the latter chapters discuss systems design specifications, file organization, and system implementation.

As part of the revision, a bimonthly newsletter containing new articles, case situations, and news related to the system development area is planned. These should be used to supplement the material in the text or the lectures, where appropriate. The newsletter will be available to instructors by writing me directly at the McIntire School of Commerce, University of Virginia, Charlottesville, VA 22903.

ACKNOWLEDGMENTS

Many individuals have contributed to the preparation of this text. For their invaluable comments and suggestions, I wish to thank my reviewers, Rod Neal, University of Arkansas; Thomas H. Weaver, Jr., Purdue University—Calumet; and Kenneth W. Veatch, San Antonio College, and my consulting editors, Robet B. Fetter, Yale University; and Claude McMillan, University of Colorado.

Special thanks to William G. Shenkir, Dean, McIntire School of Commerce, for providing microcomputer support and for his encouragement. I also recognize the constructive feedback that the MIS students at the McIntire School have given in testing various portions of the manuscript.

Elias M. Awad

Contents

Systems Analysis and Design

Part One

Overview

1 SYSTEMS CONCEPTS AND THE INFORMATION SYSTEMS ENVIRONMENT

2 THE SYSTEM DEVELOPMENT LIFE CYCLE

3 THE ROLE OF THE SYSTEMS ANALYST

Chapter 1

Systems Concepts and the Information Systems Environment

Introduction

The Systems Concept

DEFINITION

Characteristics of a System

ORGANIZATION

INTERACTION

INTERDEPENDENCE

INTEGRATION

CENTRAL OBJECTIVE

Elements of a System

OUTPUTS AND INPUTS

PROCESSOR(S)

CONTROL

FEEDBACK

At a Glance

Systems analysis is the application of the systems approach to problem solving using computers. The ingredients are systems elements, processes, and technology. This means that to do systems work, one needs to understand the systems concept and how organizations operate as a system, and then design appropriate computer-based systems that will meet an organization's requirements. It is actually a customized approach to the use of the computer for problem solving.

By the end of this chapter, you should know:

1. The primary characteristics of a system and the importance of the systems concept for developing information systems.
2. How the various elements of a system work together to interface with the end user.
3. How physical systems differ from abstract systems.
4. The unique features of formal and informal information systems.
5. The makeup of management information systems.
6. How decision support systems help in decision making.

INTRODUCTION

It's a typical day. The car starts OK, but you think with a flash of irritation that it really shouldn't take that long to get the air conditioner going. Only an hour to catch the plane, and cars are piled up on the expressway as far as the eye can see. You begin to wonder if there isn't a way to allow airport traffic to move faster. You get to the parking lot and have to walk half a mile to the plane. Where is the shuttle? Why so long a wait? Why so many obstacles?—the ticket counter, the X-ray machine, the gate attendant, etc. Each one is a system in itself, yet they are all part of the transportation system.

This book is about systems analysis and how it relates to shaping organizations, improving performance, and achieving objectives for profitability and growth. As our scenario suggests, the emphasis is on systems in action, the relationships among subsystems, and their contribution to meeting a common goal—in this case, flying passengers to destinations on time. Looking at a system and determining how adequately it functions, the changes to be made, and the quality of the output are parts of systems analysis.

Systems analysis as used in this text is the application of the systems approach to the study and solution of problems using computer-based systems. Systems thinking is integral to systems work. Organizations are complex systems that consist of interrelated and interlocking subsystems. Changes in one part of the system have both anticipated and unanticipated consequences in other parts of the system. The systems approach is a way of thinking about the analysis and design of computer-based applications. It provides a framework for visualizing the organizational and environmental factors that operate on a system. When a computer is introduced into an organization, various functions and dysfunctions operate on the user as well as the organization. Among the positive consequences are improved performance and a feeling of achievement with quality information. Among the unanticipated consequences might be (1) a possible threat to employees that their work no longer "measures up," (2) decreased morale of personnel who were not consulted about the installation, and (3) feeling of intimidation by users who have limited training in the new computer. In assessing these consequences, the analyst's role of alleviating fears and removing barriers for the user is extremely crucial for the system's success.

Systems analysis and design focus on systems, processes, and technology. Having a firm grasp of the makeup of the system in question is a prerequisite for selecting the procedure or introducing the computer for implementation. In our airport scenario, knowledge of the traffic flow, the strategic location of the airport, and how a given change will speed up airport traffic is important in deciding on improvements such as special shuttles, helicopter service, or more airport limousines to solve the problem. Thus, a background in systems concepts and a familiarity with the ways organizations function are helpful. This chapter discusses the systems concept, elaborates on the types of systems that are relevant to systems analysis, and illustrates the relationship between the knowledge of systems concepts and systems analysis.

THE SYSTEMS CONCEPT

Scholars in various disciplines who are concerned about the tendency toward the fragmentation of knowledge and the increasing complexity of phenomena have sought a unifying approach to knowledge. Ludwig von Bertalanffy, a biologist, developed a general systems theory that applies to any arrangement of elements such as cells, people, societies, or even planets.[1] Norbert Wiener, a mathematician, observed that information and communications provide connecting links for unifying fragments or elements.[2] His systems concept of information theory, which shows the parallel between the functioning of human beings and electronic systems, laid the foundation for today's computer systems. Herbert A. Simon, a political scientist, related the systems concept to the study of organizations by viewing an ongoing system as a processor of information for making decisions.[3]

Systems analysis and design for information systems were founded in general systems theory, which emphasizes a close look at all parts of a system. Too often analysts focus on only one component and overlook other equally important components. General systems theory is concerned with "developing a systematic, theoretical framework upon which to make decisions."[4] It discourages thinking in a vacuum and encourages consideration of all the activities of the organization and its external environment.[5] Pioneering work in general systems theory emphasized that organizations be viewed as total systems. The idea of systems has become most practical and necessary in conceptualizing the interrelationships and integration of operations, especially when using computers. Thus, a *system* is a way of thinking about organizations and their problems. It also involves a set of techniques that helps in solving problems.

Definition

The term *system* is derived from the Greek word *systema,* which means an organized relationship among functioning units or components. A system

---

[1] Ludwig Bertalanffy, *General Systems Theory* (New York: George Braziller, 1968).

[2] Norbert Wiener, *Cybernetics* (New York: John Wiley & Sons, 1948).

[3] Herbert A. Simon, *The Shape of Automation for Men and Management* (New York: Harper & Row, 1965).

[4] Richard A. Johnson; Fremont E. Kast; and James E. Rozensweig, *The Theory and Management of Systems* (New York: McGraw-Hill, 1973), p. 6.

[5] See Vincent P. Luchsinger, and Thomas V. Dock, *The Systems Approach: An Introduction,* 2d ed. (Dubuque, Iowa: Kendall/Hunt Publishing, 1982), p. 12.

exists because it is designed to achieve one or more objectives. We come into daily contact with the transportation system, the telephone system, the accounting system, the production system, and, for over two decades, the computer system. Similarly, we talk of the business system and of the organization as a system consisting of interrelated departments (subsystems) such as production, sales, personnel, and an information system. None of these subsystems is of much use as a single, independent unit. When they are properly coordinated, however, the firm can function effectively and profitably.

There are more than a hundred definitions of the word *system,* but most seem to have a common thread that suggests that *a system is an orderly grouping of interdependent components linked together according to a plan to achieve a specific objective.* The word *component* may refer to physical parts (engines, wings of aircraft, wheels of a car), managerial steps (planning, organizing, directing, and controlling), or a subsystem in a multilevel structure. The components may be simple or complex, basic or advanced. They may be a single computer with a keyboard, memory, and printer or a series of intelligent terminals linked to a mainframe. In either case, each component is part of the total system and has to do its share of work for the system to achieve the intended goal. This orientation requires an orderly grouping of the components for the design of a successful system.

The study of systems concepts, then, has three basic implications:

1. A system must be designed to achieve a predetermined objective.
2. Interrelationships and interdependence must exist among the components.
3. The objectives of the organization as a whole have a higher priority than the objectives of its subsystems. For example, computerizing personnel applications must conform to the organization's policy on privacy, confidentiality, and security, as well as making selected data (e.g., payroll) available to the accounting division on request.

CHARACTERISTICS OF A SYSTEM

Our definition of a system suggests some characteristics that are present in all systems: organization (order), interaction, interdependence, integration, and a central objective.

Organization

Organization implies structure and order. It is the arrangement of components that helps to achieve objectives. In the design of a business system, for example, the hierarchical relationships starting with the president on top and leading downward to the blue-collar workers represents the organization structure. Such an arrangement portrays a system-subsystem relationship, defines the authority structure, specifies the formal flow of communication, and formalizes the chain of command (see Figure 1–1). Likewise, a computer system is designed around an input device, a central processing unit, an output device, and one or more storage units. When linked together they work as a whole system for producing information.

Interaction

Interaction refers to the manner in which each component functions with other components of the system. In an organization, for example, purchasing must interact with production, advertising with sales, and payroll with personnel. In a computer system, the central processing unit must interact with the input device to solve a problem. In turn, the main memory holds programs and data that the arithmetic unit uses for computation. The interrelationship between these components enables the computer to perform.

Interdependence

Interdependence means that parts of the organization or computer system depend on one another. They are coordinated and linked together according to a plan. One subsystem depends on the input of another subsystem for proper functioning; that is, the output of one subsystem is the required

FIGURE 1–1 Organization Structure—An Example

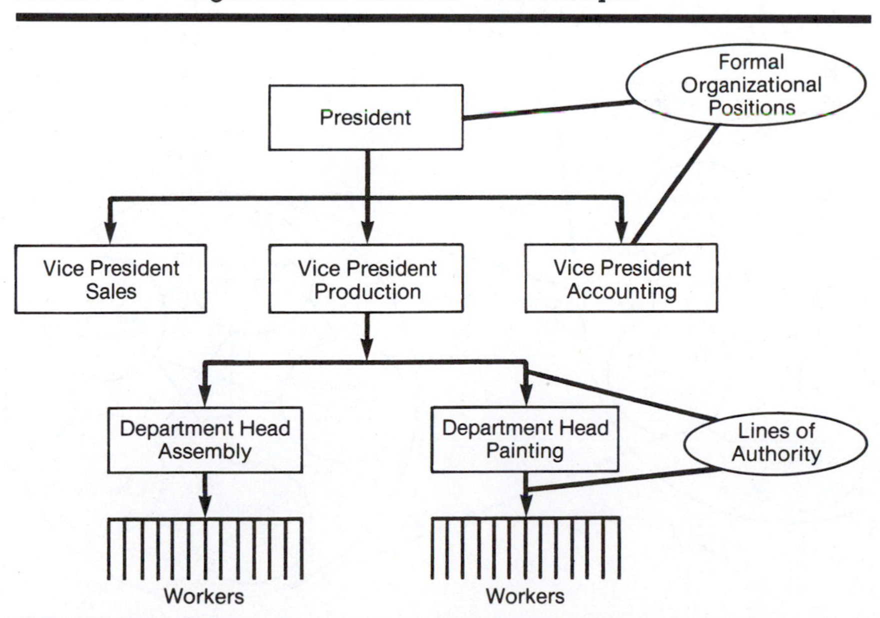

input for another subsystem. This interdependence is crucial in systems work.

To illustrate these system characteristics, Figure 1–2 shows three levels of subsystems. Each of the top inner circles represents a major subsystem of a production firm. The personnel subsystem, in turn, may be viewed as a system that consists of subsystems such as benefits, health and safety, and

FIGURE 1-2 Major Subsystems of a Production Firm

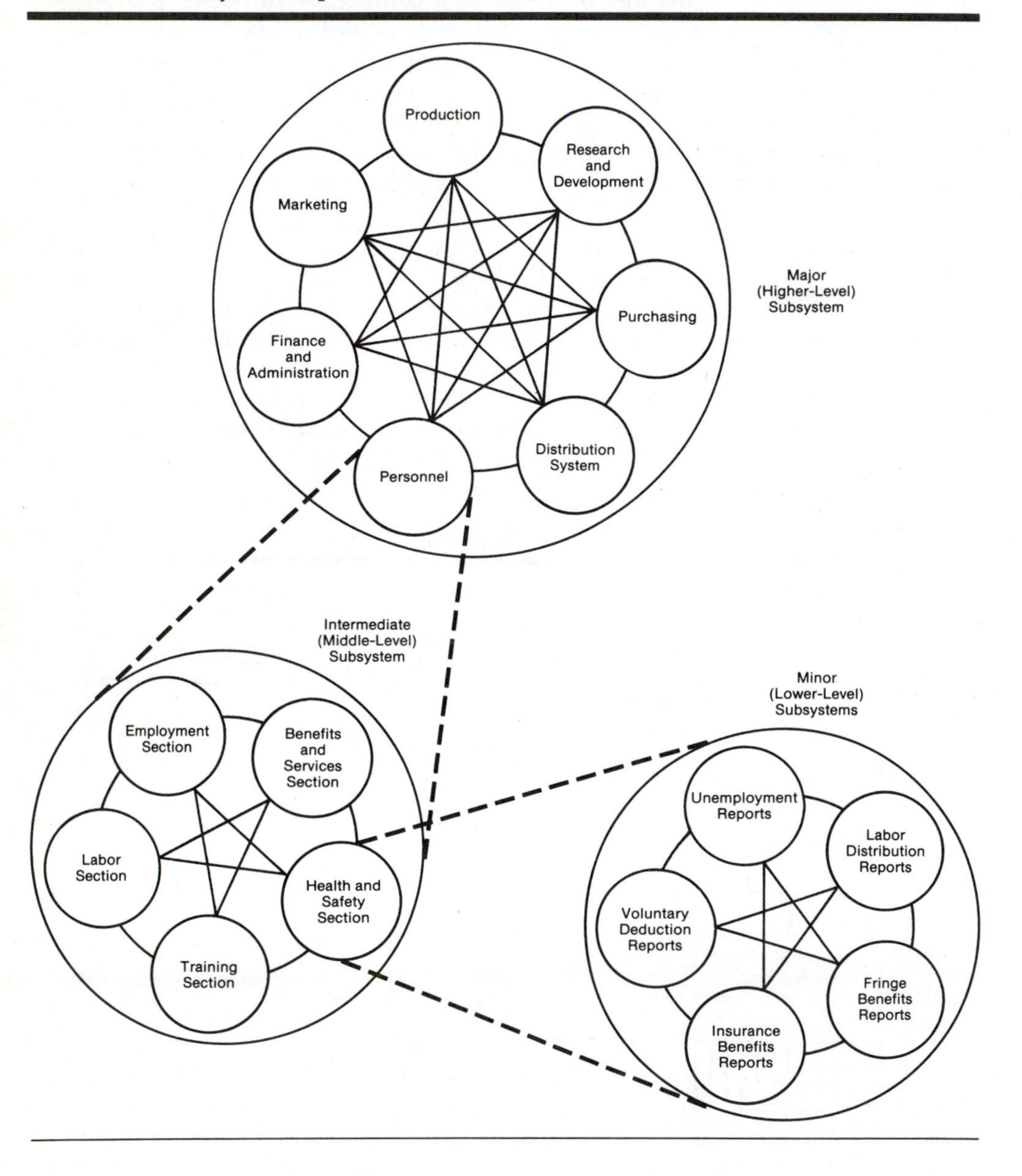

FIGURE 1–3 A Human Resources Information System

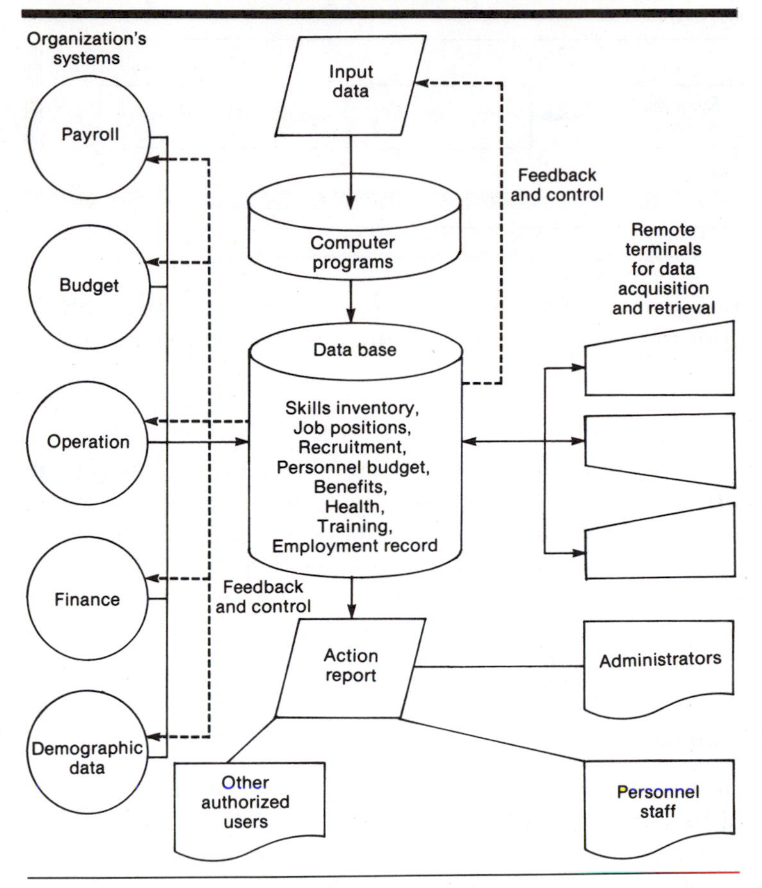

employment. Health and safety as a key personnel subsystem consists of lower-level elements that are considered vital in personnel operation. Each element may be represented by a computer-based package or is part of a human resource data base that provides information on unemployment, insurance benefits, and the like.

Figure 1–3 is an integrated information system designed to serve the needs of authorized users (department heads, managers, etc.) for quick access and retrieval via remote terminals. The interdependence between the personnel subsystem and the organization's users is obvious.

In summary, no subsystem can function in isolation because it is dependent on the data (inputs) it receives from other subsystems to perform its required tasks. Interdependence is further illustrated by the activities and support of systems analysts, programmers, and the operations staff in a computer center. A decision to computerize an application is initiated by the user, analyzed and designed by the analyst, programmed and tested by the programmer, and run by the computer operator. As

FIGURE 1–4 Task Interdependence in a Computer-Based Subsystem

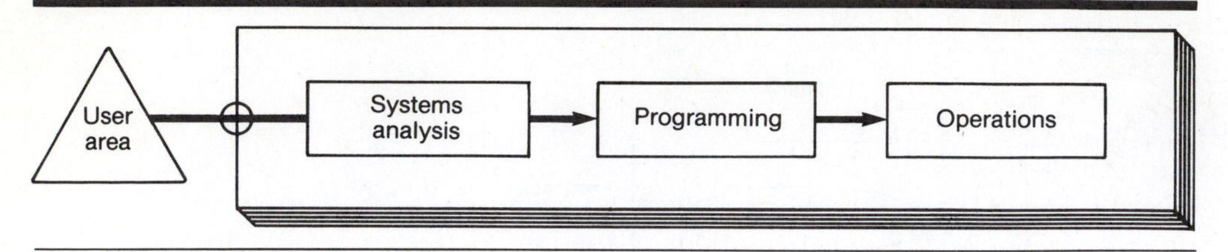

shown in Figure 1–4, none of these persons can perform properly without the required input from others in the computer center subsystem.

Integration

Integration refers to the holism of systems. Synthesis follows analysis to achieve the central objective of the organization. Integration is concerned with how a system is tied together. It is more than sharing a physical part or location. It means that parts of the system work together within the system even though each part performs a unique function. Successful integration will typically produce a synergistic effect and greater total impact than if each component works separately.

Central Objective

The last characteristic of a system is its *central objective.* Objectives may be real or stated. Although a stated objective may be the real objective, it is not uncommon for an organization to state one objective and operate to achieve another. The important point is that users must know the central objective of a computer application early in the analysis for a successful design and conversion. Later in the book, we will show that political as well as organizational considerations often cloud the real objective. This means that the analyst must work around such obstacles to identify the real objective of the proposed change.

ELEMENTS OF A SYSTEM

In most cases, systems analysts operate in a dynamic environment where change is a way of life. The environment may be a business firm, a business application, or a computer system. To reconstruct a system, the following key elements must be considered:

1. Outputs and inputs.
2. Processor(s).
3. Control.
4. Feedback.

5. Environment.
6. Boundaries and interface.

Outputs and Inputs

A major objective of a system is to produce an output that has value to its user. Whatever the nature of the output (goods, services, or information), it must be in line with the expectations of the intended user. Inputs are the elements (material, human resources, information) that enter the system for processing. Output is the outcome of processing. A system feeds on input to produce output in much the same way that a business brings in human, financial, and material resources to produce goods and services. It is important to point out here that determining the output is a first step in specifying the nature, amount, and regularity of the input needed to operate a system. For example, in systems analysis, the first concern is to determine the user's requirements of a proposed computer system—that is, specification of the output that the computer is expected to provide for meeting user requirements. Input and processing design follow (see Figure 1–5).

FIGURE 1–5 **Inputs and Outputs in a Business Operation**

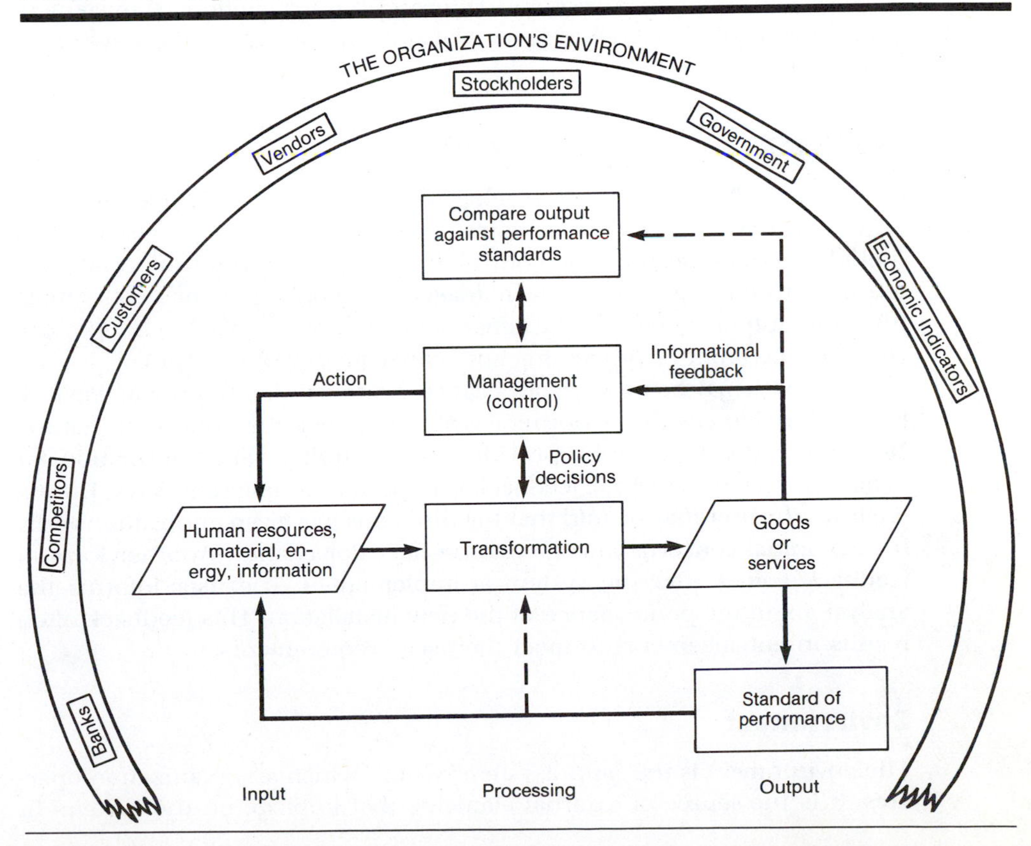

Processor(s)

The processor is the element of a system that involves the actual transformation of input into output. It is the operational component of a system. Processors may modify the input totally or partially, depending on the specifications of the output. This means that as the output specifications change, so does the processing. In some cases, input is also modified to enable the processor to handle the transformation.

Control

The control element guides the system. It is the decision-making subsystem that controls the pattern of activities governing input, processing, and output. In an organizational context, management as a decision-making body controls the inflow, handling, and outflow of activities that affect the welfare of the business. In a computer system, the operating system and accompanying software influence the behavior of the system. Output specifications determine what and how much input is needed to keep the system in balance (see Figure 1–5).

In systems analysis, knowing the attitudes of the individual who controls the area for which a computer is being considered can make a difference between the success and failure of the installation. Management support is required for securing control and supporting the objective of the proposed change.

Feedback

Control in a dynamic system is achieved by feedback. Feedback measures output against a standard in some form of cybernetic procedure that includes communication and control. In Figure 1–5, output information is fed back to the input and/or to management (controller) for deliberation. After the output is compared against performance standards, changes can result in the input or processing and, consequently, the output.

Feedback may be positive or negative, routine or informational. Positive feedback reinforces the performance of the system. It is routine in nature. Negative feedback generally provides the controller with information for action. In systems analysis, feedback is important in different ways. During analysis, the user may be told that the problems in a given application verify his/her initial concerns and justify the need for change. Another form of feedback comes after the system is implemented. The user informs the analyst about the performance of the new installation. This feedback often results in enhancements to meet the user's requirements.

Environment

The environment is the "suprasystem" within which an organization operates. It is the source of external elements that impinge on the system. In fact, it often determines how a system must function. As shown in Figure 1–5, the organization's environment, consisting of vendors, competitors, and others, may provide constraints and, consequently, influence the actual performance of the business.

Boundaries and Interface

A system should be defined by its boundaries—the limits that identify its components, processes, and interrelationships when it interfaces with another system. For example, a teller system in a commercial bank is restricted to the deposits, withdrawals, and related activities of customers' checking and savings accounts. It may exclude mortgage foreclosures, trust activities, and the like.

Each system has boundaries that determine its sphere of influence and control, although in an integrated banking-wide computer system design, a customer who has a mortgage and a checking account with the same bank may write a check through the "teller system" to pay the premium that is later processed by the "mortgage loan system." Recently, system design has been successful in allowing the automatic transfer of funds from a bank account to pay bills and other obligations to creditors, regardless of distance or location. This means that in systems analysis, knowledge of the boundaries of a given system is crucial in determining the nature of its interface with other systems for successful design.

TYPES OF SYSTEMS

The frame of reference within which one views a system is related to the use of the systems approach for analysis. Systems have been classified in different ways. Common classifications are: (1) physical or abstract, (2) open or closed, and (3) "man-made" information systems.

Physical or Abstract Systems

Physical systems are tangible entities that may be static or dynamic in operation. For example, the physical parts of the computer center are the offices, desks, and chairs that facilitate operation of the computer. They can be seen and counted; they are static. In contrast, a programmed computer is a dynamic system. Data, programs, output, and applications change as the user's demands or the priority of the information requested changes.

Abstract systems are conceptual or nonphysical entities. They may be as straightforward as formulas of relationships among sets of variables or models—the abstract conceptualization of physical situations. A model is a representation of a real or a planned system. The use of models makes it easier for the analyst to visualize relationships in the system under study. The objective is to point out the significant elements and the key interrelationships of a complex system.

Systems Models

In no field are models used more widely and with greater variety than in systems analysis. The analyst begins by creating a model of the reality (facts, relationships, procedures, etc.) with which the system is concerned. Every computer system deals with the real world, a problem area, or a reality outside itself. For example, a telephone switching system is made up of subscribers, telephone handsets, dialing, conference calls, and the like. The analyst begins by modeling this reality before considering the functions that the system is to perform.

Various business system models are used to show the benefits of abstracting complex systems to model form. The major models discussed here are schematic, flow, static, and dynamic system models.

Schematic Models. A schematic model is a two-dimensional chart depicting system elements and their linkages. Figure 1–6 shows the major elements of a personnel information system together with material and information flow.

Flow System Models. A flow system model shows the flow of the material, energy, and information that hold the system together. There is an orderly flow of logic in such models. A widely known example is PERT (Program Evaluation and Review Technique). It is used to abstract a real-world system in model form, manipulate specific values to determine the critical path, interpret the relationships, and relay them back as a control. The probability of completion within a time period is considered in connection with time, resources, and performance specifications (see Figure 1–7). PERT is discussed in detail in Chapter 15.

Static System Models. This type of model exhibits one pair of relationships such as activity-time or cost-quantity. The Gantt chart, for example, gives a static picture of an activity-time relationship. In Figure 1–8, planned activities (stamping, sanding, etc.) are plotted in relation to time. The date column has light lines that indicate the amount of time it takes to complete a given activity. The heavy line represents the cumulative time schedule for each activity. The stamping department, for example, is scheduled to start working on order number 25 Wednesday morning and complete the job by the same evening. One day is also scheduled for order number 28, two days for order number 22, and two days (May 10–11) for order number 29. The total of six days is represented by the heavy line opposite the stamping department. The broken line indicates that the department is two days behind schedule. The arrowhead indicates the date when the chart is to be in effect. The Gantt chart is discussed in more detail in Chapter 15.

Dynamic System Models. Business organizations are dynamic systems. A dynamic model approximates the type of organization or applica-

FIGURE 1–6 Personnel Information Flow in a Banking Environment

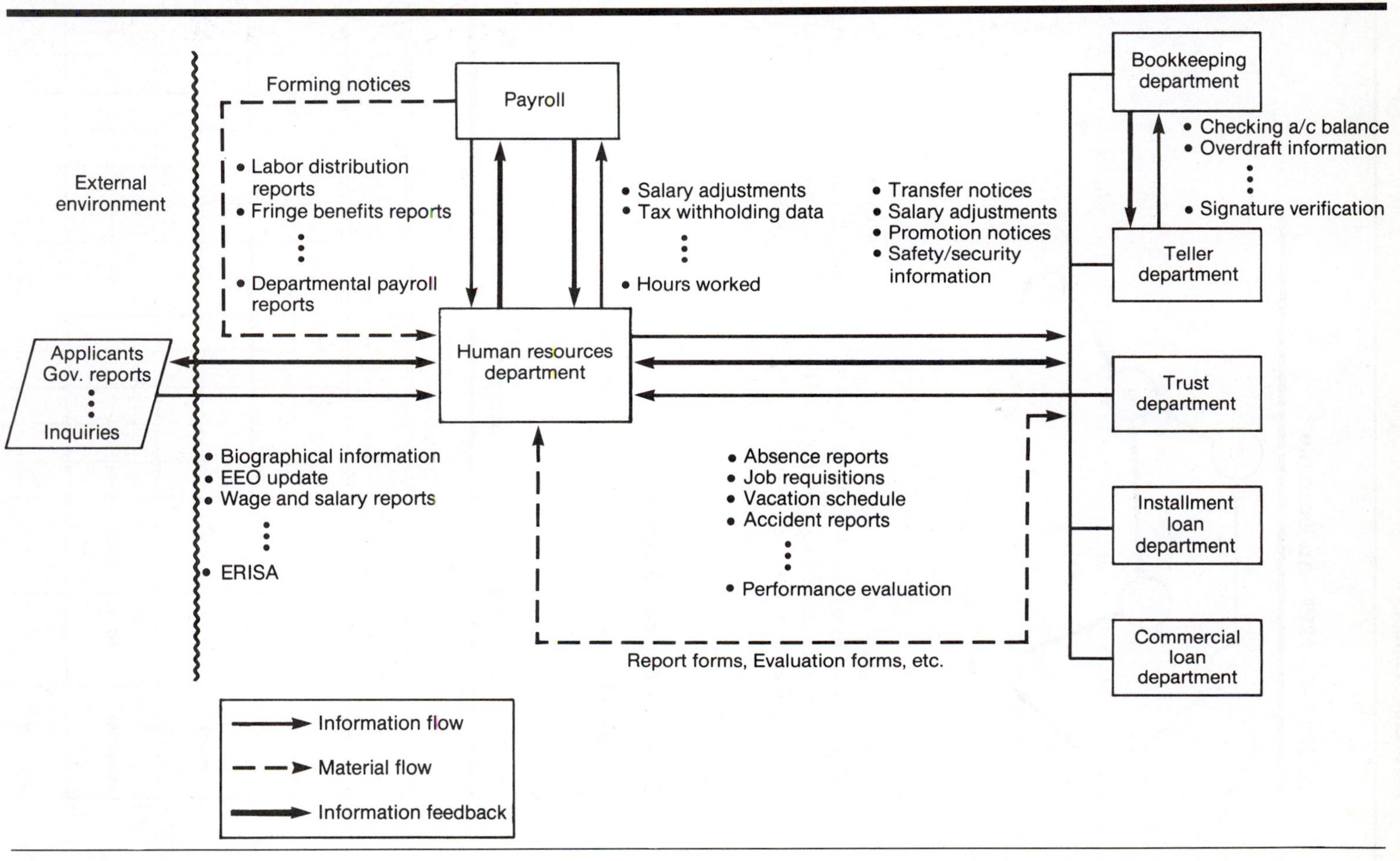

Source: W. Cascio and E. M. Awad, *Human Resources Management: An Information Systems Approach* (Reston, Va.: Reston Publishing, 1981), p. 46.

FIGURE 1–7 PERT—An Example

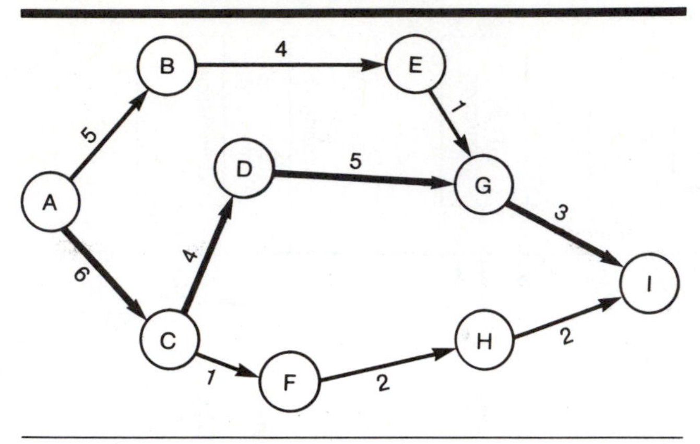

tions that analysts deal with. It depicts an ongoing, constantly changing system. As mentioned earlier, it consists of (1) inputs that enter the system, (2) the processor through which transformation takes place, (3) the program(s) required for processing, and (4) the output(s) that result from processing (see Figure 1–5).

Open or Closed Systems

Another classification of systems is based on their degree of independence. An *open* system has many interfaces with its environment. It permits inter-

FIGURE 1–8 Gantt Chart—An Example

Gantt Chart														
Departments	Number of workers	Capacity per week	May				5	6						12
Stamping	75	3,000			25		28		22			29		
Sanding	10	400			21				25					
Assembly	60	2,400	19						20					
Painting	8	320	13					1	4					

action across its boundary; it receives inputs from and delivers outputs to the outside. An information system falls into this category, since it must adapt to the changing demands of the user. In contrast, a *closed* system is isolated from environmental influences. In reality, a completely closed system is rare. In systems analysis, organizations, applications, and computers are invariably open, dynamic systems influenced by their environment.

A focus on the characteristics of an open system is particularly timely in the light of present-day business concerns with computer fraud, invasion of privacy, security controls, and ethics in computing. Whereas the technical aspects of systems analysis deal with internal routines within the user's application area, systems analysis as an open system tends to expand the scope of analysis to relationships between the user area and other users and to environmental factors that must be considered before a new system is finally approved. Furthermore, being open to suggestions implies that the analyst has to be flexible and the system being designed has to be responsive to the changing needs of the user and the environment.

Five important characteristics of open systems can be identified.

1. *Input from outside.* Open systems are self-adjusting and self-regulating. When functioning properly, an open system reaches a *steady state* or *equilibrium*. In a retail firm, for example, a steady state exists when goods are purchased and sold without being either out of stock or overstocked. An increase in the cost of goods forces a comparable increase in prices or decrease in operating costs. This response gives the firm its steady state.

2. *Entropy.* All dynamic systems tend to run down over time, resulting in entropy or loss of energy. Open systems resist entropy by seeking new inputs or modifying the processes to return to a steady state. In our example, no reaction to increase in cost of merchandise makes the business unprofitable which could force it into insolvency—a state of disorganization.

3. *Process, output, and cycles.* Open systems produce useful output and operate in cycles, following a continuous flow path.

4. *Differentiation.* Open systems have a tendency toward an increasing specialization of functions and a greater differentiation of their components. In business, the roles of people and machines tend toward greater specialization and greater interaction. This characteristic offers a compelling reason for the increasing value of the concept of systems in the systems analyst's thinking.

5. *Equifinality.* The term implies that goals are achieved through differing courses of action and a variety of paths. In most systems, there is more of a consensus on goals than on paths to reach the goals.

Understanding system characteristics helps analysts to identify their role and relate their activities to the attainment of the firm's objectives as they undertake a system project. Analysts are themselves part of the organization. They have opportunities to adapt the organization to changes through computerized applications so that the system does not "run down." A key to this process is information feedback from the prime user of the new system as well as from top management. The design and maintenance of management information systems will be discussed later in the chapter.

Table 1–1 summarizes the main elements and reinforces the theme that the process of designing information systems borrows heavily from a general knowledge of systems theory. The objective is to make a system more efficient by modifying its goals or changing the outputs.

Man-Made Information Systems

Ideally, information reduces uncertainty about a state or event. For example, information that the wind is calm reduces the uncertainty that the boat trip will be pleasant. An information system is the basis for interaction between the user and the analyst. It provides instructions, commands, and feedback. It determines the nature of the relationships among decision makers. In fact, it may be viewed as a decision center for personnel at all levels. From this basis, an *information system* may be defined as a set of devices, procedures, and operating systems designed around user-based criteria to produce information and communicate it to the user for planning, control, and performance. In systems analysis, it is important to keep in mind that considering an alternative system means improving one or more of these criteria.

TABLE 1–1 Systems Theory and Information System Design

Primary System Element	Relevance for Information System Design
1. A system is an entity	Define system under study before evaluating its subsystems
2. A system has components	Determine the primary components of the present system and how they relate to one another before deciding on design changes
3. A system is goal oriented	Specify the goal(s) of the information system under study
4. A system comprises input/processing/output	Specify the inputs, processing, and outputs of the information system being analyzed
5. A system increases entropy	Designing an information system is critical to the organization's performance and growth
6. A system exhibits equifinality	An information system has many components and there are several ways of designing it to achieve the same goal
7. A system maintains a steady state	An information system should be designed to stabilize and maintain the informational needs of management to reduce uncertainty in decision making

Many practitioners fail to recognize that a business has several information systems; each is designed for a purpose and works to accommodate data flow, communications, decision making, control, and effectiveness. The major information systems are formal, informal, and computer based.

Formal Information Systems

A formal information system is based on the organization represented by the *organization chart*. The chart is a map of positions and their authority relationships, indicated by boxes and connected by straight lines.[6] It is concerned with the pattern of authority, communication, and work flow. Information is formally disseminated in instructions, memos, or reports from top management to the intended user in the organization. This structure also allows feedback up the chain of command for follow-up. In Figure 1–1 input from the environment provides impetus for policy decisions by top management. Policies are generalizations that specify what an organization ought to do. Policies are translated into directives, rules, and regulations and transmitted to lower-level management for implementation. The output represents employee performance.

Categories of Information. There are three categories of information related to managerial levels and the decisions managers make. The first level is *strategic* information, which relates to long-range planning policies that are of direct interest to upper management. Information such as population growth, trends in financial invesetment, and human resources changes would be of interest to top company officials who are responsible for developing policies and determining long-range goals. This type of information is achieved with the aid of *decision support systems* (DSS), which will be explained later in the chapter.

The second level of information is *managerial* information. It is of direct use to middle management and department heads for implementation and control. Examples are sales analysis, cash flow projections, and annual financial statements. This information is of use in short- and intermediate-range planning—that is, months rather than years. It is maintained with the aid of *management information systems* (MIS), which also will be covered later in the chapter.

The third information level is *operational* information, which is short-term, daily information used to operate departments and enforce the day-to-day rules and regulations of the business. Examples are daily employee absence sheets, overdue purchase orders, and current stocks available for sale. Operational information is established by *data processing systems* (DPS) (see Figure 1–9).

The nature of the information and managerial levels is also related to the major types of decision making: structured and unstructured decision making. An organizational process that is closed, stable, and mechanistic

[6] J. L. Massie and John Douglas, *Managing,* 3d ed. (Englewood Cliffs, N.J.: Prentice-Hall, 1982), p. 270.

FIGURE 1–9 Management and Information Levels in a Typical Organization

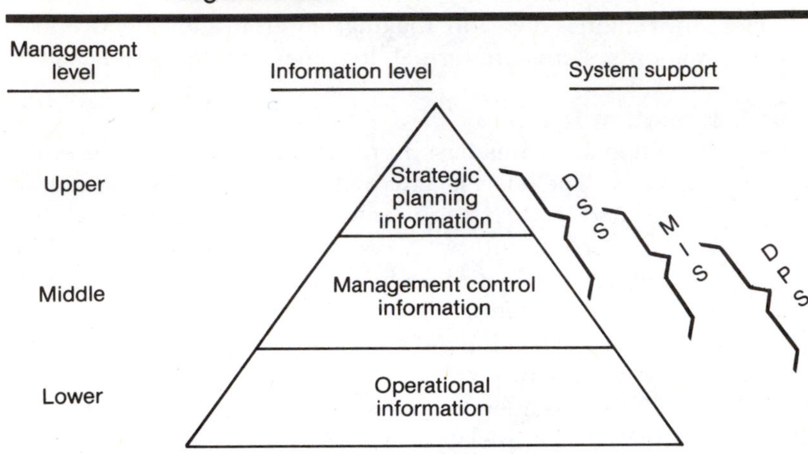

tends to be more structured, computational, and relies on routine decision making for planning and control. Such decision making is related to lower-level management and is readily supported with computer systems. In contrast, open, adaptive, dynamic processes increase the uncertainty associated with decision making and are generally evidenced by a lack of structure in the decision-making process. Lack of structure as well as extraorganizational and incomplete information make it difficult to secure computer support. Table 1–2 summarizes the characteristics of decision making and the information required at different managerial levels.

Therefore, in designing an information system, the analyst needs to determine the type of information needed, the level of the information, how it is structured, and in what format it is before deciding on the system

TABLE 1–2 Characteristics of Decision Making and Information Required at Different Managerial Levels

	Lower Management	Middle Management	Top Management
Characteristics of decision making	(Structured) Computational/ routine →		(Unstructured) Judgmental
Organizational process	Closed/stable mechanistic →		Open/adaptive/ dynamic
Examples	Production scheduling	Capacity planning	New product planning
Characteristics of information	Organizational →		Environmental
Examples	Sales order	Sales analysis	Industry forecasts

needed to produce it. This is another reason for having a background in systems theory and organizations.

Informal Information Systems

The formal information system is a power structure designed to achieve company goals. An organization's emphasis on control to ensure performance tends to restrict the communication flow among employees, however. As a result, an informal information system develops. It is an *employee-based* system designed to meet personnel and vocational needs and to help solve work-related problems. It also funnels information upward through indirect channels. In this respect, it is a useful system because it works within the framework of the business and its stated policies.

In doing a systems study, the analyst should have a knowledge of the chain of command, the power-authority-influence network, and how decisions are made to get a feel for how much support can be expected for a prospective installation. Furthermore, knowledge about the inner workings of the employee-based system is useful during the exploratory phase of analysis. Employee cooperation and participation are crucial in preventing sabotage and training users. Since computers cannot provide reliable information without user staff support, a proper interface with the informal communication channels could mean the difference between the success and failure of new systems.

Computer-Based Information Systems

A third class of information system relies on the computer for handling business applications. The computer is now a required source of information. Systems analysis relies heavily on computers for problem solving. This suggests that the analyst must be familiar with computer technology and have experience in handling people in an organizational context. The role of the analyst is discussed further in Chapter 3.

Management Information Systems (MIS). The computer has had a significant impact on the techniques used by management to operate a business. The level of the manager in the organization is also a factor in determining the kind of information needed to solve a problem (see Figure 1–10). *Lower-level* management needs detailed internal information to make day-to-day, relatively structured control decisions. *Higher-level* management, for whom long-range objectives are the primary concerns, requires summarized information from a variety of sources to attain goals. In either case, management action is based on information that is accurate, relevant, complete, concise, and timely. MIS has been successful in meeting these information criteria quickly and responsively.

MIS is a person-machine system and a highly integrated grouping of information-processing functions designed to provide management with a comprehensive picture of specific operations. It is actually a combination of information systems. To do the job, it should operate in real time, handling

FIGURE 1-10 Management and Information Levels in an Organization

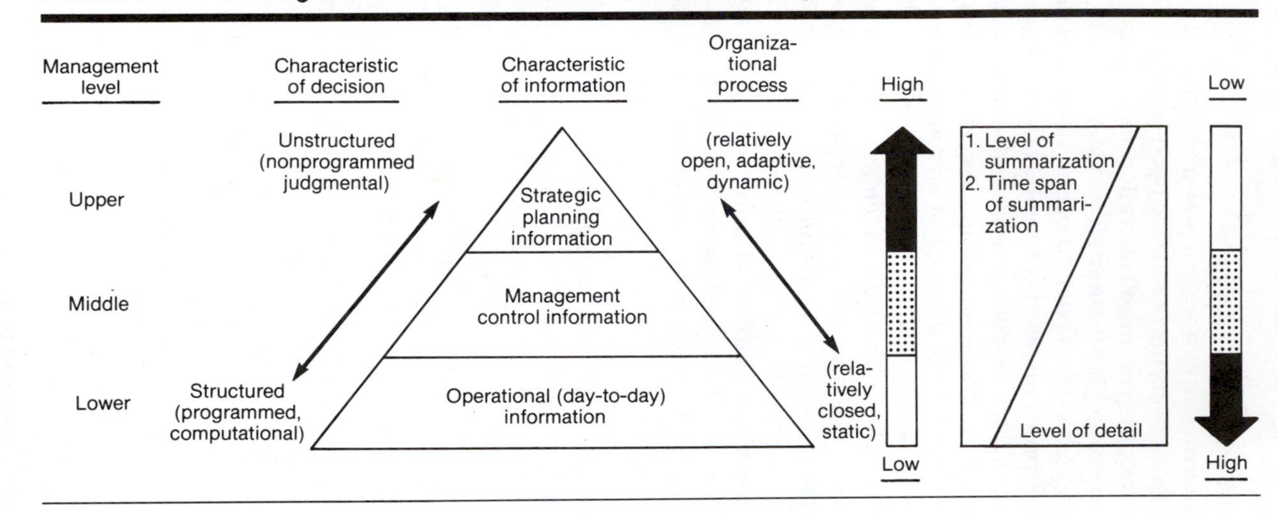

inquiries as quickly as they are received. Management information must also be available early enough to affect a decision. Operationally, MIS should provide for file definition, file maintenance and updating, transaction and inquiry processing, and one or more data bases linked to an organizational data base. Within an MIS, a single transaction can simultaneously update all related data files in the system. In so doing, data redundancy (duplication) and the time it takes to duplicate data are kept to a minimum, thus insuring that data are kept current at all times (see Figure 1–11).

A key element of MIS is the data base—a nonredundant collection of interrelated data items that can be processed through application programs and available to many users. All records must be related in some way. Sharing common data means that many programs can use the same files or records. Information is accessed through a data base management system

FIGURE 1–11

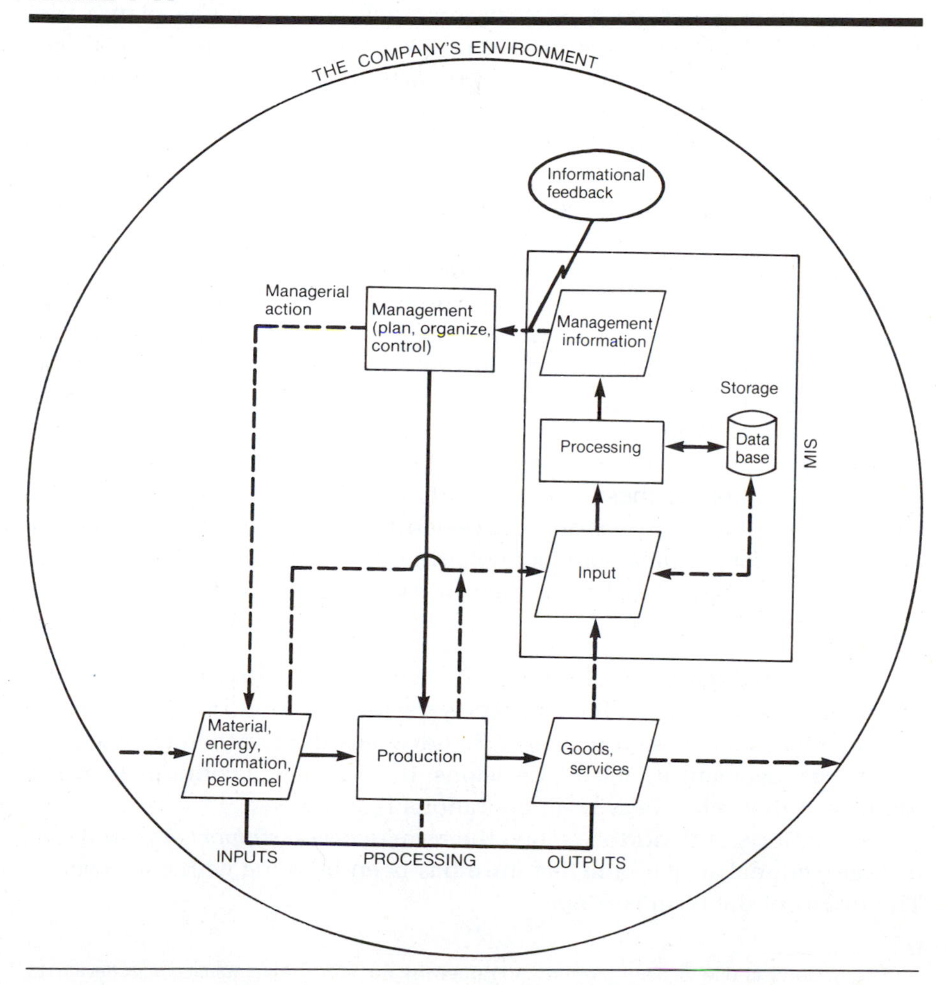

(DBMS). It is a part of the software that handles virtually every activity involving the physical data base.

There are several advantages to a data base system:

1. Processing time and the number of programs written are substantially reduced.
2. All applications share centralized files.
3. Storage space duplication is eliminated.
4. Data are stored once in the data base and are easily accessible when needed.

The two primary drawbacks of a data base are the cost of specialized personnel and the need to protect sensitive data from unauthorized access. There is more on data bases in Chapter 12.

The primary users of MIS are middle and top management, operational managers, and support staff. Middle and top management use MIS for preparing forecasts, special requests for analysis, long-range plans, and periodic reports. Operational managers use MIS primarily for short-range planning and periodic and exception reports. The support staff finds MIS useful for the special analysis of information and reports to help management in planning and control. Providing data for use in MIS is the function of most levels of personnel in the organization. Once entered into the system, the information is no longer owned by the initiating user but becomes available to all authorized users.

Today's typical MIS poses several problems. Most MIS reports are historical and tend to be dated. Another problem is that many installations have data bases that are not in line with user requirements. This means that many MIS environments have not been congruent with the real world of the user. Finally, an inadequate or incomplete update of the data base jeopardizes the reliability for all users.

A major problem encountered in MIS design is obtaining the acceptance and support of those who will interface with the system. Personnel who perceive that their jobs are threatened may resist the implementation of MIS. In understanding both technology and human behavior, the analyst faces the challenge of selling change to the right people for a successful installation.

Decision Support Systems (DSS). One reason cited in the literature for management's frustration with MIS is the limited support it provides top management for decision making. DSS advances the capabilities of MIS. It assists management in making decisions. It is actually a continually evolving model that relies heavily on operations research.

Since Gorry and Morton coined the term *decision support system* (DDS) in their seminal article,[7] the literature has been bursting with controversy. The origin of the term is simple:

---

[7] G. A. Gorry and M. S. Scott Morton, "A Framework for MIS," *Sloan Management Review* 13 (Fall 1971), pp. 55–70.

- *Decision*—emphasizes decision making in problem situations, not information processing, retrieval, or reporting.
- *Support*—requires computer-aided decision situations with enough "structure" to permit computer support.
- *System*—accentuates the integrated nature of problem solving, suggesting a combined "man," machine, and decision environment.

Beginning with management decision systems in the early 1970s, the concept of interactive computer-based systems supporting unstructured decision making has been expanded to include everything but transaction processing systems. A typical early definition required an interactive computer-based system to help users use data and models to solve unstructured problems. There are authors today who view DSS as an extension of MIS, DSS as independent of MIS, or MIS as a subset of DSS. The commonly accepted view in the literature views DSS as a second-generation MIS. MIS is generated when we add predefined managerial reports that are spun out of the transaction processing, report generation, and online inquiry capabilities—all integrated with a given functional area such as production MIS or personnel MIS. DSS results from adding external data sources, accounting and statistical models, and interactive query capabilities. The outcome is a system designed to serve all levels of management, and top management in particular, in dealing with "what if" unstructured problem situations. It is a system with the intrinsic capability to support ad hoc data analysis as well as decision-modeling activities.[8]

The field of DSS is young and evolving. The analyst's present role is to look at existing DSS packages, their attributes, capabilities, and design considerations, consider how they differ from MIS design, and learn more about the methodology needed to provide a proper fit between DSS as a future-oriented system and management requirements for various levels of decision making. Since DSS appears to be the wave of the future (evidenced by the surge of DSS software packages and the technology that supports them), systems analysts will need to upgrade their knowledge to meet the changing demands of users and organizations alike.

Herbert Simon described decision making as a three-phase continuous process model beginning with *intelligence* and moving toward *design* and *choice* (see Figure 1–12). The process is invoked by the recognition of a problem. The resulting decision is then directed at solving the problem.

The *intelligence* phase of decision making involves the awareness of a problem at a symptomatic level; it requires a closer look at the problem and a thorough evaluation of the variables and their relationships. For example, the symptom of a problem is a large number of auto accidents, but the actual cause of the problem turns out to be that the state discontinued auto inspections, lowered the minimum drinking age, has an inadequate police force, or a combination of all these factors. The more intelligence manage-

---

[8] See Peter Keen, and M. S. Scott Morton, *Decision Support Systems: An Organizational Perspective* (Reading, Mass.: Addison-Wesley Publishing, 1978).

FIGURE 1–12 Simon's Decision-Making Process

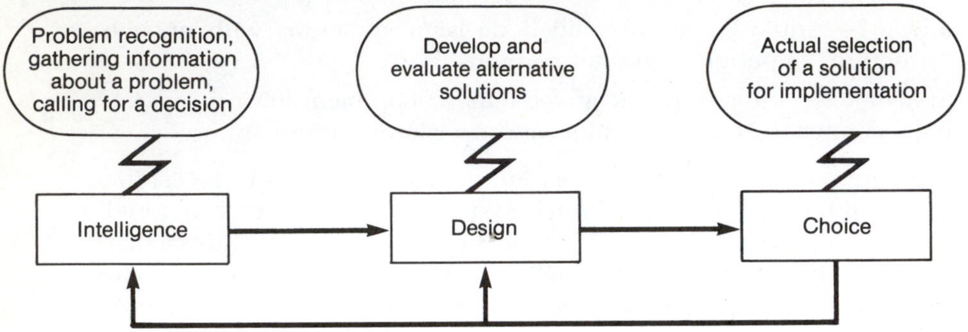

Source: Herbert A. Simon, *The New Science of Management Decisions* (Englewood Cliffs, N.J.: Prentice- Hall, 1960), pp. 54–5.

ment has about the cause of a problem, the better is the likelihood of designing a good decision. A DSS can provide intelligence through information retrieval and statistical packages.

The *design* phase of decision making focuses on the evaluation of decision alternatives. During this phase, computer-based deterministic or stochastic models may be used for decision design. DSS plays a major role in decision design under uncertainty. The output of the model(s) is the basis for the *choice* phase of decision making.

ILLUSTRATION—A DYNAMIC PERSONNEL INFORMATION SYSTEM MODEL

Understanding the systems concept and the role of the computer in generating information for decision making is a prerequisite for designing computer-based systems to serve the needs of business users. To illustrate the multifaceted nature of systems thinking, the master model in Figure 1–13 shows a human resource information system (HRIS) that operates within and is influenced by environmental and organizational factors. The environmental factors are the actions of competitors; geographic, sociocultural, political, and economic factors; and government regulations. The organizational factors include company goals; the existing structure, tools, and technology; company policies; and physical and financial resources.

Once these factors are defined, the next step for analysis is to determine the outputs, inputs, processes, and feedback mechanisms. The output should measure employee performance and how well performance meets prescribed standards. Input is defined in terms of the type and caliber of employees to be hired. This involves job analysis, human resources planning, recruitment, and selection. Each input becomes a part of a human resource management data base that is accessible for future reference.

FIGURE 1–13. Personnel Information Systems Model

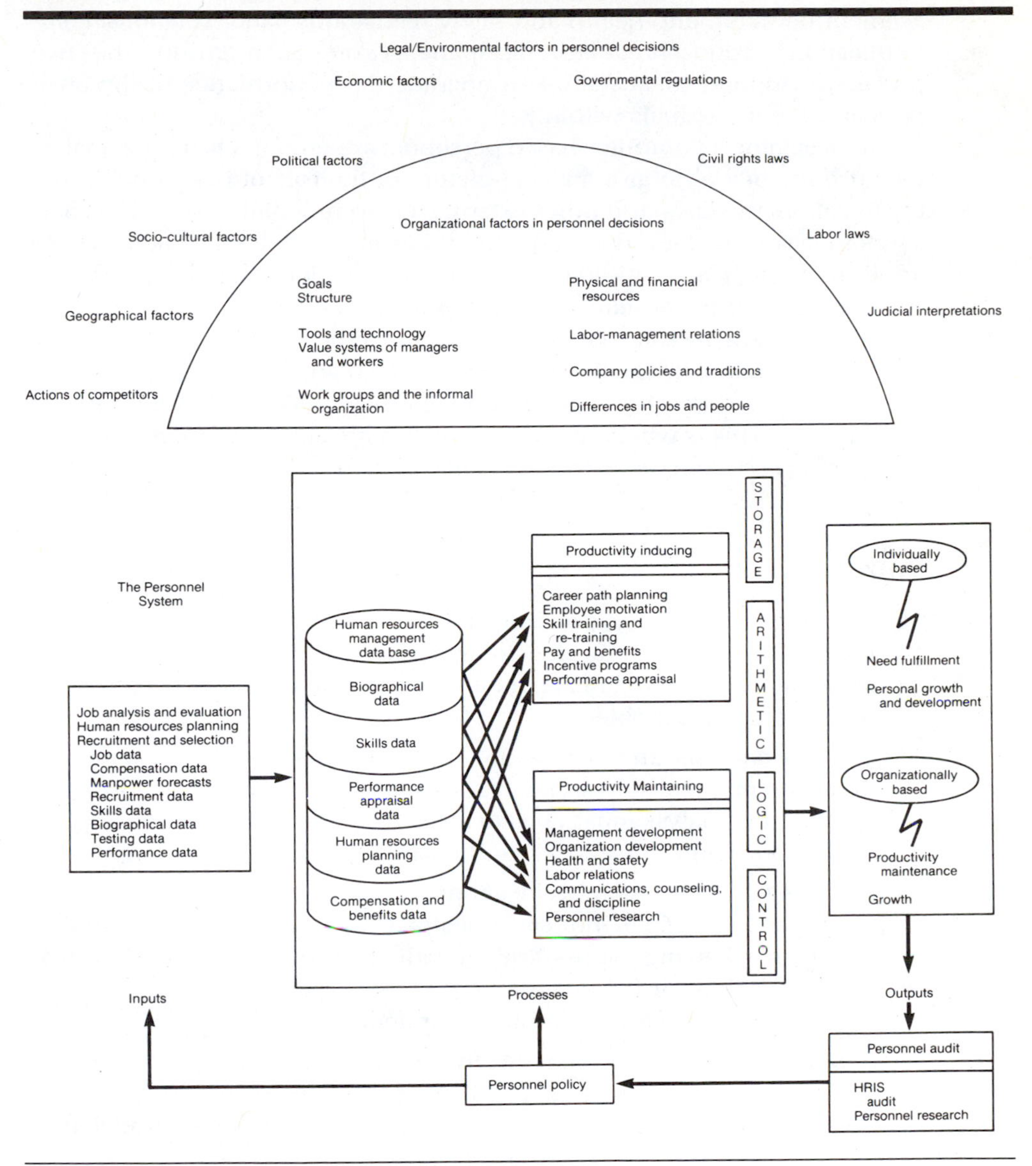

Source: W. Cascio, and E. M. Awad, *Human Resources Management: An Information Systems Approach.* (Reston, Va.: Reston Publishing, 1981), p. 63.

As processes, the model first specifies *productivity-inducing* processes, which include career path planning, employee motivation programs, skill training, pay and benefits, incentive programs, and performance appraisal. The second process is called a *productivity-maintaining* process, which is designed to examine organizational tasks and establish job requirements to channel emloyee talent toward goal attainment. This is done through management development, health and safety regulations, labor relations, communications, counseling, and discipline. Taken as a group, the two processes respond to, sustain, and upgrade the performance of the business and the individuals within it.

In designing a computer-based personnel system for a firm, the analyst may use the model to give a global picture of the role of personnel in the organization, the laws and other external (environmental) factors that personnel faces in making decisions, the types and numbers of activities or applications that are candidates for computerization, and the priority of each application for design and implementation. By following such a procedure and knowing how personnel "hangs together" within the framework of the organization, the analyst can more easily plan for change, sell change to the users, and provide a good fit for the new personnel system in the organization. This is what analysis and design are about: *awareness, understanding, observation, assessment, justification, design, testing, and implementation.*

Summary

1. A system is an orderly grouping of interdependent components linked together according to a plan to achieve a specific objective. Its main characteristics are organization, interaction, interdependence, integration, and a central objective.
2. Systems analysis and design are the application of the systems approach to problem solving, generally using computers. To reconstruct a system, the analyst must consider its elements—outputs and inputs, processors, control, feedback, and environment.
3. Systems fall into three classifications:
   a. Physical (tangible entities) or abstract, such as models or formulas.
   b. Open (allowing inputs and providing outputs) or closed, which isolates it from the environment.
   c. Man-made such as information systems.
4. A system may be a schematic (two-dimensional), a flow system to abstract the real world (e.g., PERT chart), a static (exhibiting one relationship such as a Gantt chart), or a dynamic system model that approximates the business organization.
5. An information system is an open system that allows inputs and facilitates interaction with the user. The main characteristics of an open system are input from outside, processing, output, operating in cycles through feedback, differentiation, and equifinality.
6. There are three levels of information in organizations that require a special type of information system:
   a. *Strategic* information relates to long-range planning policies and

upper management. It is achieved with the aid of decision support systems.

b. *Managerial* information helps middle management and department heads in policy implementation and control. It is maintained with the aid of management information systems.

c. *Operational* information is daily information needed to operate the business. It is established by data processing systems.

7. The nature of information and managerial levels is related to whether decision making is structured or unstructured. A relatively closed, stable, and mechanistic business tends to develop a more structured information process for planning and control. Open, dynamic organizations are associated with uncertain environments where more unstructured rather than structured decisions are made to cope with uncertainty.
8. The informal communications network of employees in the organization is a useful source for examining systems, since computers cannot provide all the information. User staff support is important.
9. Managerial levels determine the kind of information needed to solve a problem. Lower-level management needs detailed information to make day-to-day structured decisions. Higher-level management requires summarized information to attain goals. MIS is a combination of information for various managerial levels.
10. The key element of MIS is the data base—ideally, a nonredundant collection of interrelated data items that are processed through application programs. The data base offers several advantages to end users and is the wave of the future.
11. The historical nature of MIS reports and the need to deal with unstructured problem situations prompted the introduction of DSS. The concept is future oriented, emphasizing decision making in problem situations, not information processing; it requires a computer-aided environment and accentuates a combined "man," machine, and decision environment.

Key Words

Abstract System
Closed System
Control
Data Base
Data Base Management System (DBMS)
Decision Support System (DSS)
Differentiation
Dynamic System Model
Entropy
Equifinality
Equilibrium
Feedback
Flow System Model
Gantt Chart
General Systems Theory
Information
Information System

Input
Integration
Interdependence
Management Information System (MIS)
Model
Open System
Organization
Organization Chart
Output
PERT
Physical System
Policy
Processing
Schematic Model
Static System Model
Steady State
System
Systems Analysis

Review Questions

1. You have heard people discuss systems. What is a system? What is systems analysis?
2. From your understanding of this chapter, do you need a computer to do systems analysis? Discuss.
3. Take an organization with which you are familiar and examine the following:
   a. Primary subsystems.
   b. Characteristics.
   c. Elements.
   d. Purpose.
4. List the parts and functions of the following systems:
   a. Microcomputer.
   b. Stapler.
   c. Your business school or department.
5. Consider an automobile and a hospital as two systems. Identify the following as an input and/or output for each system:
   a. Batteries.
   b. Cured patients.
   c. Doctors.
   d. Driver's performance.
   e. Drugs.
   f. Gasoline.
   g. Information.
   h. Motion.
   i. A patient who died.
   j. Tires.
   k. X-ray machine.
6. What are the elements of a system? Can you have a viable system without feedback? Explain.
7. Distinguish between:
   a. Interaction and interdependence.

   b. Physical and abstract systems.
   c. Open and closed systems.
   d. Schematic and static systems models.

8. Discuss the primary characteristics of open systems. In what way is a system entropic?
9. How important is the informal information system in systems analysis? Explain.
10. What categories of information are relevant to decision making in business? Relate each category to the managerial level and an information system.
11. Discuss the concepts of MIS and DSS. How are they related? How do they differ?
12. Why is a data base important in MIS? Explain.
13. Write a short essay on the concept and uses of DSS. Include a brief discussion on the relationship between DSS and decision making.

Application Problems

1 Touhy Lumber is a large wholesale and retail distributor of building supplies. It purchases products in truckloads and sells them to builders and homeowners in smaller orders, depending on the item. The firm is very successful. It has two other stores—all located in a large midwestern city with a population of 6 million.

The organization structure of the firm is as follows:

President and Owner.

General Manager.

Three supervisors in charge of contractor, lumber, and hardware departments, respectively.

Three employees in the contractor's department, 14 in the lumber area, and 9 in hardware.

The president's main concern is to control the huge inventory and many supplies that arrive by truck and train three times a week. The firm has no prior computer experience. No expansion of business activities is planned, although the firm advertises heavily in the local newspapers and on radio to maintain a high sales volume. Products such as roofing material, lumber, bathroom fixtures, and paint continue to be the main lines of the firm.

You have been asked by the firm's president to examine the firm's structure, market share, and overall performance as a basis for discussing a possible computer system for inventory control. All you know about the company is what is described here. You are to prepare for the first meeting with the president.

Assignment

a. What questions will you ask to do the following?
   i. Develop an organization chart.
   ii. Understand the organization structure and the type of system the firm is.
   iii. Learn about the current inventory system.
   iv. Assess the relationship between the organization and the community, customers, vendors, and others.

   Remember that your purpose is to find out whether there is a need for a system study to install computer-based inventory control.

b. Using the case situation of Touhy Lumber, how important are the following systems concepts for systems analysis? Explain.
   i. Data base.
   ii. Feedback.
   iii. Interdependence.
   iv. Open and closed systems.
   v. Organization chart.
   vi. System-subsystem interface.

---

2 You are in a coffee shop across the street from school having lunch. A customer walks up to the counter. You observe the following:

Customer: Hi Jane, I'd like a burger to go. Everything but onions.

Jane (waitress): Anything else?

Customer: Yes, a small order of fries and a root beer.

Jane: That'll be $2.35.

She collects the cash and places the order through an electronic cash register that automatically displays the order on a TV screen in the back room where orders are prepared. When the order is ready, Jane puts it in a bag and hands it to the customer.

Assignment

a. Explain the pattern of this system in action. Specifically discuss the following:
   i. The organization system's characteristics.
   ii. The subsystems, information flow, and interfaces.

iii. The types of interdependence in the organization structure and the nature of feedback.
iv. Inputs and outputs and environment.
v. Formal and informal information systems.

b. If you were to improve the performance of this establishment, what would you do? How? Explain.

---

3 You are waiting in line to register for this semester's courses. A student ahead of you is talking to a registration counselor. You overhear the following conversation.

Student: Good morning, I'd like to register. [*He hands the counselor an IBM card listing the courses he wants to take for the semester.*]

Counselor: [*She looks over the class enrollment sheets and finds EDP 320 (systems analysis) closed.*] I'm sorry, but the systems class is closed. I can't add you on. You'll have to take it next term. Check with your advisor for another course. Management 415 is also closed. If you have to take it, you need permission from the instructor. Accounting 410 is OK, but your advisor did not initial it on the card. The other three courses are OK. I'll put your name down.

Student: You've got to be joking. This is my last semester. I have to take the systems course or I won't be able to graduate.

Counselor: Sorry, there is little I can do. Why don't you see your advisor about that?

Student: I saw her yesterday. She said she was going to be out of town till school starts next Wednesday.

Counselor: Then see the department chairman. If I were you, I'd go right away, because the classes are filling fast.

Student: [*looking frustrated*] I'll see. [*He dashes out the door.*]

Assignment

a. What type of system is it (open versus closed)? Why? Specify the system-subsystem linkages, interfaces, and interdependence.

b. In what way is the registration process a subsystem? Explain.

Selected References

Bertalanffy, Ludwig. *General Systems Theory.* New York: George Braziller, 1968.

Garry, G. A., and M. S. Scott Morton. "A Framework for MIS." *Sloan Management Review* 13 (Fall 1971), pp. 55–70.

Johnson, Richard A.; Fremont E. Kast; and James E. Rozensweig. *The Theory and Management of Systems.* New York: McGraw-Hill, 1973.

Keen, Peter, and M. S. Scott Morton. *Decision Support Systems: An Organizational Perspective.* Reading, Mass.: Addison-Wesley Publishing, 1978.

Kerola, P., and A. Jarvinen. "The PSC-Systemeering Model and Its Influence on the Basic Concept Structure of Data System Development." In *Tempere Report (1978): Summary Report of the Systemeering Research Seminar of Tampere 21,* ed. P. Kerola; M. Klemola; H. Kamarainen; and L. Lyytineh. University of Oulu, Institute of Data Processing, Tampere, Finland 31.12.1978 (1978).

Luchsinger, Vincent P., and Thomas V. Dock. *The Systems Approach: An Introduction.* 2d ed. Dubuque, Iowa: Kendal/Hunt Publishing, 1982.

Massie, J. L., and John Douglas. *Managing* 3d ed. Englewood Cliffs, N.J.: Prentice-Hall, 1982, p. 270.

Simon, Herbert A. *The Shape of Automation for Men and Management.* New York: Harper & Row, 1965.

Welke, R. J., and B. P. Konsynski. "Technology, Methodology & Information Systems: A Tripartite View." *Data Base,* Fall 1982, pp. 41–57.

Wetherbe, James C. *Systems Analysis and Design: Traditional, Structured, and Advanced Concepts and Techniques.* 2d ed. Minneapolis, Minn.: West Publishing, 1984, pp. 21–38.

Wiener, Norbert. *Cybernetics.* New York: John Wiley & Sons, 1948.

Chapter 2

The System Development Life Cycle

At a Glance

Systems analysts work with users to identify goals and build systems to achieve them. System development revolves around a life cycle that begins with the recognition of user needs. Following a feasibility study, the key stages of the cycle are evaluation of the present system, information gathering, cost/benefit analysis, detailed design, and implementation of the candidate system. The life cycle is not a procedure that deals with hardware and software. It is building computer-based systems to help the user operate a business or make decisions effectively and manage an enterprise successfully. This is the basis for learning systems analysis.

By the end of this chapter, you should know:

1. The makeup of the system development life cycle.
2. What prompts users to request change.
3. The components of a feasibility study.
4. The factors to consider in a candidate system.
5. How to plan and control for system success.

PLANNING AND CONTROL FOR SYSTEM SUCCESS

Prototyping

INTRODUCTION

In Chapter 1 we discussed the importance of systems concepts for developing business information systems. Developing such systems expedites problem solving and improves the quality of decision making. This is where the role of systems analysts becomes crucial. They are confronted with the challenging task of creating new systems and planning major changes in the organization. Like architects, they work with users to identify the goal(s), agree on a procedure and a timetable, and deliver a system that meets the user's requirements. It is a job that requires much personal contact between the analyst and members of the organization.

This chapter focuses on the stages of the system development life cycle, sometimes referred to as a *system study*. The systems analyst gives a system development project meaning and direction. A candidate system is approached after the analyst has a thorough understanding of user needs and problems, has developed a viable solution to these problems, and then communicates the solution(s) through the installation of a candidate system. Candidate systems often cut across the boundaries of users in the organization. For example, a billing system may involve users in the sales order department, the credit department, the warehouse, and the accounting department. To make sure that all users' needs are met, a *project team* that represents each user works with the analyst to carry out a system development project. In complex projects, representatives from other user areas influenced by the candidate system as well as information systems specialists may also be included.

THE SYSTEM DEVELOPMENT LIFE CYCLE

To understand system development, we need to recognize that a candidate system has a life cycle, just like a living system or a new product. Systems analysis and design are keyed to the system life cycle. The stages are shown in Figure 2-1. The analyst must progress from one stage to another methodically, answering key questions and achieving results in each stage.

A word of caution regarding life cycle activities: We isolate and sequence these activities for learning purposes, but in real life they overlap and are highly interrelated. For example, when the analyst is evaluating an existing operation, he/she is probably thinking about an alternative way that would improve the system or wondering whether a given piece of hardware would be a critical cost item to consider for a candidate system. Therefore, there can easily be overlap during any phase of the cycle. In fact, it may act as a basis for modifying earlier steps taken. We now describe each of these steps.

Recognition of Need—What Is the Problem?

One must know what the problem is before it can be solved. The basis for a candidate system is recognition of a need for improving an information

FIGURE 2–1 System Development Life Cycle

Stage	Key Question	Result
1. Recognition of need		
Preliminary survey/ initial investigation	What is the problem or opportunity?	Statement of scope and objectives Performance criteria
2. Feasibility study		
Evaluation of existing system and procedures Analysis of alternative candidate systems Cost estimates	What are the user's demonstrable needs? Is the problem worth solving? How can the problem be redefined?	Technical/behavioral feasibility Cost/benefit analysis System scope and objectives Statement of new scope and objectives
3. Analysis		
Detailed evaluation of present system Data collection	What must be done to solve the problem? What are the facts?	Logical model of system—e.g., data dictionary, data flow diagram Pertinent data
4. Design		
General design specifications Detailed design specifications Output Input Files Procedures	In general, how must the problem be solved? Specifically, how must the problem be solved? What is the system (processing) flow? Does the user approve the system?	Design of alternative solutions Final cost/benefit analysis Hardware specifications Cost estimates Implementation specifications Implementation schedule Approval of systems by user
Program construction Testing Unit testing Combined module testing User acceptance testing	How well do individual programs/modules test out? How ready are programs for acceptance test?	Programs Test plans Security, audit, and operating procedures Actual hardware use Formal system test
5. Implementation		
User training File/system conversion	What is the actual operation? Are user manuals ready? Are there delays in loading files?	Training program User-friendly documentation
6. Post-implementation and maintenance		
Evaluation Maintenance Enhancements	Is the key system running? Should the system be modified?	User requirements met User standards met Satisfied user

system or a procedure. For example, a supervisor may want to investigate the system flow in purchasing, or a bank president has been getting complaints about the long lines in the drive-in. This need leads to a preliminary survey or an *initial investigation* to determine whether an alternative system can solve the problem. It entails looking into the duplication of effort, bottlenecks, inefficient existing procedures, or whether parts of the existing system would be candidates for computerization.

If the problem is serious enough, management may want to have an analyst look at it. Such an assignment implies a commitment, especially if the analyst is hired from the outside. In larger environments, where formal procedures are the norm, the analyst's first task is to prepare a statement specifying the scope and objective of the problem. He/she then reviews it with the user for accuracy. At this stage, only a rough "ball park" estimate of the development cost of the project may be reached. However, an accurate cost of the next phase—the feasibility study—can be produced.

Impetus for System Change

The idea for change originates in the environment or from within the firm (see Figure 2–2). Environment-based ideas originate from customers, vendors, government sources, and the like. For example, new unemployment compensation regulations may make it necessary to change the reporting procedure, format, and content of various reports, as well as file structures. Customer complaints about the delivery of orders may prompt an investigation of the delivery schedule, the experience of truck drivers, or the volume of orders to be delivered. When investigated, each of these ideas may lead to a problem definition as a first step in the system life cycle process.

Ideas for change may also come from within the organization—top management, the user, the analyst (see Figure 2–2). As an organization changes its operations or faces advances in computer technology, someone within the organization may feel the need to update existing applications or improve procedures. Here are some examples:

- An organization acquires another organization.
- A local bank branches into the suburbs.
- A department spends 80 percent of its budget in one month.
- Two departments are doing essentially the same work, and each department head insists the other department should be eliminated.
- A request for a new form discloses the use of bootleg (unauthorized) forms.

Serious problems in operations, a high rate of labor turnover, labor-intensive activities, and high reject rates of finished goods, also prompt *top management* to initiate an investigation. Other examples are:

- A report reaches a senior vice president and she suspects the figures.

FIGURE 2-2 Major Sources of Change

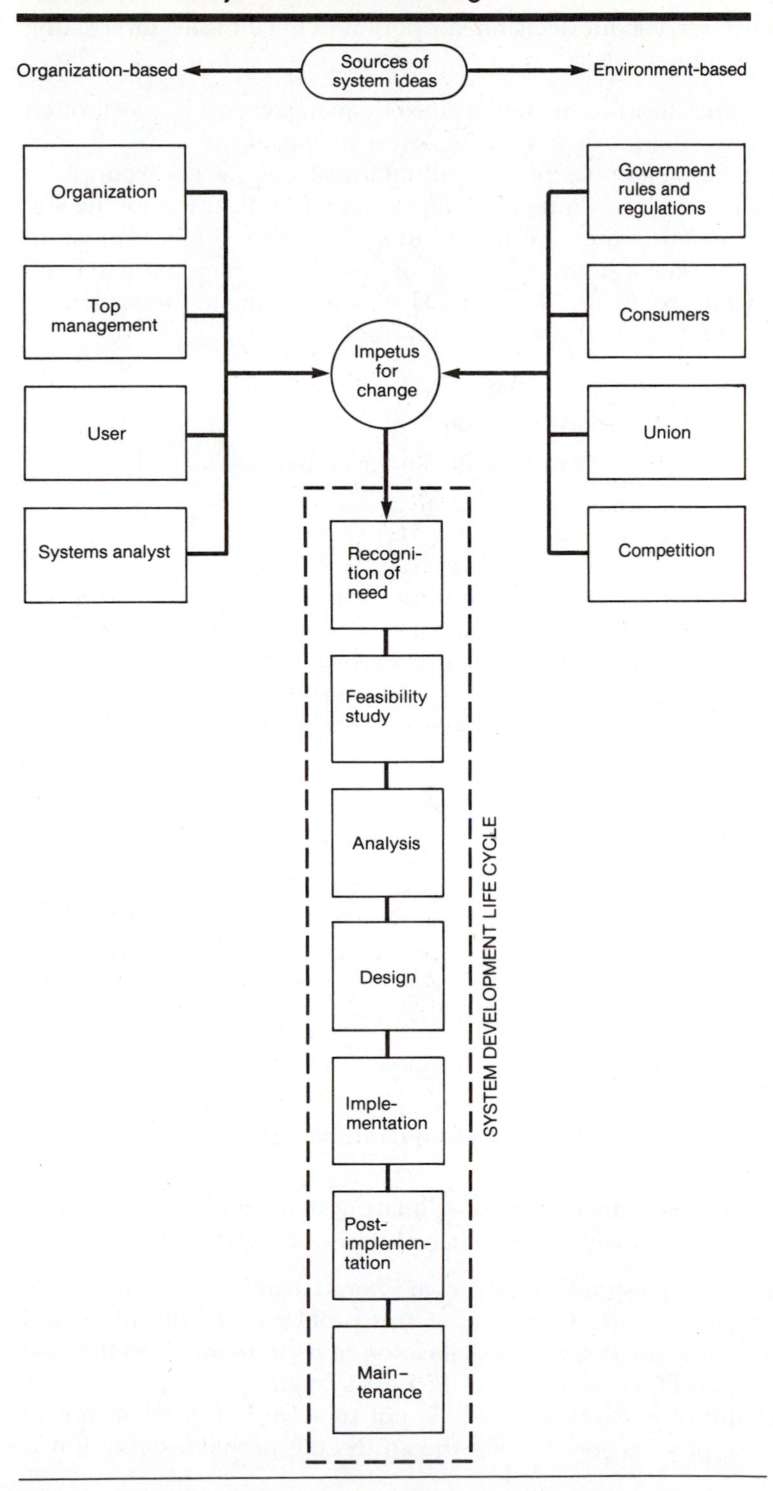

- The company comptroller reads an IRS audit report and starts thinking.
- An executive read about decision support systems for sales forecasting and it gives him an idea.

Many of these ideas lead to further studies by management request, often funneled downward and carried out by lower management.

User-originated ideas also prompt initial investigations. For example, a bank's head teller has been noticing long customer lines in the lobby. She wants to know whether they are due to the computer's slow response to inquiries, the new tellers' limited training, or just a sudden increase in bank business. To what extent and how quickly a user-originated idea is converted to a feasibility study depend on several factors:

- The risks and potential returns.
- Management's bias toward the user.
- Financial costs and the funds available for system work.
- Priorities of other projects in the firm.
- The persuasive ability of the user.

All these factors are crucial for a prompt response to a user request for change. A systems analyst is in a unique position to detect and even recommend change. Experience and previous involvement in the user's area of operations make him/her a convenient resource for ideas. The role and status of the analyst as a professional add credibility to the suggestions made.

Feasibility Study

Depending on the results of the initial investigation, the survey is expanded to a more detailed feasibility study. As we shall learn in Chapter 7, a feasibility study is a test of a system proposal according to its workability, impact on the organization, ability to meet user needs, and effective use of resources. It focuses on three major questions:

1. What are the user's demonstrable needs and how does a candidate system meet them?
2. What resources are available for given candidate systems? Is the problem worth solving?
3. What are the likely impact of the candidate system on the organization? How well does it fit within the organization's master MIS plan?

Each of these questions must be answered carefully. They revolve around investigation and evaluation of the problem, identification and description of candidate systems, specification of performance and the cost of each system, and final selection of the best system.

The objective of a feasibility study is not to solve the problem but to acquire a sense of its scope. During the study, the problem definition is crystallized and aspects of the problem to be included in the system are determined. Consequently, costs and benefits are estimated with greater accuracy at this stage. Cost/benefit analysis is described in Chapter 8.

The result of the feasibility study is a formal proposal. This is simply a report—a formal document detailing the nature and scope of the proposed solution. The proposal summarizes what is known and what is going to be done. It consists of the following:

1. *Statement of the problem*—a carefully worded statement of the problem that led to analysis.
2. *Summary of findings and recommendations*—a list of the major findings and recommendations of the study. It is ideal for the user who requires quick access to the results of the analysis of the system under study. Conclusions are stated, followed by a list of the recommendations and a justification for them.
3. *Details of findings*—an outline of the methods and procedures undertaken by the existing system, followed by coverage of the objectives and procedures of the candidate system. Included are also discussions of output reports, file structures, and costs and benefits of the candidate system.
4. *Recommendations and conclusions*—specific recommendations regarding the candidate system, including personnel assignments, costs, project schedules, and target dates.

After the proposal is reviewed by management, it becomes a formal agreement that paves the way for actual design and implementation. This is a crucial decision point in the life cycle. Many projects die here, whereas the more promising ones continue through implementation. Changes in the proposal are made in writing, depending on the complexity, size, and cost of the project. It is simply common sense to verify changes before committing the project to design.

Analysis

Analysis is a detailed study of the various operations performed by a system and their relationships within and outside of the system. A key question is: What must be done to solve the problem? One aspect of analysis is defining the boundaries of the system and determining whether or not a candidate system should consider other related systems. During analysis, data are collected on the available files, decision points, and transactions handled by the present system. We shall learn in Chapters 6 and 8 about some logical system models and tools that are used in analysis. Data flow diagrams, interviews, on-site observations, and questionnaires are examples. The interview is a commonly used tool in analysis. It requires special skills and sensitivity to the subjects being interviewed. Bias in data collection and interpretation can be a problem. Training, experience, and common sense are required for collection of the information needed to do the analysis.

Once analysis is completed, the analyst has a firm understanding of what is to be done. The next step is to decide how the problem might be solved. Thus, in systems design, we move from the logical to the physical aspects of the life cycle.

Design

The most creative and challenging phase of the system life cycle is system design. The term *design* describes a final system and the process by which it is developed. It refers to the technical specifications (analogous to the engineer's blueprints) that will be applied in implementing the candidate system. It also includes the construction of programs and program testing. The key question here is: How should the problem be solved? The major steps in design are shown in Figure 2–3.

The first step is to determine how the output is to be produced and in what format. Samples of the output (and input) are also presented. Second, input data and master files (data base) have to be designed to meet the requirements of the proposed output. The operational (processing) phases are handled through program construction and testing, including a list of the programs needed to meet the system's objectives and complete documentation. Finally, details related to justification of the system and an estimate of the impact of the candidate system on the user and the organization are documented and evaluated by management as a step toward implementation.

The final report prior to the implementation phase includes procedural flowcharts, record layouts, report layouts, and a workable plan for implementing the candidate system. Information on personnel, money, hardware, facilities, and their estimated cost must also be available. At this point, projected costs must be close to actual costs of implementation.

In some firms, separate groups of programmers do the programming, whereas other firms employ analyst-programmers who do analysis and design as well as code programs. For this discussion, we assume that analysis and programming are carried out by two separate persons. There are certain functions, though, that the analyst must perform while programs are being written. Operating procedures and documentation must be completed. Security and auditing procedures must also be developed. A detailed discussion of security and audit is presented in Chapter 16.

Implementation

The implementation phase is less creative than system design. It is primarily concerned with user training, site preparation, and file conversion. When the candidate system is linked to terminals or remote sites, the telecommunication network and tests of the network along with the system are also included under implementation.

During the final testing, user acceptance is tested, followed by user

FIGURE 2-3 Steps in System Design

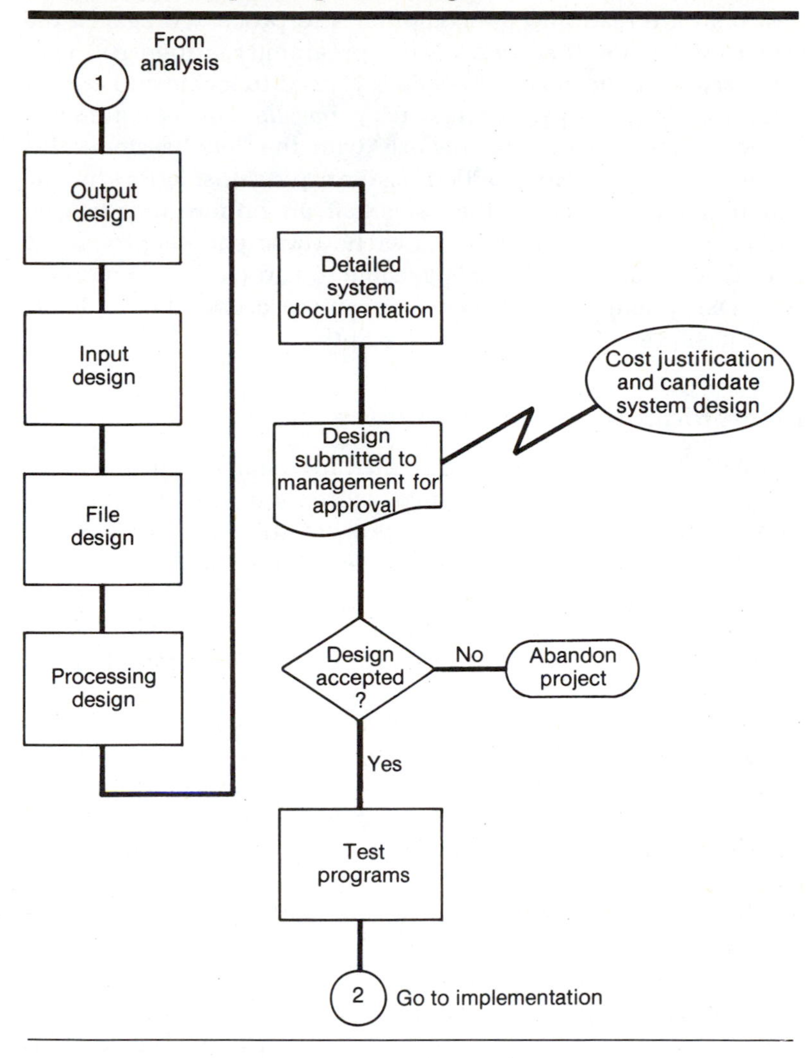

training. Depending on the nature of the system, extensive user training may be required. Conversion usually takes place at about the same time the user is being trained or later.

In the extreme, the programmer is falsely viewed as someone who ought to be isolated from other aspects of system development. Programming is itself design work, however. The initial parameters of the candidate system should be modified as a result of programming efforts. Programming provides a "reality test" for the assumptions made by the analyst. It is therefore a mistake to exclude programmers from the initial system design.

System testing checks the readiness and accuracy of the system to access, update, and retrieve data from new files. Once the programs become available, test data are read into the computer and processed against the file(s) provided for testing. If successful, the program(s) is then run with "live" data. Otherwise, a diagnostic procedure is used to locate and correct errors in the program. In most conversions, a *parallel run* is conducted where the new system runs simultaneously with the "old" system. This method, though costly, provides added assurance against errors in the candidate system and also gives the user staff an opportunity to gain experience through operation. In some cases, however, parallel processing is not practical. For example, it is not plausible to run parallel two online point-of-sale (POS) systems for a retail chain. In any case, after the candidate system proves itself, the old system is phased out.

Post-Implementation and Maintenance

After the installation phase is completed and the user staff is adjusted to the changes created by the candidate system, evaluation and maintenance begin. Like any system, there is an aging process that requires periodic *maintenance* of hardware and software. If the new information is inconsistent with the design specifications, then changes have to be made. Hardware also requires periodic maintenance to keep in tune with design specifications. The importance of maintenance is to continue to bring the new system to standards.

User priorities, changes in organizational requirements, or environmental factors also call for system *enhancements*. To contrast maintenance with enhancement, if a bank decided to increase its service charges on checking accounts from $3.00 to $4.50 for a minimum balance of $300, it is maintenance. However, if the same bank decided to create a personal loan on negative balances when customers overdraw their account, it is enhancement. This change requires evaluation, program modifications, and further testing. Software maintenance is covered in Chapter 14.

Project Termination

A system project may be dropped at any time prior to implementation, although it becomes more difficult (and costly) when it goes past the design phase. Generally, projects are dropped if, after a review process, it is learned that:

- Changing objectives or requirements of the user cannot be met by the existing design.
- Benefits realized from the candidate system do not justify commitment to implementation.
- There is a sudden change in the user's budget or an increase in design costs beyond the estimate made during the feasibility study.
- The project greatly exceeds the time and cost schedule.

In each case, a system project may be terminated at the user's request.

In contrast to project termination is new system failure. There are many reasons a new system does not meet user requirements:

- User requirements were not clearly defined or understood. Figure 2–4 illustrates this point.
- The user was not directly involved in the crucial phases of system development.
- The analyst, programmer, or both were inexperienced.
- The systems analyst (or the project team) had to do the work under stringent time constraints. Consequently, not enough thought went into the feasibility study and system design.
- User training was poor.
- Existing hardware proved deficient to handle the new application.
- The new system left users in other departments out of touch with information that the old system had provided.
- The new system was not user-friendly.
- Users changed their requirements.
- The user staff was hostile.

The list can be expanded to include many more causes. The important point is that although advances in computer systems and software make life easier for the analyst, the success of a system project depends on the

FIGURE 2–4 The Systems Design Procedure

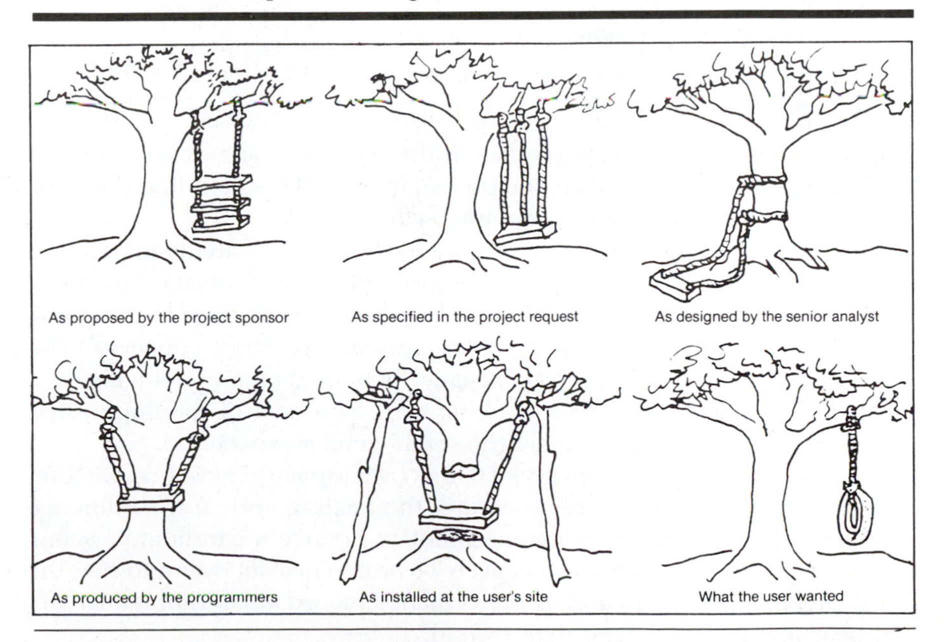

Source: Adapted from the Educational Exploration Center *Newsletter*, Minneapolis, Minnesota.

experience, creative ability, and knowledge of the analyst and the support from the user staff. This suggests that the analyst be skilled in the state of the art (hardware and software) as well as in dealing with people. The role of the analyst is covered in Chapter 3.

CONSIDERATIONS FOR CANDIDATE SYSTEMS

In today's business, there is more demand for computer services than there are resources available to meet the demand. The demand is made up of the following:

1. Operations of existing systems.
2. Maintenance that focuses on "patching" programs—often representing over 50 percent of maintenance.
3. Enhancements that involve major modifications in program structure or equipment.
4. Requests for candidate systems.

All these demands require resources—human, financial, and technological. On the human side, the computer department has to provide the following:

- Computer operators to run equipment.
- Data entry personnel.
- Systems analysts to define and design specifications.
- Application programmers to convert system specifications to computer programs.
- Maintenance programmers to repair errors.
- Supervisors, project leaders, and managers to coordinate the jobs with the users.

Thus, the basic problem is to match the demands for services with the available resources. How much one project is favored over another depends on technical, behavioral, and economic factors.

The *technical* factor involves the system department's ability to handle a project. Much depends on the availability of qualified analysts, designers, and software specialists to do the work. This is especially true in designing data bases and implementing complex systems for large concerns. The alternative to abandoning a project because of limited talent on the inside is free-lancing it to an outside consulting firm. The cost of developing the project has to be weighed against the total benefits expected.

The *behavioral* factor involves (1) the user's past experience with an existing system, (2) the success record of the analyst, and (3) the influence the user can exert on upper management to finance a candidate system. Political considerations that subjectively favor one project over another, the status of the department, and its performance record are additional factors that bear on funding a candidate system.

Perhaps the most important criterion in selecting a project is the *economic* factor. It focuses on the system's potential return on investment. What is considered an acceptable rate varies with different formulas, the variables chosen, and the like. System consultants suggest an annual rate of return of just over 20 percent.

Political Considerations

In conjunction with the preceding considerations is the political factor, which is partly behavioral. Imagine this setting: Managers in a production firm are considering two office automation proposals: proposal A—a teleconferencing system designed to reduce travel costs, and proposal B—a sales support system. Proposal A is justified by hard figures, but it was turned down. Instead, proposal B (poorly presented and justified) was sponsored by an influential executive and had the support of the committee. It passed because the right people were convinced it should.

Politics is the art of using influence and building coalitions when routine procedures do not achieve the right results. When system projects are developed, a collaborative relationship with the end user is helpful. A user who participated in building a system rarely criticizes it. If such a participative relationship comes too late, resistance can crop up and politics comes into play. The trick is to anticipate resistance early and turn it into support.

Planning and Control for System Success

What can the analyst do to ensure the success of a system? First, a plan must be devised, detailing the procedure, some methodology, activities, resources, costs, and timetable for completing the system. Second, in larger projects, a project team must be formed of analysts, programmers, a system consultant, and user representatives. Shared knowledge, interaction, and the coordination realized through team effort can be extremely effective in contrast with individual analysts doing the same work. Finally, the project should be divided into manageable modules to reflect the phases of system development—analysis, design, and implementation.

Most of this work falls under project management and control. The main idea behind the system development life cycle is to formalize a means of establishing control over a complex process. Work units have to be structured at three major levels for effective control of the project (see Figure 2–5). At the lowest level, work assignments are broken down into small manageable tasks. A *task* is usually a well-defined, structured work unit that can be carried out by one individual. The task can be easily budgeted and scheduled and its quality measured. It can be easily completed independent of other tasks and other project team members. If rework is necessary, there is minimal loss or impact on other tasks, except where time is critical.

FIGURE 2–5 Phases, Activities, and Tasks of a System Project

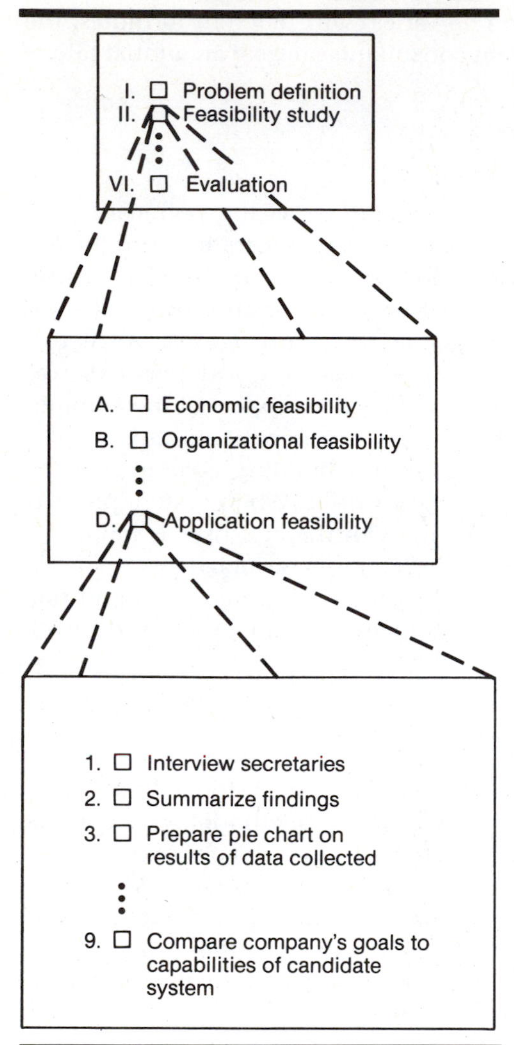

The second level at which work units are structured involves activities that have a larger scope and are designed to produce substantial results. An *activity* is a group of logically related tasks that serve one phase of the system development life cycle.

A *phase,* a third level of control, is a set of activities that bring the project to a critical milestone. *Milestones* are steppingstones that make up the entire project.

To illustrate these levels, Figure 2–6 shows six major phases of a system master plan for a given client (user). Each phase (numbered by a roman

FIGURE 2–6 Major Phases of a System Master Plan—An Example

(client name)
OVERVIEW OF SYSTEMS MASTER PLAN

Page______ of ______
Date____________

Phase No.	TASK/ACTIVITY	REF	DELEGATE TO			IMPLEMENTATON DATES				REVIEW		GANTT CHART					
			Company IND	Company DEPT	Con IND	MAN HRS	START DATE	FINISH DATE	VAR	Co.	Con.						
I	DEFINITION PHASE																
	A. Definition of organizational responsibilities																
	B. Definition of industry/environment																
	C. Definition of current business																
	D. Definition of system goals																
II	FEASIBILITY PHASE																
	A. Economic feasibility																
	B. Organizational feasibility																
	C. Alternative systems feasibility																
	D. Application feasibility																
III	RESEARCH PHASE																
	A. System Configuration audit																
	B. Hardware audit																
	C. Software audit																
IV	SELECTION PHASE																
	A. Requests for Proposals																
	B. Selection Criteria																
	C. Evaluation of Vendors																
V	IMPLEMENTATION PHASE																
	A. Physical Site Preparation																
	B. Training Plan																
	C. Installation of Hardware																
	D. Installation of System Software																
	E. Custom Programming (optional)																
	F. Installation of Application Software																
	G. Hardware/Software Testing (vendor data)																
	H. Documentation Plan (optional)																
	I. Systems Development																
	J. Parallel Testing (user data)																
	K. Conversion Cutover																
VI	EVALUATION PHASE																
	A. Project Evaluation																
	B. Internal Control Review																
	C. Documentation Review																
	D. Recommendations																

numeral) has the activities that must be carried out. The table includes specifications regarding the area to which each activity should be delegated, implementation dates, and a Gantt chart to show the range of completion dates for each activity. Figure 2–7 expands on Phase I (definition phase). Under each of the four activities (represented by A, B, C, and D) are individual tasks that must be completed. Similar to the organization in Figure 2–6, each task is delegated to an area or a team member and implementation dates are defined. Note that each activity falls within a specific phase of the system master plan. Phase boundaries do not overlap, although activities from different phases may be completed simultaneously.

In planning a project, the following steps should be taken:

1. Identify the activities in each phase and the tasks within each activity.
2. Calculate the budget for each phase and obtain agreement to proceed.
3. Review, record, and summarize progress on activities periodically.
4. Prepare a project progress report at the end of a reporting month.

FIGURE 2-7 Activities and Tasks of the Definition Phase

(client name)
SYSTEMS MASTER PLAN
DEFINITION PHASE

Page______ of ______
Date____________

Phase No.	TASK/ACTIVITY	N/A	SCOPE/COMMENT	DOCUMENT	REF	DELEGATE TO			IMPLEMENTATON DATES				REVIEW	
						Company		Con	MAN	START	FINISH	VAR		
						IND	DEPT	IND	HRS	DATE	DATE		Co.	Con.
I	DEFINITION PHASE													
	A. ORGANIZATIONAL RESPONSIBILITIES													
	• Receive approval from upper management to perform the systems study													
	• Select a qualified person(s) to undertake and supervise the systems study													
	• Decide whether to use consultants or vendors in the systems study													
	• Solicit recommendations and cooperation from all departments													
	• Delegate responsibility to appropriate department personnel													
	• Orient the staff as to the general goals, purpose, techniques, and extent of their involvement in the study													
	• Draw up a time schedule and work program by phases to control the completion of tasks													
I	B. INDUSTRY/ENVIRONMENT													
	• Document significant national and local economic conditions													
	• Document the effect of economic conditions on client industry and business													
	• Document the nature and competition of the client's industry													
	• Document system problems unique to the client's industry													
I	C. CURRENT BUSINESS													
	• Prepare a description of the business													
	• Prepare a description of major products or services													
	• Document the corporate structure													
	• Prepare an organizational chart													
	• Prepare a list of all current and proposed employees and their functions													
	• Review the most recent financial statements and budgets													
	• Review the long-term and short-term financing and budgeting of the current and proposed facilities													
	• Document any significant or relevant matters which affect systems planning													

In summary, system development should not be regarded merely as some procedure that deals with hardware and software. The original assumptions upon which system specifications were based should be tested and reevaluated with the user in mind. Managing system projects includes the important responsibility of seeing to it that all features of the candidate system—technological, logical, and behavioral—are considered before implementation and maintenance.

PROTOTYPING

As can be deduced from the discussion on system development, there are two major problems with building information systems: (1) the system development life cycle takes too long and (2) the right system is rarely developed the first time. Lengthy development frustrates the user. Analysts seem to get bogged down with tedious methodologies for developing systems. The reason they often come up with the wrong system is that they

FIGURE 2–7 *(concluded)*

(client name)
SYSTEMS MASTER PLAN
DEFINITION PHASE

Page ______ of ______
Date ______

Phase No.	TASK/ACTIVITY	N/A	SCOPE/COMMENT	DOCUMENT		DELEGATE TO			IMPLEMENTATON DATES				REVIEW	
						Company		Con	MAN	START	FINISH	VAR		
					REF	IND	DEPT	IND	HRS	DATE	DATE		Co.	Con.
I	DEFINITION PHASE D. SYSTEM GOALS • Document the general long-term and short-term business and systems objectives • Document areas to improve by department and/or function • Document the proposed hardware capabilities/objectives • Document the proposed system software capabilities/objectives • Document the proposed data processing application capabilities/objectives • Document the proposed word processing application capabilities/objectives • Document the proposed systems personnel capabilities/objectives													

expect users to define their information requirements. It usually turns out that what users ask for is not what they want, and what they want is not what they need.[1]

An alternative to this "paralysis by analysis" is an advanced technique called *prototyping*. Prototyping recognizes problems of cognitive style and uses advanced computer technology (see Figure 2–8). It advocates building a simple system through trial and error and refining it through an iterative process. The most extensive research on prototyping has been conducted by Naumann and Jenkins.[2] The basic steps are:

1. Identify the user's information and operating requirements.

[1] James C. Wetherbe, "Advanced System Development Techniques Avoid 'Analysis by Paralysis,'" *Data Management*, February 1984, p. 49.

[2] J. D. Naumann and M. A. Jenkins, "Prototyping: The New Paradigm for System Development," *MIS Quarterly*, September 1982, pp. 15–21.

FIGURE 2-8 System Development Life Cycle with Prototyping

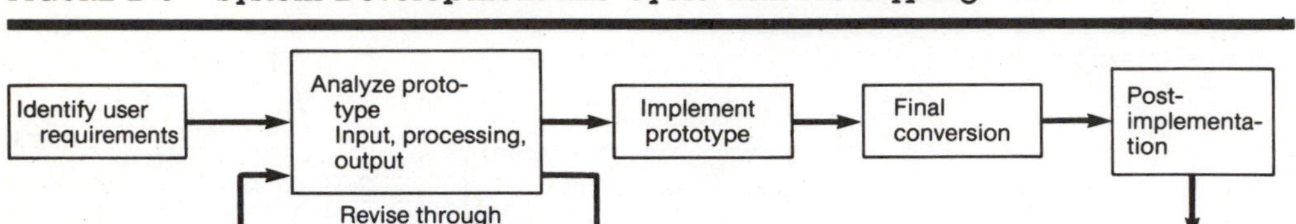

Source: Adapted from J. C. Wetherbe, "Advanced System Development Techniques Avoid 'Analysis by Paralysis,'" *Data Management,* February 1984, p. 51.

2. Develop a working prototype that focuses on only the most important functions, using a basic data base.
3. Allow the user to use the prototype, discuss requested changes, and implement the most important changes.
4. Repeat the next version of the prototype with further changes incorporated until the system fully meets user requirements.

Prototyping and advanced system development techniques have been successful in a wide variety of applications. The benefits include shorter development time, more accurate user requirements, and greater user participation and support. Prototyping as a strategy for determining user requirements is discussed in Chapter 5.

Summary

1. Systems analysis and design are keyed to the life cycle. The stages are:
   a. Recognition of the need for change.
   b. Feasibility study.
   c. Analysis of the present system.
   d. Design of a candidate system.
   e. Testing and implementation of the system.
   f. Post-implementation.
2. The idea for change originates in the environment (government, consumers, union, etc.) or from within the firm (user, analyst, etc.). Once the problem is verified, an initial investigation is conducted to determine whether change is feasible. If the answer is yes, a feasibility study is authorized.
3. Analysis is a detailed study of the various operations performed by a system. This involves gathering information and using structured tools for analysis.
4. System design refers to the technical specifications that will be applied in implementing the candidate system. This involves input/output, file, and processing design.

5. Implementation is concerned with detail—the physical creation of the candidate system. The key point is actual operation and user acceptance testing before the system is released to the user.
6. After implementation, maintenance begins. This includes enhancements, modifications, or any change from the original specifications. This phase terminates the system development life cycle.
7. In deciding on the project to design, a number of factors are considered: technical (availability of qualified specialists), operational (user's experience with similar projects), and economic (cost effectiveness of the proposed system). Political considerations also play a role in the final selection.
8. To ensure the success of the system, careful and often extensive planning is required. This work falls under project management, where a project is organized into work unit levels involving tasks (lowest level), activities, phases, and milestones. The overall management process is crucial to the successful completion of systems.

Key Words

Activity
Analysis
Candidate System
Design
Feasibility Study
Implementation
Initial Investigation
Milestone
Parallel Run
Phase
Prototyping
System Development
Task

Review Questions

1. What is the system development life cycle? How does it relate to systems analysis?
2. How would an analysis determine the user's needs for a system? Explain.
3. Where do ideas for a proposed system originate? To what extent does the analyst assist in this regard?
4. Distinguish between initial investigation and feasibility study. In what way are they related?
5. Why is a system proposal so crucial for system design? Explain.
6. What is the difference between analysis and design? Can one begin to design without analysis? Why?
7. What activities make up system design? How does system design simplify implementation?
8. A number of activities are carried out under implementation. Elaborate.
9. When does an analyst terminate a project? How does it tie in with post-implementation? Explain.

10. There are several considerations in deciding on a candidate system. What are they? Why are they important? Be specific.
11. Explain briefly the levels of structuring work units in system development.

Application Problems

1 Refer to Application Problem 3 in Chapter 1 and determine the steps you would take to develop the registration cycle.

2 An analyst was asked by a bank president to look into installing online terminals for all tellers. The bank has 15 tellers in the lobby and 9 tellers in the drive-in. The teller department reports to operations. The bank is organized with the following departments:

a. Operations department, which includes bookkeeping, tellers, check processing, microfilm, customer service, and accounting. This department has 58 employees and 4 officers.

b. Commercial and installment loan department, which handles auto loans, home improvement loans, secured loans, and commercial loans for small businesses. The department has eight employees and three officers.

c. Audit department, which verifies the entire bank's operations, audits the books, and investigates cash shortages or overages.

d. Safe deposit department, which is in charge of 6,000 safe deposit boxes housed in a vault off the lobby. Three employees handle customer service and billing.

e. Personnel department, which performs services such as health insurance administration, recruitment, testing, training, and human resources planning. The department is run by a director and one assistant.

The present teller system has electronic terminals that use paper tape to capture deposits, withdrawals, and other transactions. At the end of the day, the tape goes to data processing, where the information is entered on tape for final processing. If a teller needs an account balance, she dials bookkeeping, which looks up the balance in a computer-generated report. The teller then writes the balance on a slip and hands it to the customer. An online teller terminal performs the same function in a matter of seconds.

Assignment

a. If you were the analyst, what detailed plan would you lay out to represent the system life cycle for a possible installation? Explain each step in detail.

b. In this particular case, where would you start the cycle? What step do you consider most crucial? Explain.

---

3 The vice president of a large retail store wants to modify is order entry system. He states the problem as follows: "I need a report that gives me information on the aging of back orders."

Assignment

a. What questions would you ask?

b. Outline the procedure to follow.

Selected References

Brooks, Cyril; Phillip Grouse; D. Ross Jeffrey; and Michael Lawrence. *Information System Design.* Englewood Cliffs, N.J.: Prentice-Hall, 1982, pp. 86–105.

Naumann, J. D., and M. A. Jenkins. "Prototyping: The New Paradigm for System Development." *MIS Quarterly,* September 1982, pp. 49–53.

Senn, James A. *Analysis and Design of Information Systems.* New York: McGraw-Hill, 1984, pp. 17–23.

Wetherbe, James C. "Advanced System Development Techniques Avoid 'Analysis by Paralysis.'" *Data Management,* February 1984, pp. 49–52.

Chapter 3

The Role of the Systems Analyst

At a Glance

A feasibility study and the eventual design and implementation of a computer system are the functions of the systems analyst. The role requires a combination of skills, experience, personality, and common sense. Many individuals in business can benefit by learning the duties and responsibilities of the analyst. Users, managers, and accountants must understand how the analyst relates to them and how interface should be maintained.

Systems analysis has a history dating to the late 1890s. The role of the analyst has been emerging with the changing technology. This means that technical skills are required for systems work. The fact that a system is designed for a specific user also means that the analyst must have interpersonal skills. The multifaceted role of the analyst, then, warrants detailed discussion.

By the end of this chapter, you should know:

1. What it takes to do systems analysis.
2. The academic and personal qualifications of systems analysts.
3. The multifaceted role of the analyst.
4. How to maintain proper interface with the user.
5. The behavioral issues that contribute to system success.
6. The place of the analyst in the MIS organization.

INTRODUCTION

Designing and implementing systems to suit organizational needs are the functions of the systems analyst. He/she plays a major role in seeing business benefit from computer technology. The analyst is a person with unique skills. The job is not confined to data processing as such, because it deals heavily with people, procedures, and technology. Common sense, a structured framework, and a disciplined approach to solving problems are a part of analysis. Many individuals in business can benefit by learning the functions of and techniques used by the analyst. Users, managers, accountants, and auditors must understand how the analyst relates to them. This chapter, then, describes the role of the analyst in system development and the interface maintained between the analyst and the user.

DEFINITION

The role of the analyst has been emerging with changing technology. The literature suggests several definitions of *analyst,* but there seems to be a common thread. A representative definition is the *Random House Dictionary:* "a person who conducts a methodical study and evaluation of an activity such as a business to identify its desired objectives in order to determine procedures by which these objectives can be gained."[1] A similar definition is offered by Nicholas: "The task of the systems analyst is to elicit needs and resource constraints and to translate these into a viable operation."[2]

HISTORICAL PERSPECTIVE

The Early Years

Systems analysis dates back to the late 1890s to the principles of Frederick Taylor. Taylor's approach focused on the effective use of humans at work."[3] The four key steps in scientific management are:

1. Develop an ideal method of doing a task and establish a standard for it. In turn, the worker should be paid an incentive for exceeding the standard.
2. Select the best person for the job and train him/her accordingly.
3. Incorporate the scientific method with well-trained people.

---

[1] *Random House Dictionary,* 1979.

[2] John M. Nicholas, "Transactional Analysis for Systems Professionals," *Journal of System Management,* October 1978, p. 6.

[3] Henry C. Lucas, Jr., *The Analysis, Design, and Implementation of Information Systems* (New York: McGraw-Hill, 1981), p. 14.

4. Establish cooperation between manager and worker based on the division of labor.

Taylorism had problems, however. It was basically physiological in emphasis and ignored behavior. It applied to lower-level, repetitive tasks and said nothing about decision making in organizations. It assumed that money is a primary motivator and that humans always act rationally. It also suggested that people are lazy. They do not want to work, so close supervision is required.

Later in the 1940s, Taylor's scientific approach was modified by theorists such as Maslow,[4] McGregor,[5] Likert, and Schein. They focused on the psychological needs of people at work. Maslow, for example, developed the needs hierarchy as a way of understanding people in organizations. According to the hierarchy, a lower-level need (e.g., survival) must be satisfied before a person is interested in a higher-level need, such as independence or self-actualization.

This modified approach to scientific management helped analysts realize that people make the system, not just the hardware.[6] Early analysts are credited with working in factories, specializing in the improvement of work methods, and setting time standards for production.

The War Effort

Prior to World War I, factories were labor-intensive, but the war brought about automation because of labor shortages. Automation produced goods faster, cheaper, and quicker than before. The creation of unions and higher wages and production costs precipitated a demand for analysts to improve working conditions and establish time standards and incentive wages.[7] With the large number of untrained workers, analysts also developed methods and programs for training and development.

The same picture continued throughout the 1950s, but the 1960s marked the beginning of a new era in systems analysis and operations research. The federal government was stockpiling weapons in huge quantities. Analysts performed cost/benefit analyses of military weapon systems and used inventory-control techniques to determine where to store weapons and in what quantities.

With the advent of the commercial computer, organizations began to look for ways to approach problems that might be solved by the computer. The demand for systems analysts took a new direction. The analyst began to assume the role of a problem solver and a specialist in computer-based applications.

---

[4] E. Schein, *Organizational Psychology,* 2d ed. (Englewood Cliffs, N.J.: Prentice-Hall, 1970).

[5] D. McGregor, *The Human Side of Enterprise* (New York: McGraw-Hill, 1960).

[6] Stephen A. Dimino, "Return to Basics," *Journal of Systems Management,* December 1982, p. 28.

[7] Alfred Hunt, *The Management Consultant* (New York: John Wiley & Sons, 1977), pp. 13–14.

WHAT DOES IT TAKE TO DO SYSTEMS ANALYSIS?

An analyst must possess various skills to effectively carry out the job. Specifically, they may be divided into two categories: interpersonal and technical skills. Both are required for system development. *Interpersonal* skills deal with relationships and the interface of the analyst with people in business. They are useful in establishing trust, resolving conflict, and communicating information.[8] *Technical* skills, on the other hand, focus on procedures and techniques for operations analysis, systems analysis, and computer science.

The interpersonal skills relevant to systems work include the following:

1. *Communication*—having the ability to articulate and speak the language of the user, a "flare" for mediation, and a knack for working with virtually all managerial levels in the organization. Communication is not just reports, telephone conversations, and interviews. It is people talking, listening, feeling, and reacting to one another, their experience and reactions. Some indicators of a climate of closed communication are defensive memos, excessive correspondence, and a failure to speak up for fear of being identified. Therefore, opening communication channels are a must for system development.
2. *Understanding*—identifying problems and assessing their ramifications, having a grasp of company goals and objectives, and showing sensitivity to the impact of the system on people at work.
3. *Teaching*—educating people in use of computer systems, selling the system to the user, and giving support when needed.
4. *Selling*—selling ideas and promoting innovations in problem solving using computers.

Technical skills include:

1. *Creativity*—helping users model ideas into concrete plans and developing candidate systems to match user requirements.
2. *Problem solving*—reducing problems to their elemental levels for analysis, developing alternative solutions to a given problem, and delineating the pros and cons of candidate systems.
3. *Project management*—scheduling, performing well under time constraints, coordinating team efforts, and managing costs and expenditures.
4. *Dynamic interface*—blending technical and nontechnical considerations in functional specifications and general design[9]

---

[8] Nicholas, "Transactional Analysis, pp. 6–11.

[9] Laura Scharer, "Systems Analysts Performance: Criteria and Priorities," *Journal of Systems Management*, February 1982, pp. 10–15.

5. *Questioning attitude and inquiring mind*—knowing the what, when, why, where, who, and how a system works.
6. *Knowledge of the basics of the computer and the business function.*

Systems analysts require interpersonal as well as technical skills, although the necessity for both skills depends on the stages of system development. Figure 3–1 illustrates the skills expected across the phases of system development: analysis, design, implementation, and maintenance. During analysis, there is greater need for interpersonal skills—working with the user to determine requirements and translate them into design criteria. During design, the major thrust is to develop a detailed design of the candidate system—highly technical procedures and methodologies. Even then, there is some emphasis on the interpersonal factor—the analyst/user interface and user participation as a step toward training and implementation. During program construction, coding and testing are carried out with some user participation.

During system implementation, technical and interpersonal skills converge. The technical aspects focus on "proving" the software and preparing for the final conversion of files and documentation. The interpersonal aspects deal with user training and selling the user on the benefits and potential of the candidate system. During the maintenance stage the role of the analyst drops off, except when unanticipated problems develop.

Overall, it can be seen that a successful analyst blends the realities of

FIGURE 3–1 Interpersonal and Technical Skills Necessary in System Development

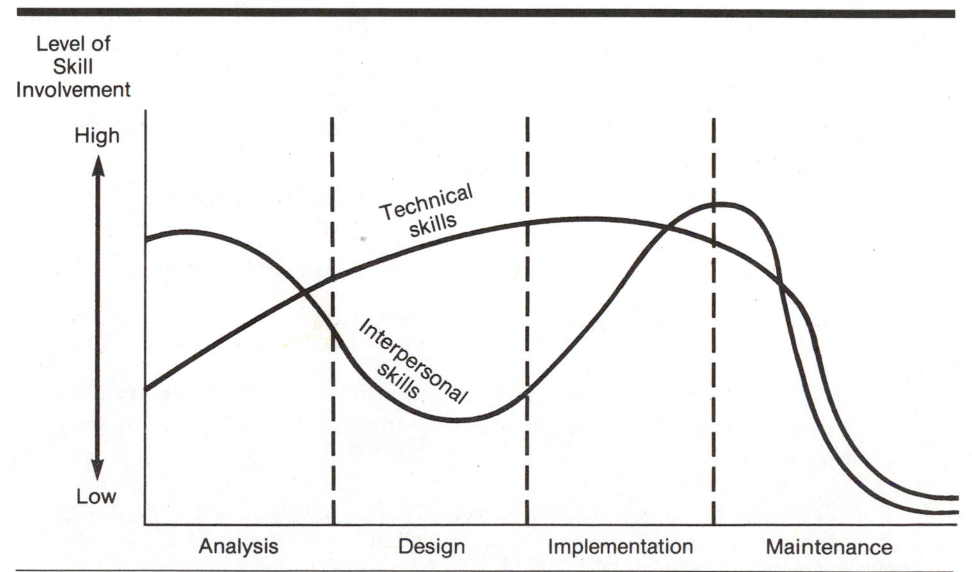

the human factor with the structured techniques and procedures that permit problem solution through the computer. Both skills are required for a lasting interface.

Academic and Personal Qualifications

How does the analyst acquire these skills? The answer is in education, experience, and personality. Most of today's analysts are college graduates with majors in accounting, management, or information systems. The latter major is becoming so popular that more and more universities have programs that include courses in systems analysis, project management, and data base design. More than 30 universities offer doctorates in the field.

The background and experience of analysts include:

1. A background in systems theory and organization behavior.
2. Familiarity with the makeup and inner workings of major application areas such as financial accounting, personnel administration, marketing and sales, operations management, model building, and production control.
3. Competence in system tools and methodologies and a practical knowledge of one or more programming and data base languages.
4. Experience in hardware and software specifications, which is important for selection.

Awad conducted a study to determine the personal attributes of analysts and what attracts them to systems analysis.[10] The attributes are:

1. *Authority*—the confidence to "tell" people what to do. Much of this quality shows in project management and team work to meet deadlines.
2. *Communication skills*—ability to articulate and focus on a problem area for logical solution.
3. *Creativity*—trying one's own ideas, developing candidate systems using unique tools or methods.
4. *Responsibility*—making decisions on one's own and accepting the consequences of these decisions.
5. *Varied skills*—doing different projects and handling change.

These academic and personal qualifications highlight the role of the analyst and distinguish between analysis and traditional programming.

---

[10] E. M. Awad, "Vocational Needs, Reinforcers, and Job Satisfaction of Analysts and Programmers in a Banking Environment," from the 10th Australian Computer Conference, Melbourne, Australia, September 1983.

THE MULTIFACETED ROLE OF THE ANALYST

Among the roles an analyst performs are change agent, monitor, architect, psychologist, salesperson, motivator, and politician. Let's briefly describe each role.

Change Agent

The analyst may be viewed as an agent of change. A candidate system is designed to introduce change and reorientation in how the user organization handles information or makes decisions. It is important, then, that change be accepted by the user. The way to secure user acceptance is through user participation during design and implementation.[11] The knowledge that people inherently resist change and can become ineffective because of excessive change should alert us to carefully plan, monitor, and implement change into the user domain.

In the role of a change agent, the systems analyst may select various styles to introduce change to the user organization. The styles range from that of persuader (the mildest form of intervention) to imposer (the most severe intervention). In between, there are the catalyst and the confronter roles.[12] When the user appears to have a tolerance for change, the persuader or catalyst (helper) style is appropriate. On the other hand, when drastic changes are required, it may be necessary to adopt the confronter or even the imposer style. No matter what style is used, however, the goal is the same: to achieve acceptance of the candidate system with a minimum of resistance.

Investigator and Monitor

In defining a problem, the analyst pieces together the information gathered to determine why the present system does not work well and what changes will correct the problem. In one respect, this work is similar to that of an investigator—extracting the real problems from existing systems and creating information structures that uncover previously unknown trends that may have a direct impact on the organization.[13]

Related to the role of investigator is that of monitor. To undertake and successfully complete a project, the analyst must monitor programs in relation to time, cost, and quality. Of these resources, time is the most

---

[11] William Feeney and Frea Sladek, "The System Analyst As a Change Agent," *Datamation,* November 1977, p. 85.

[12] Ibid, p. 86

[13] Robb Ware, "From Technician to Problem-solver: Training the Systems Analyst," *Data Management,* March 1983, p. 21.

important. If time "gets away," the project suffers from increased costs and wasted human resources.[14] Implementation delays also mean the system will not be ready on time, which frustrates users and customers alike.

Architect

The architect's primary function as liaison between the client's abstract design requirements and the contractor's detailed building plan may be compared to the analyst's role as liaison between the user's logical design requirements and the detailed physical system design. As architect, the analyst also creates a detailed physical design of candidate systems. He/she aids users in formalizing abstract ideas and provides details to build the end product—the candidate system.[15]

Psychologist

In systems development, systems are built around people. This is perhaps a bit exaggerated, but the analyst plays the role of a psychologist in the way he/she reaches people, interprets their thoughts, assesses their behavior, and draws conclusions from these interactions. Understanding interfunctional relationships is important. Here is an illustration of the complexities that may arise in such relationships:

> Harry, the new analyst at C. O. Ball Insurance, had discovered an obvious flaw in the information flow. Betty in premium accounting was processing premium payments and then mailing the checks to Marie for deposit. Marie sat only 40 feet away, but the mail system in the company took two days. At their next meeting, Harry suggested to Betty, "Why don't you carry the checks over to Marie? She is only a few steps away. We can improve cash flow tremendously in. . . ." He never got to finish. Betty had begun to sniffle. She jumped up and ran into the ladies room. Harry planned to discuss the situation with his boss the next morning. When he returned to his office, however, he found a note to see Mr. Carlisle, the general agent, immediately. Harry's fifth sense detected trouble as he walked into Carlisle's office. Carlisle said, "Harry, why are you upsetting my sister-in-law with some nonsense about her husband's first wife?[16]

As can be seen, it is important that the analyst be aware of people's feelings and be prepared to get around things in a graceful way. The art of listening is important in evaluating responses and feedback.

---

[14] Leslie Matthies, "Time: The Analyst's Basic Resource," *Systems Management,* November 73, pp. 18–19.

[15] Edward M. Canavan, "The Mysterious Systems Analyst," *Journal of Systems Management,* May 1980, p. 34.

[16] Wm. C. Ramsgard, "The Systems Analyst—Doctor of Business," *Journal of Systems Management,* July 1974, p. 10.

Salesperson

Selling change can be as crucial as initiating change. As we shall learn in Chapter 7, the oral presentation of the system proposal has one objective—selling the user on the system. Selling the system actually takes place at each step in the system life cycle, however. Sales skills and persuasiveness, then, are crucial to the success of the system.

Motivator

A candidate system must be well designed and acceptable to the user. System acceptance is achieved through user participation in its development, effective user training, and proper motivation to use the system. The analyst's role as a motivator becomes obvious during the first few weeks after implementation and during times when turnover results in new people being trained to work with the candidate system. The amount of dedication it takes to motivate users often taxes the analyst's abilities to maintain the pace. What was once viewed as a challenge can easily become a frustration if the user's staff continues to resist the system.

Politician

Related to the role of motivator is that of politician. In implementing a candidate system, the analyst tries to appease all parties involved. Diplomacy and finesse in dealing with people can improve acceptance of the system. Inasmuch as a politician must have the support of his/her constituency, so is the analyst's goal to have the support of the users' staff. He/she represents their thinking and tries to achieve their goals through computerization.

In summary, these multiple roles require analysts to be orderly, approach a problem in a logical, methodical way, and pay attention to details. They are usually factual and "down to earth" and see the world as logical, orderly, and predictable. They prefer to concentrate on objective data, seek the best method, and be highly prescriptive. They appear to be cool and studious. During casual conversation, the analyst might preface remarks with "It stands to reason . . . ," "If you look at it logically . . . ," or "If we just go about it methodically. . . ."[17] These characteristics point out the strengths of analysts: They focus on method and plan, point out details, are good at model building, perform best in structured situations, and seek stability and order. Their primary liabilities are having a tendency toward "tunnel vision," being overly inflexible and cautious, and overplanning.

---

[17] Robert M. Bramson and Allen F. Harrison, "The Orderly Ways of Analysts," *Computer Decisions*, November 1983, p. 112.

THE ANALYST/USER INTERFACE

An important aspect of system development is a viable interface between the analyst and the user. Analysts must devote as much skill and effort to achieve a productive relationship with the user as they devote to the technical requirements of the system. Most research indicates that as the number of users increases, the probability of system failure increases without close analyst/user interface.

The interface problem emerged in the 1960s when there was an adversary relationship between the analyst and the user.[18] The analyst represented management and essentially imposed change on the user. Analysis was a defensive business with the overriding concern to avoid failure rather than achieve success. User involvement was limited. The analyst was "kingpin." Later, growth in user-friendly technology, improved knowledge of the user in information systems, and maturity of the analyst paved the way for greater user participation in system development. It also narrowed the cultural gap between the user and the analyst.

Behavioral Issues

Much research has been done to study users and their relationships with systems analysts. Increasing reports of system failures that were not caused by technical problems made it necessary to seek a better understanding of the analyst/user interface. Let's examine user motivation, analyst/user differences, and the political factor.

User Motivation

The motivational approach in system development states that the candidate system should satisfy the users' needs if they are going to use it. Several models of user behavior attempted to look at the motivation behind system acceptance. For example, Lucas's descriptive model of user behavior identifies attitudinal, personal, and situational factors that affect system use. Use depends on both positive attitudes toward the systems's features and the decision-making style of the user.[19]

The expectancy theory of user motivation stresses two important relationships that have a bearing on user acceptance. The first relationship is between effort and performance. The user determines the probability that a certain level of motivation or effort will improve job performance. So, a user who perceives a system to be of low quality will put forth limited effort to use it. According to Zmud, the perceived link between effort and performance is formulated before the system design is completed.[20] The second

---

[18] Martin Lasdin, "Games Played Between Users and Providers," *Computer Decisions*, October 1980, pp. 72–73.

[19] Henry C. Lucas, *The Analysis, Design, and Implementation*, op. cit, pp. 35–38.

[20] Robert W. Zmud, "The Role of Individual Differences in MIS Development, unpublished paper, Georgia State University, 1980.

important relationship is between performance and rewards. To the extent that rewards (intrinsic and extrinsic) are contingent upon performance, motivation may be enhanced. Also, the value that the user places on the perceived rewards from a candidate system determine the motivation to use the system.

One conclusion from the motivational perspective is that the impact of the computer on the user's daily work is important. Systems design is essentially task design. A candidate system must be tailored to the user's tasks to be accepted.[21]

Analyst/User Differences

On the surface, differences in education, experience, and language are quite obvious. The analyst's impatience with the user's ignorance about terminology like *chip* and CRT and the user's impatience with the analyst's limited understanding of the business, however, often lead to conflict during system development. The user also tends to take for granted the analyst's knowledge and expects the computer to solve virtually all problems. These unrealistic expectations are barriers to the interface.[22] Much of it is defensive behavior: The user does not want to seem dumb about the technology.[23]

On the other hand, most systems analysts feel limited responsibility for the effects of new systems they implement. A study by Bostrom and Heiner showed that analysts are preoccupied with low costs and technical sophistication. They tend to ignore the trade-offs that are essential to a design that makes the user comfortable. They also seem to view themselves as creators rather than participants in the development process. The result is ignoring user suggestions and producing analyst-oriented rather than user-oriented systems.[24]

Beneath the surface lies the fundamental difference in the way analysts and users process information. Cognitive styles, for example, produce interesting insights into the problem of designing computer systems. If the dominant style of the analyst is analytical and that of the user is heuristic, then the prospect of designing a system that meets the user's expectations is remote. If the analysis (and therefore the system) is presented in a format that fits the cognitive style of the user, however, then the candidate system is expected to be successful.[25]

---

[21] Daniel Robey, "Perspectives of the User Interface," in *Proceedings of the 17th Annual Conference of the SIGCPR,* 1980, p. 23.

[22] Larry G. Faerber and Richard L. Ratcliff, "People Problems Behind MIS Failures," *Financial Executive,* April 1980, pp. 18–24.

[23] Pamela McGhee, "What Do Users Want?" *Computerworld,* March 23, 1981, p. 84.

[24] R. P. Bostrom and J. S. Heiner, "MIS Problems and Failures," *MIS Quarterly,* December 1977, pp. 48–50.

[25] I. Benbasat and R. N. Taylor, "The Impact of Cognitive Styles on Information System Design," *MIS Quarterly,* June 1978, pp. 143–48.

Two implications may be drawn for system design. First, there is a need for mutual understanding between the analyst and the user. Second, once differences are understood and accepted, alleviating them may be possible through a deeper involvement of the user and support of the analyst.

The Political Factor

Since information is a source of organizational power, the process of system development may be viewed as a contest for power where analysts have the initial advantage. System development is often viewed as a bargaining process, where analysts and users attempt to enhance their power positions and self-interests. To achieve political rationality, the analyst should not brag about the technical aspects of the candidate system. A system that is simple to explain and easy to understand is more readily accepted than a technical presentation; it also makes the user feel less vulnerable. In essence, the political factor prompts the analyst to honestly assess the motives of all parties involved and attempt to remove barriers that lead to system failure.

Conflict Resolution

The role of the analyst requires coordinating a vast network of people's ideas and integrating them into a "systemeering" process design. Taggart suggests that the concerns of the user be integrated into the system life cycle and recommends *systemeering*—a term used in Nordic countries to describe a comprehensive set of activities associated with system development.[26] Analysts are expected to adapt their own personal style to user personality factors to improve communication and promote system success.

Related to the success of the system is the extent to which a commitment can be made to avoid future conflict. The complexity of technical and behavioral factors in system development has caused several large firms to develop approaches to secure system success. For example, Boeing Aircraft Corporation developed a Detailed Definition of Requirements (DDR), which is a binding agreement between analysts and users. The analyst agrees to design a system to meet the user's requirements, and the user, in turn, agrees to accept the system based on what has been specified.[27] In effect, DDR forces communication between the analyst and the user and bridges the knowledge gap by requiring all parties to be explicit.

---

[26] Wm. Taggart, "Human Information Processing Styles and the Information System Architect in the PSC Systemeering Model," in *Proceedings of the 17th Annual Conference of the Special Interest Group on Computer Personnel Research,* 1980, pp. 62–76.

[27] Hillard McLamore, "How to Avoid Hassles and Headaches When Systems People and Users Meet," *SAM Advanced Management Journal,* Summer 1978, pp. 5–12.

THE PLACE OF THE ANALYST IN THE MIS ORGANIZATION

The MIS Organization

The organizing function is the task of grouping and assigning work elements to appropriate areas. An MIS manager organizes by assigning tasks, dividing work into specific jobs, and defining the relationship among them. The MIS organization structure encompasses supervisory levels, authority relationships, and the general pattern of activities carried out by employees at each level. Figure 3-2 shows a general organization structure of an MIS facility. The chain of command extends from manager to subordinate. There is a direct line of authority from the director of MIS services to each of the supervisors in charge of analysis and design, programming, and opera-

FIGURE 3-2 General Organization Chart of an MIS Division

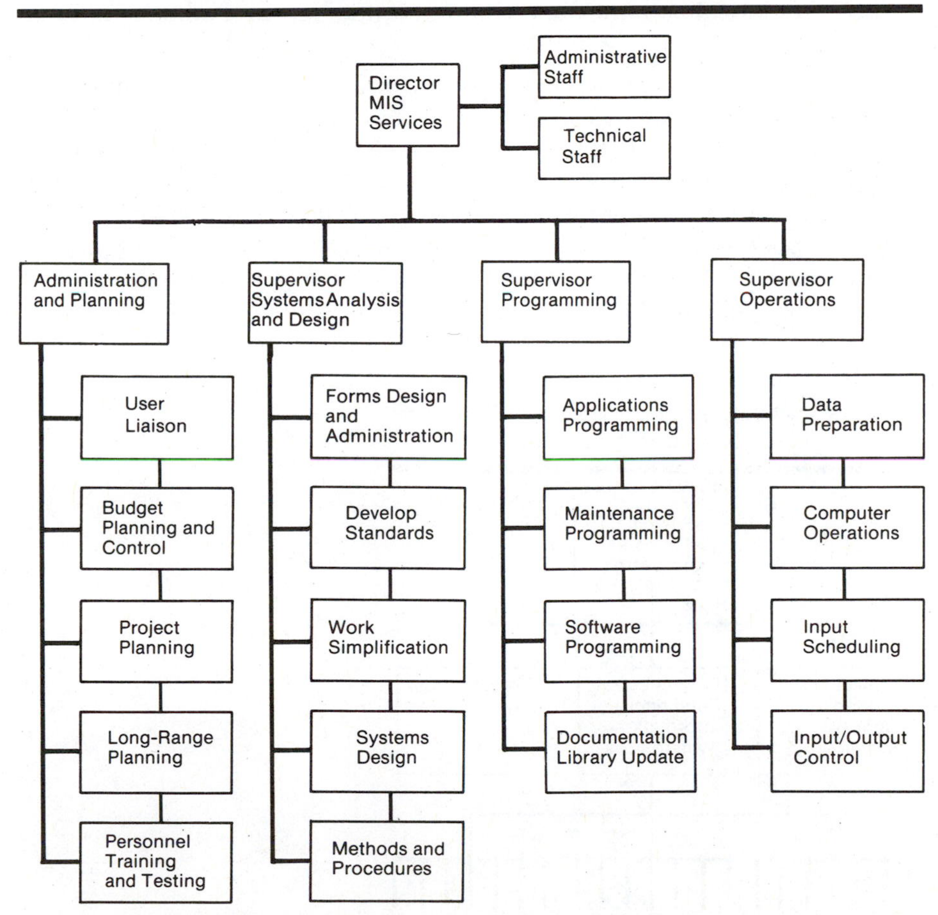

tions. This line of authority permits a supervisor to exercise direct command over subordinates to carry out their tasks.

Primary Functions of an MIS Facility

The structure of an MIS facility is organized around the primary functions to be performed. The functional requirements of an MIS facility center around the following areas: administration, systems analysis and design, programming, and operations.

1. *Administration* is represented by four activities:
   - *a.* User liaison which handles the changing needs of the user and user-systems relationships.
   - *b.* Long-range planning, which includes personnel selection, recruitment, application development, and planning for anticipated changes in hardware and software.
   - *c.* Budget planning and control of the entire MIS division.
   - *d.* Personnel administration and training for upgrading employee skills.

2. *Systems analysis and design* may be organized as project-oriented, pool-oriented, or functional. In a *project-oriented* arrangement a team of analysts is formed to work on one project. As shown in Figure 3–3, each team has a project leader who reports directly to the systems manager. This arrangement is typical of smaller installations that handle limited projects.

In a *pool-oriented* arrangement, analysts work on any system assignment within the firm. Once the job is completed, they return to the pool for another assignment. In Figure 3–4, a system team is on loan to report

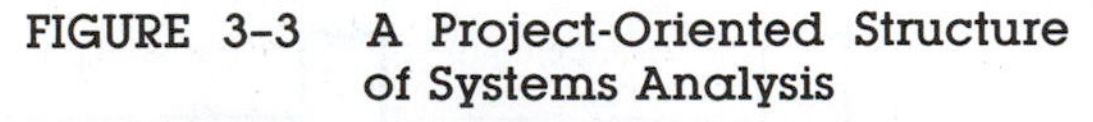

FIGURE 3–3 A Project-Oriented Structure of Systems Analysis

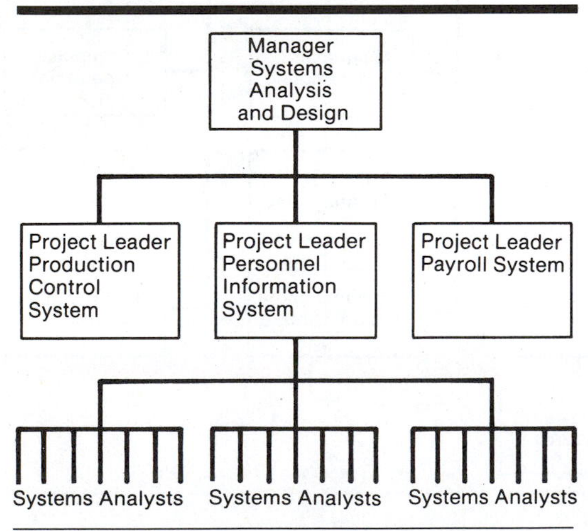

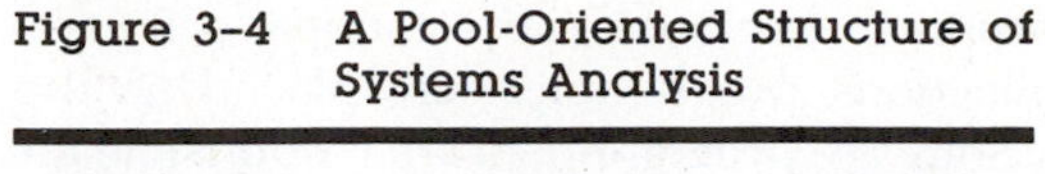

Figure 3–4 A Pool-Oriented Structure of Systems Analysis

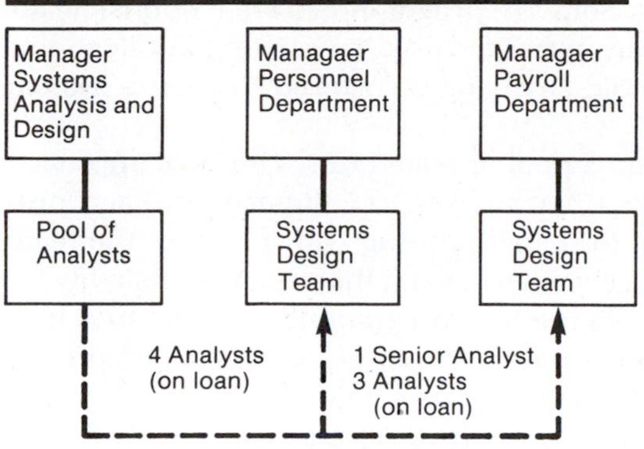

directly to the manager of the operating department where the application design is requested. This arrangement gives the department some control over its own application. The outcome is greater user participation and support in system development.

The *functional* structure of analysis assigns a group of analysts to serve a specific system. For example, there may be a personnel systems team, a production systems team, and a marketing systems team. Each team has a manager who reports directly to the director of system development (see Figure 3–5). This arrangement is suitable for a large computer facility.

FIGURE 3–5 A Functional Structure of Systems Analysis

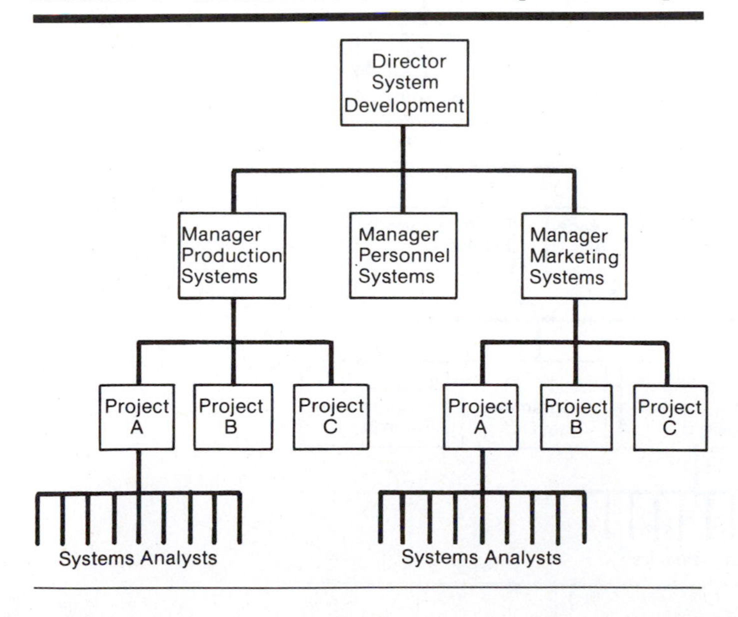

3. *Progamming* is structured around three areas: applications, software, and maintenance. In many applications, programmers are placed on the same project team with analysts. Software programmers are responsible for modifying compilers and software packages used by existing applications. Maintenance programmers handle all changes required to keep a system operating.

Programming is organized on a pool or team basis. The pool approach clusters programmers into a pool that reports to a supervisor. Each programmer is assigned a program to prepare through all stages. In the team approach, each programmer is assigned to a team that has responsibility for a specific project. As shown in Figure 3–6, programmers report to a lead programmers (team leader), who reports to the project leader. The project leader is usually a section head who supervises a number of teams in the section and reports to the manager of the programming area.

4. *Operations* handles job scheduling and supportive services such as supplies inventory, programming, and data library. These functions are coordinated with systems design and programming under a master plan (see Figure 3–7).

FIGURE 3–6 A Team-Oriented Structure of Programmers

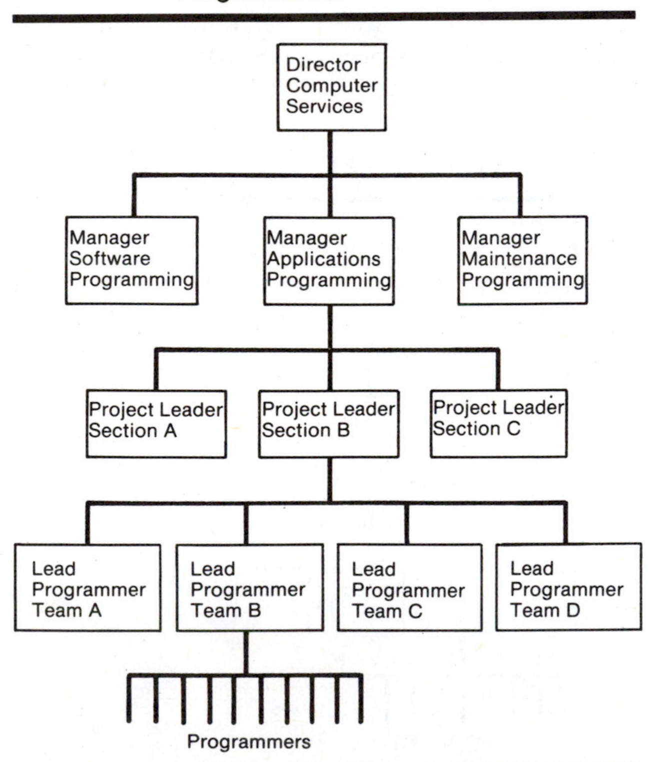

FIGURE 3-7 General Structure of Computer Operations

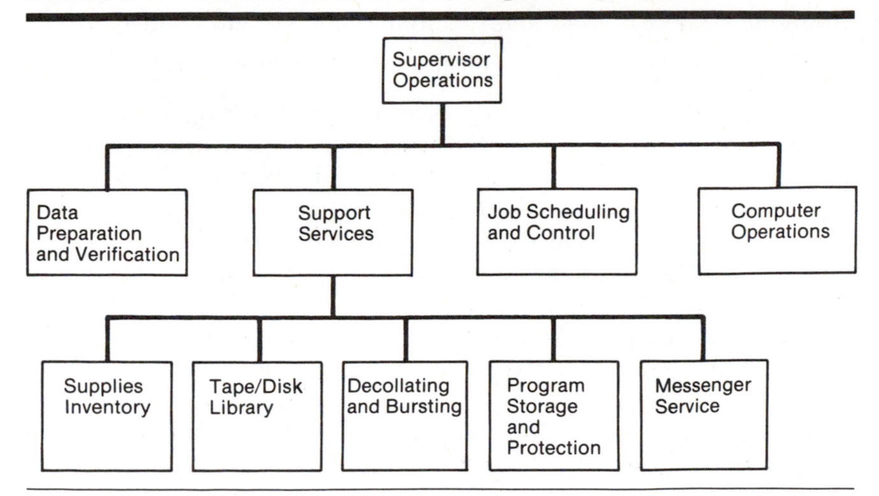

Functions of Key System Personnel

Manager—MIS Services. The MIS manager is responsible for planning, organizing, coordinating, and directing the activities for the entire division. Typically, managers have good technical and managerial skills. The job requirements depend on the size of the firm and the sophistication of the MIS facility. Typically, a college degree or even a master's degree is required. A degree in business administration with emphasis on information systems would be suitable for work in a commercial installation.

In addition to managing analysts, programmers, and other specialists, the MIS manager now has to work with executives at all levels. In essence, the manager has become an executive who makes decisions that can affect the functioning of the entire organization. The key qualities are:

1. Skill in planning, organizing, and controlling the work of the division.
2. Ability to deal logically with difficult problems and cope with new situations.
3. Technical knowledge of hardware, software packages, and networking.
4. Ability to relate to others.
5. Broad knowledge of the business of the employing organization.

Manager—Systems Department. A second-level managerial position in the MIS division is the systems supervisor. The primary functions are:

1. Preparing long-range plans for system projects.
2. Authorizing system projects.
3. Organizing and staffing project teams.

4. Preparing and maintaining system procedures.
5. Conducting system surveys and recommending system changes.
6. Establishing standards and specifications for proposed hardware.

Systems management requires a project leader. The project team includes analysts, programmers, a user representative, and an outside consultant. The project leader reports to the manager of the systems department. Most design development work is pursued in a matrix organization with user representatives, systems analysts, programmers, and others reporting to their administrative leader as well as the team leader for the duration of the job.

Project management deals with the system development life cycle, including feasibility studies, design, system specifications, programming, testing, and implementation. It is a detailed procedure for evaluating project resources and managing allocated costs. The master plan includes a description of the various tasks, the estimated project start-up and termination dates, and the time, cost, and work force required for each aspect of the project. A detailed discussion of project management is in Chapter 15.

From these activities, it is obvious that the project leader's job is unique. The leader must work with analysts and users. To do the job well, a technical and administrative background is important. The primary attributes are:

1. Ability to design systems and write procedures.
2. Ability to sell new ideas or new equipment to users.
3. Imagination in exploring new and better ways of designing systems.
4. Integrity in dealing with vendors and safeguarding company records.
5. Perceptiveness in understanding subordinates and how they react to a work situation.

The Systems Analyst. A key member of the project team is the systems analyst. This job carries considerable responsibility, high status, and attractive pay. As shown in Figure 3-8, analysts have the highest perceived status in an MIS facility.

RISING POSITIONS IN SYSTEM DEVELOPMENT

The Paraprofessional

The tasks that make up the system development process are changing. With an increase in the use of structured tools, there are emerging tasks that are less technical or creative than the traditional ones. Rather than the analyst ignoring them or tying up valuable time, they are carried out by less experienced paraprofessionals. These tasks are categorized as follows:

FIGURE 3-8 Relative Status and Job Levels in an MIS Facility

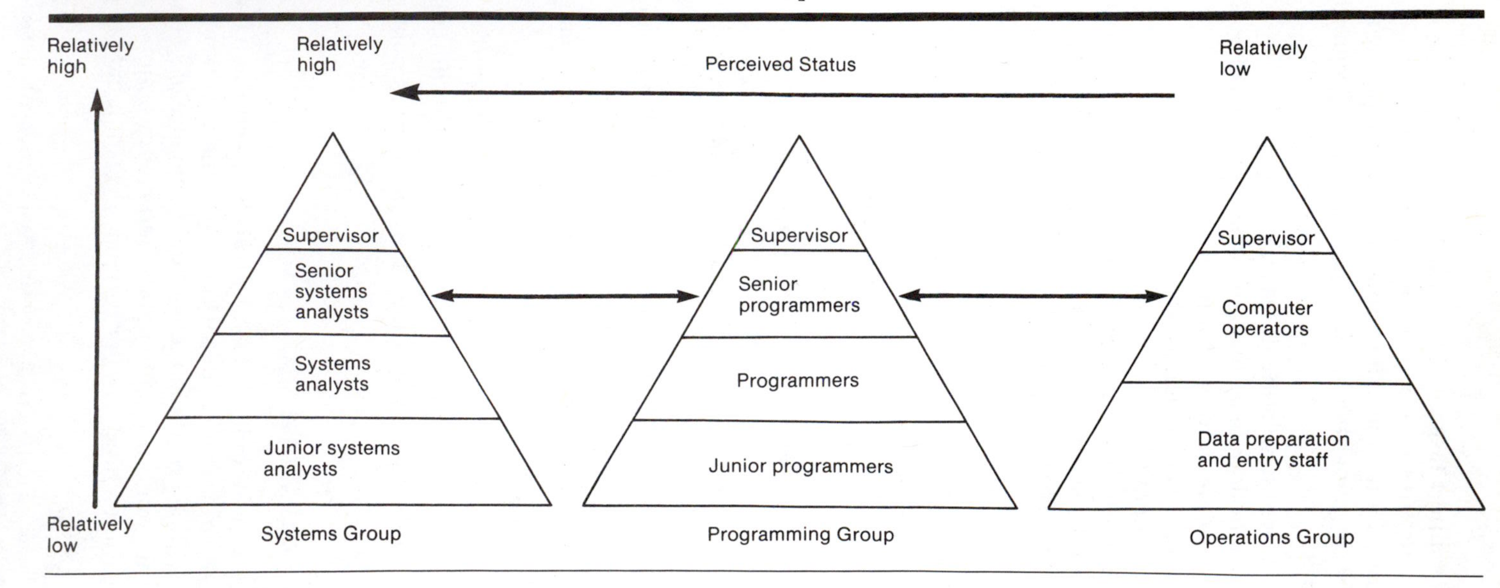

1. General support tasks.
   a. Maintain current documentation on a daily basis.
   b. Maintain a technical literature and information retrieval service.
   c. Assist in constructing and maintaining a program development library.
2. Specific system development tasks.
   a. Draft data flow diagrams and structure charts based on specifications from system development personnel.
   b. Maintain data dictionaries.
   c. Code programs.
   d. Prepare and maintain test data files, check test results, and modify programs based on a predetermined procedure.
   e. Prepare user manuals and other documentation.
   f. Prepare supportive training materials when needed.

A shortage of qualified analysts and their high salaries suggest the value of separating essential tasks and assigning some to trained personnel who have not reached the level of education, skill, experience, and pay of the seasoned analyst. According to Harris and Hoffman, paraprofessionals perform tasks of the following types:

1. The task is not significantly complex or does not require "high-level" ability, education, skill, or experience.
2. The task is repetitious and there is a relatively well-defined process for performing it.
3. The general support tasks require limited time for instruction relative to the time it takes to perform them.[28]

Some characteristics desirable in a paraprofessional are:

1. Communication skills.
2. Ability to think like the system person being supported.
3. Ability to work independently and perform with minimal supervision.
4. Creative problem-solving ability and an attitude conducive to trying things out.

The Technical Writer

Documentation is one of the best liaisons between the technical and user worlds. As the number of user-driven systems increases, their use and acceptance will be greatly influenced by the quality of system documentation. From the rough draft, omissions, and contradictions of the programmer's notes, the technical writer extracts a clean, lucid set of instructions

[28] James N. Harris and Thomas R. Hoffman, "Para-professionals in System Development," *Journal of Systems Management,* November 1983, p. 27.

for the candidate system. The writer seeks to uncover the hidden logic of the program and convert it into an operational pattern. This is where the technical writer makes major contributions.

Technical writers are usually brought in when the candidate system nears completion. Problems often result from the analyst's lack of interest in a system that is already history. When errors are uncovered in a system and reflect on the analyst, irritation turns into hostility. This makes the job of the technical writer a real challenge. There is evidence of the emergence of technical writing as a computer-based specialty in the proliferation of writing subspecialties and the consulting positions they have spawned. There are methods analysts who develop standards and procedures manuals, indexers, hardware, and operations writers. Experienced writers in high-tech areas such as expert systems earn more than $75 per hour.[29]

CONCLUSIONS

In this chapter, an attempt was made to discuss the role of the analyst and the importance of the analyst/user interface for successful systems. Although the interface may be vulnerable because of advancing technology, the basic foundations of human interaction and organization remain important for system development. In a speech before the National Association of Secondary Principals, Birnbaum pointed out that "man-made complications that surround the creation of systems are the principal barriers to our success today."[30] This illustrates the need for maintaining a viable interface throughout the system life cycle.

Communication between the user and the analyst is probably the most important aspect of the interface. Analysts need to explain and users need to understand the framework surrounding the candidate system before there is total acceptance. George Bundy, former Kennedy administration aide, once explained the relationship between the White House (analysts) and the people of the United States (users) as follows:

> I think we're a little like the Harlem Globetrotters in the performance they put on before the game; they pass the ball under their legs and under their arms and throw it down the field (court). It is all very dazzling, but at some point, the whistle blows, the game begins, and then what matters is whether you make baskets.[31]

In other words, the analyst cannot treat users like an audience waiting to be impressed. The user has to get involved. User participation is important.

---

[29] Gene Knauer, "The Rise of the Technical Writer," *Computerworld*, January 9, 1984, p. 1ff.

[30] Joel S. Birnbaum, "Microelectronics and Education," *What the Next Decade Will Bring* Moscone Center, San Francisco, March 19, 1983.

[31] Edward A. Tomeski and Harold Lazarus, *People-Oriented Computer Systems* (New York: Van Nostrand Reinhold, 1975), p. 238.

Summary

1. A systems analyst is a person who conducts a study, identifies activities and objectives, and determines a procedure to achieve the objectives. Systems analysis has a history dating back to Taylor. Early analysts worked in factories, specializing in improving work methods and setting time standards for production. With the advent of the computer, the analyst assumed the role of a problem solver and a specialist in developing computer applications.

2. Success in systems analysis requires interpersonal and technical skills. Interpersonal skills emphasize communication and interface with the user, whereas technical skills include creativity, problem solving, and managing the overall project. During analysis, there is greater need for interpersonal skills, but during design there is greater emphasis on technical skills. During implementation, both skills are needed.

3. A career in systems analysis requires academic preparation, experience, and good interpersonal relations. The person must be familiar with the inner-workings of business and competent in system tools and methodologies. The personal qualities include creativity and communications skills and being systematic and sensitive.

4. Analysts perform a multitude of roles—as change agent, investigator, architect, psychologist, salesperson, and motivator. They also need to understand politics.

5. An important aspect of system development is the viability of the relationship between the analyst and the user. Research indicates that the probability of system failure increases without a good interface.

6. Three behavioral issues are related to system success: establishing user motivation, narrowing analyst/user differences, and neutralizing political factors. Related to system success also is the extent of the commitment to avoid future conflict. Therefore, coordination of people's ideas for system design is important for system development.

7. The analyst is the key member of the MIS organization. An MIS facility consists of four major areas:

   a. *Administration,* which handles user liaison, long-range planning, budget planning and control, and personnel.

   b. *Systems analysis and design,* which is organized as project oriented, pool oriented, or functional.

   c. *Programming,* which is structured around applications, software, and maintenance. It is organized on a pool or a team basis.

   d. *Operations,* which deals with job scheduling, supplies inventory, and the data library.

8. Of all the managerial and administration positions in the MIS area, the position of the analyst is perhaps the most crucial. The job carries considerable responsibility, high status, and attractive pay.

9. The tasks that make up system development are changing. To perform them well, assistance is becoming available from paraprofessionals. Although less experienced than the systems analyst, they perform general support tasks and specific system development tasks such as drafting data flow diagrams and maintaining data dictionaries. Care must be taken to determine what tasks should be assigned to the paraprofessional.
10. Another new position in systems work is the technical writer, who is brought in when the candidate system is near completion. The main contribution is quality documentation.

Key Words

Conflict Resolution
Expectancy Theory
Functional Structure
Paraprofessional
Pool-Oriented Structure
Project-Oriented Structure
Systems Analyst
Team-Oriented Structure
Technical Writer

Review Questions

1. Trace the history of systems analysis and compare the major changes or differences in today's role of the analyst.
2. What were the main contributions of Taylor, Maslow, and McGregor to systems analysis?
3. Based on the material in the chapter, what does it take to do systems analysis? Do you agree? Explain.
4. Elaborate on the technical and interpersonal skills required of systems analysts. When is one skill favored over the other? Why?
5. What academic qualifications are important for systems work? What about the personal attributes? Explain.
6. The chapter approaches systems analysis as a multifaceted role. Explain and illustrate situations where the roles might be best applied in the system development life cycle.
7. The political factor has been brought up in the literature on several occasions. In what respect should the analyst be a politician? What would be an example where political considerations are used in systems work?
8. What is meant by the analyst/user interface? Why is it a problem?
9. Discuss the behavioral issues involved in understanding the analyst/user interface.
10. When we discuss the analyst/user interface, does it imply differences? If so, how pronounced are the differences? How are they resolved? Discuss.

11. Explain the makeup of and the activities undertaken by the MIS organization. Where does the analyst fit in?
12. Distinguish between the following:
    a. Project-oriented and pool-oriented arrangements.
    b. Pool approach and team approach in programming.
    c. The jobs of the manager (systems department) and the systems analyst.
13. Visit the MIS department of a local firm and ask the manager his/her opinion of the role of the paraprofessional and the technical writer in system development.

Application Problems

1 The systems department of a large, Chicago-based insurance company has an opening for a junior systems analyst. The job requires a college degree in management information systems, business administration, or computer science. Eighty percent of the work is assisting a senior systems analyst in data collection, documentation, and some cost analysis. The remaining 20 percent involves programming and program maintenance, using Cobol. The firm has an ongoing training program at all levels.

The personnel department placed an ad in the Sunday paper, and after screening, selected two candidates for the position. As the systems manager, you have their resumes and have had a chance to interview both applicants. A summary of the resumes is as follows.

Name: James E. Hart, Jr.

Age: 21—single

Education: B.S. in computer science. GPA = 3.64 out of 4.00.

Experience: Two years (part-time) as Fortran programmer in a public utility; one year (part-time) as RPG programmer in the school's computer center.

Test scores: Verbal = 61
Math = 92

Interview results: Left a good impression regarding technical ability, very personable, but lacked knowledge of business world—how organizations are managed, etc.

Name: Jane Foresight

Age: 28—divorced, two children (4 and 6)

Education: Two years (junior college). Major in data processing. Finishing college in four months. Major in accounting; minor in business administration. GPA = 3.0 out of 4.0.

Experience: One year (full-time) with a local accounting firm as Cobol programmer; two years with a retail chain as computer operator; 1.5 years as a keypunch operator at a bank.

Test scores: Verbal = 87
Math = 75

Interview results: Personable and assertive, demonstrated proficiency in Cobol programming, has been holding two jobs to support family, limited knowledge of business organizations.

Assignment

a. What additional information would you like to gather during a second interview?

b. Which applicant would you choose? Why? Justify your answer.

---

2 An outside systems analyst was asked by the bank's president to contact Mrs. Mandelbaum about some problem in the accounting department. During the first meeting, Mrs. Mandelbaum, the supervisor, said there have been problems reconstructing the Christmas Club account three out of five days each week. The tape shows figures that are either more or less than actual cash receipts. The head teller blames it on the old teller terminals, which are "down" most of the time. The accounting department counters by emphasizing that the teller department is accountable for the difference. Everything points to human error.

With this in mind, the analyst went straight to the teller in charge of the club account and pretty much accused her of stealing money. Following this meeting, he conducted interviews with the head teller and the reconciling clerk in charge of the club accounts. Late that day, he phoned the president and told him it would take a week to correct the problem.

The next day, the analyst studied the coupons and the procedure followed by the clerk. Although counting coupons was manual, there was nothing in particular to explain the difficulty in balancing the receipts.

During the next three days, the analyst observed the teller and how she handled the coupons. Everything seemed normal, although there were occasional errors in keying the amounts or code. The last day was Friday and the analyst had no answers. The president was disappointed with the lack of feedback. When confronted, the analyst admitted that he had little knowledge of accounting, auditing, or how accounts are reconciled, but if he had another week, he could learn enough to computerize the whole operation.

Toward the end of the day, a customer's passbook was posted with the wrong amount. Upon closer inspection, it was found that the teller machine always printed 9 instead of 0. When the machine was replaced, the Christmas Club account balanced with no difficulty.

Assignment

a. What mistakes did the analyst make in handling the project? Did he use a correct procedure? Explain.

b. What type of personality would it take to deal with a problem like the one just described? Why?

c. What qualifications do you suggest the analyst possess to handle this case? Why? Explain.

d. In deciding on an analyst, what criteria do you suggest the bank consider for future invitations? Elaborate.

Selected References

Anderson, Wm. S. "The Expectation Gap." *Journal of Systems Management,* June 1978, pp. 6–10.

Awad, E. M. "Vocational Needs, Reinforcers, and Job Satisfaction of Analysts and Programmers in a Banking Environment." *Tenth Australian Computer Conference,* Melbourne, Australia, September 26, 1983.

Benbasat, I., and R. Taylor IV. "The Impact of Cognitive Styles on Information System Design." *MIS Quarterly,* June 1978, pp. 143–48.

Birnbaum, Joel S. "Microelectronics and Education: What the Next Decade Will Bring" Moscone Center, San Francisco, March 19, 1983.

Bostrom, R. P., and J. S. Heiner. "MIS Problems and Failures. A Sociotechnical Perspective, Part I: The Causes." *MIS Quarterly* 1, no. 3 (September 1977), pp. 17–32.

________. "Part II: The Application of Socio-technical Theory." *MIS Quarterly* 1, no. 4 (December 1977), pp. 11–28.

Bramson, Albert M., and Allen F. Harrison. "The Orderly Ways of Analysts." *Computer Decisions,* November 1983, p. 112ff.

Campbell, Robert Braun. "Analyzing Systems Analysts." *Journal of Systems Management,* March 1983, pp. 22–23.

Canavan, Edward M. "The Mysterious Systems Analyst." *Journal of Systems Management,* May 1980, pp. 34–37.

Cochran, Terry L. "Know Thy User." *Datamation,* January 1979, pp. 46–49.

Dimino, Stephen A. "Return to Basics." *Journal of Systems Management,* December 1982, pp. 28–31.

Faeber, Larry G., and Richard L. Ratcliff. "People Problems Behind MIS Failures." *Financial Executive,* April 1980, pp. 18–24.

Feeney, Wm., and Frea Sladek. "The Systems Analyst As a Change Agent." *Datamation,* November 1977, pp. 85–88.

Frankel, Mark S. "In Search of Professionalism." *Data Management,* August 1982, pp. 21–27.

Gore, Marvin, and John Stubbe. *Elements of Systems Analysis.* 3d ed. Dubuque, Iowa: Wm. C. Brown, 1983, pp. 30–48.

Harris, James N., and Thomas R. Hoffman. "Paraprofessionals in System Development." *Journal of Systems Management,* November 1983, pp. 25–29.

Hunt, Alfred. *The Management Consultant.* New York: John Wiley & Sons, 1977, pp. 13–14.

Knauer, Gene. "The Rise of the Technical Writer." *Computerworld (In-Depth),* January 9, 1984, p. 1Dff.

Kneitel, Arnold M. "Futility of User Involvement." *Infosystems,* November 1978, pp. 92–96.

Lasdin, Martin. "Games Played Between Users and Providers." *Computer Decisions,* October 1980, pp. 72–73.

Lucas, Henry C. Jr. *The Analysis, Design, and Implementation of Information Systems.* New York: McGraw-Hill, 1981.

Matthies, Leslie. "Time: The Analyst's Basic Resource." *Journal of Systems Management,* November 1973, pp. 18–19.

McGhee, Pamela. "What Do Users Want?" *Computerworld,* March 23, 1981, p. 84.

McGregor, D. *The Human Side of Enterprise.* New York: McGraw-Hill, 1960.

McLamore, Hillard. "How to Avoid Hassles and Headaches When Systems People and Users Meet." *SAM Advanced Management Journal,* Summer 1979, pp. 5–12.

Miller, M. S. "Problem Avoidance in the User/Analyst Relationship." *Journal of Systems Management,* May 1981, pp. 34–39.

Mollen, Dave. "Narrowing the Gap." *Datamation,* May 1980, pp. 195–198.

Moran, Thomas P. "An Applied Psychology of the User." *Computing Surveys* 13, no. 1 (March 1981), pp. 1–11.

Moriarty, Robert F., and Joseph C. Yeager. "Human Resistance a Priority in Overcoming Barriers to Automation." *Information & Records Management,* August 1983, pp. 18–19ff.

Moynihan, John A. "What Users Want." *Datamation,* March 1982, pp. 116–118.

Munro, M. C., and G. B. Davis. "Determining Management Information Needs—A Comparison of Methods." *MIS Quarterly* 1, no. 2 (June 1977), pp. 55–67.

Nicholos, John M. "Transactional Analysis for Systems Professionals." *Journal of Systems Management,* October 1978, pp. 6–11.

Ramsgard, Wm. C. "The Systems Analyst—Doctor of Business." *Journal of Systems Management,* July 1974, pp. 10–14.

Robey, Daniel. "Perspectives of the User Interface." In *Proceedings of the 17th Annual Conference of the Special Interest Group of Computer Personnel Research (SIGCPR),* June 1980, p. 23.

__________. "User Attitudes and Management Information Systems." *Academy of Management,* September 1979, pp. 527–38.

Scharer, Laura. "Systems Analysts Performance: Criteria and Priorities." *Journal of Systems Management,* February 1982, pp. 10–15.

Schein, E. *Organizational Psychology.* 2d ed. Englewood Cliffs, N.J.: Prentice-Hall, 1970.

Shapin, Paul G. "System Selection Needs Users Involvement for Success." *Infosystems,* September 1980, p. 104ff.

Synnot, Wm. R., and **Wm. H. Gruber.** "The Care and Feeding of Users." *Datamation,* March 1982, pp. 191–204.

Taggart, Wm. "Human Information Processing Styles and the Information System Architect in the PSC Systemeering Model." In *Proceedings of the 17th Annual Conference of the Special Interest Group of Computer Personnel Research (SIGCPR),* June 1980, pp. 62–76.

Tomeski, Edward A., and **Harold Lazarus.** *People-Oriented Computer Systems.* New York: Van Nostrand Reinhold, 1975, pp. 238–45.

"Too Many Professionals?" Editorial, *Newsweek on Campus,* University of Virginia, March 1984, pp. 8ff.

Vitalari, N. P. "An Investigation of the Problem Solving Behavior of Systems Analysts." Unpublished Ph.D. dissertation, University of Minnesota, 1981.

Ware, Robb. "From Technician to Problem-solver: Training the Systems Analyst." *Data Management,* March 1983, pp. 20–21.

Weinberg, Victor. "Structured Analysis and the Analyst/User Relationship." *Infosystems* 24 (1977), pp. 58–70.

Welke, Richard J. "User-Oriented Approaches to MIS." *Canada Magazine,* August 1979, pp. 62–68.

Zmud, Robert W. "The Role of Individual Differences in MIS Departments." Unpublished paper, Georgia State University, 1980.

Part Two

Systems Analysis

Chapter 4

Systems Planning and the Initial Investigation

INTRODUCTION

It is always wise to look ahead, but it is difficult to look further than you can see.

Winston Churchill's remark points out that planning future endeavors, though difficult, is important in managing operations. System development is no exception. Identifying the need for a new information system and launching an investigation and a feasibility study must be based on an MIS master plan that has management support. Planning cuts across all phases of the system life cycle. It is the first step in developing and managing systems. Understanding MIS planning functions and their relationships to systems analysis and design, then, is crucial to successful computer installations. This chapter reviews the functions and levels of planning and discusses the steps in the initial investigation. The role and importance of project management are emphasized later in the chapter.

BASES FOR PLANNING IN SYSTEMS ANALYSIS

Planning information systems in business has become increasingly important during the past decade. First, information is now recognized as a vital resource and must be managed. It is equal in importance to cash, physical facilities, and personnel. Second, more and more financial resources are committed to information systems. As computer systems are becoming integral to business operations, top management is paying more attention to their development. Third, there is a growing need for formal long-range planning with information systems that are complex, require months or years to build, use common data bases, or have a greater competitive edge.[1] The objectives are to map out the development of major systems and reduce the number of small, isolated systems to be developed and maintained. Proper planning for information systems ensures that the role played by the system will be congruent with that of the organization.

Dimensions of Planning

The following conditions dictate today's business strategies:

1. High interest rates make it more important that business realizes a good return on investment.
2. Inflation puts pressure on profit when it occurs.
3. The growing trend toward guaranteed employment suggests that costs are becoming fixed and the commitment to business expansion may not be easily changed.
4. Resource shortages impede expansion.

---

[1] Ephraim R. McLean and John V. Soden, *Strategic Planning for MIS* (New York: John Wiley & Sons, 1977), p. 6.

At a Glance

The most critical phase of managing system projects is planning. To launch a system investigation, we need a master plan detailing the steps to be taken, the people to be questioned, and the outcome expected. The initial investigation has the objective of determining whether the user's request has potential merit. The major steps are defining user requirements, studying the present system to verify the problem, and defining the performance expected by the candidate system to meet user requirements. When the initial investigation is completed, the user receives a system proposal summarizing the findings and recommendations of the analyst.

By the end of this chapter, you should know:

1. Why planning is important in systems analysis.
2. What planning dimensions govern information system development.
3. How to determine the user's information requirements.
4. How prototyping is used in determining information requirements.
5. What factors determine the need for a feasibility study.

BACKGROUND ANALYSIS

FACT-FINDING

Review of Written Documents
On-Site Observations
Interviews and Questionnaires

FACT ANALYSIS

DETERMINATION OF FEASIBILITY

FIGURE 4–1 Technology Cost/Benefit Curve

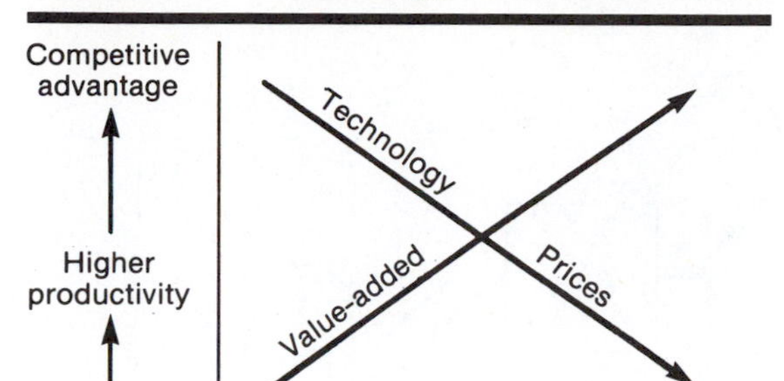

5. Regulatory constraints slow entry into the market.
6. Increased productivity paves the way for expansion.

Information systems embedded in an organization provide users with the opportunity to *add value* to products and business operations at lower costs. Therefore, they must be carefully planned[2] (see Figure 4–1).

Strategic MIS Planning

Planning for information system development must be done within the framework of the organization's overall MIS plan. It may be viewed from two dimensions: (1) The *time horizon* dimension specifies whether it is short range (usually less than two years), which is tantamount to the MIS yearly plan, medium term (two to five years), or long range (more than five years). (2) The *focus* dimension tells whether the primary concern is strategic, managerial, or operational. Strategic (MIS) planning is an orderly approach that determines the basic objectives for the user to achieve, the strategies and policies needed to achieve the objectives, and the tactical plans to implement the strategies.[3] The first task in strategic planning is to set the MIS objectives and the results expected. Consideration of these objectives must deal with their fit with the organization's strategic plan, the types of systems and services to be offered, the role of users in system development, and the technology to be used. Once the MIS objectives are set, MIS policies are defined as a guideline to be used in carrying out strategy. MIS policies, in turn, are translated into long-range (conceptual), medium-range (managerial), and short-range (operational) plans for implementation (see Figure 4–2).

---

[2] Harvey Poppel, *Strategic Impact of Information Technology* (New York: Deltak, Inc., 1982), pp. 5, 9.

[3] George A. Steiner, "Formal Strategic Planning in the United States Today," *Long Range Planning* 16, no. 3 (1983), pp. 13–17.

FIGURE 4–2 MIS Strategic Planning—Conceptual Model

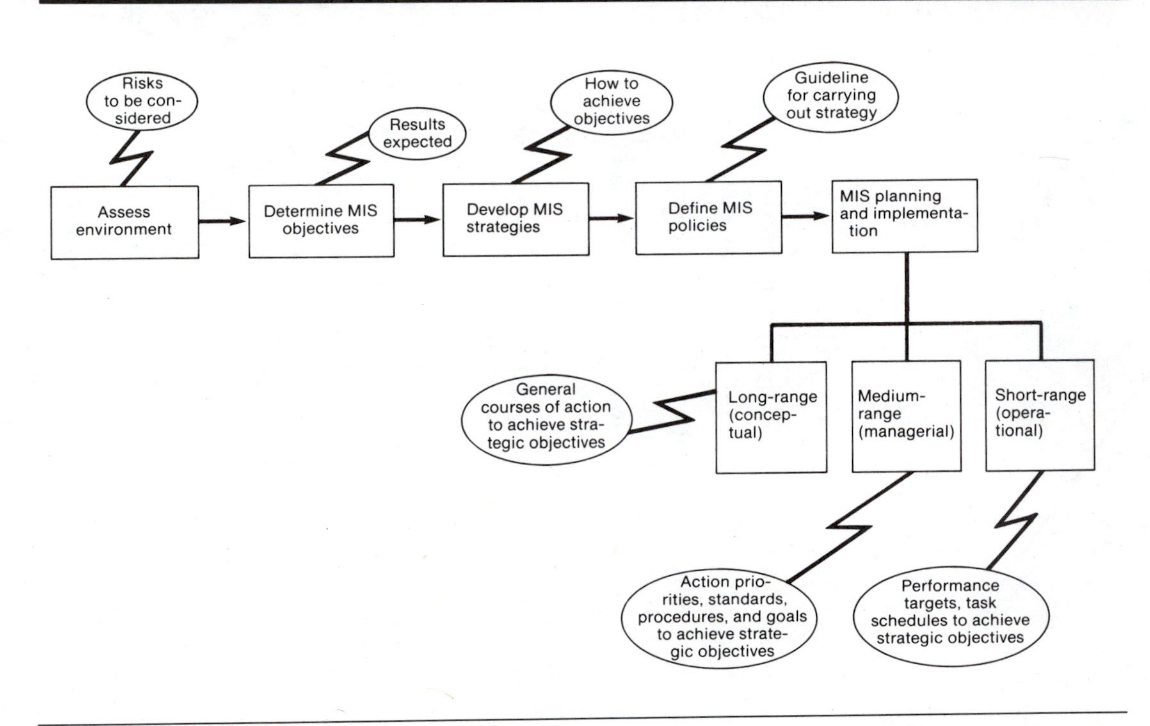

Source: Adapted from Ephraim R. McLean and John V. Sodon, *Strategic Planning for MIS* (New York: John Wiley & Sons, 1977), pp. 24–25.

In determining the MIS strategic plan, several questions need to be asked at the start:

1. What MIS objectives and strategies can be derived from the corporate strategic plan?
2. Who will review and approve the plan?
3. How long is the planning horizon? What will the plan contain?
4. What will the plan focus on (e.g., new technology, computer security, new application development)?

In most cases, the answers depend on the structure and complexity of the MIS organization, the level of computerization in the firm, the "hit rate" of the MIS division, and the influence of MIS in getting projects approved by top management.

Managerial and Operational MIS Planning

Managerial MIS planning integrates strategic with operational plans. It is a process in which specific functional plans are related to a specific number of years to show how strategies are to be carried out to achieve long-range plans. The next step is to devise short-range plans that spell out the day-to-day activities of the system. They are programmed plans requiring a year's commitment. For example, the operating expense budget, the human resource budget of each computer application, and timetables for implementing a new system are all short-range plans designed to implement the organization's master plan by computerizing the labor-intensive areas of the business.

The MIS operating plan requires the heaviest user involvement to define fully the system's requirements. System development must support organizational MIS objectives as laid out in the corporate plan and identify and select applications that are the organization's priorities. This important link has been described by Bowman, Davis, and Wetherbe in a three-stage model:[4]

1. Strategic system planning—establishing relationships between the organization plan and the plan for a candidate system.
2. Information requirements analysis—identifying organization requirements to direct the specific application of system development projects.
3. Resource allocation—determining hardware, software, telecommunication, facilities, personnel, and financial resources to execute the development of the system.

Planning for system development activities is a major aspect of the overall planning job in terms of its importance and commitment of planning resources. Broad corporate strategic objectives should be the basis for

[4] Brent Bowman, Gordon Davis, and James Wetherbe, "Modeling for MIS," *Datamation*, July 1981, pp. 155–64.

FIGURE 4–3 A Top-Down Approach to System Planning

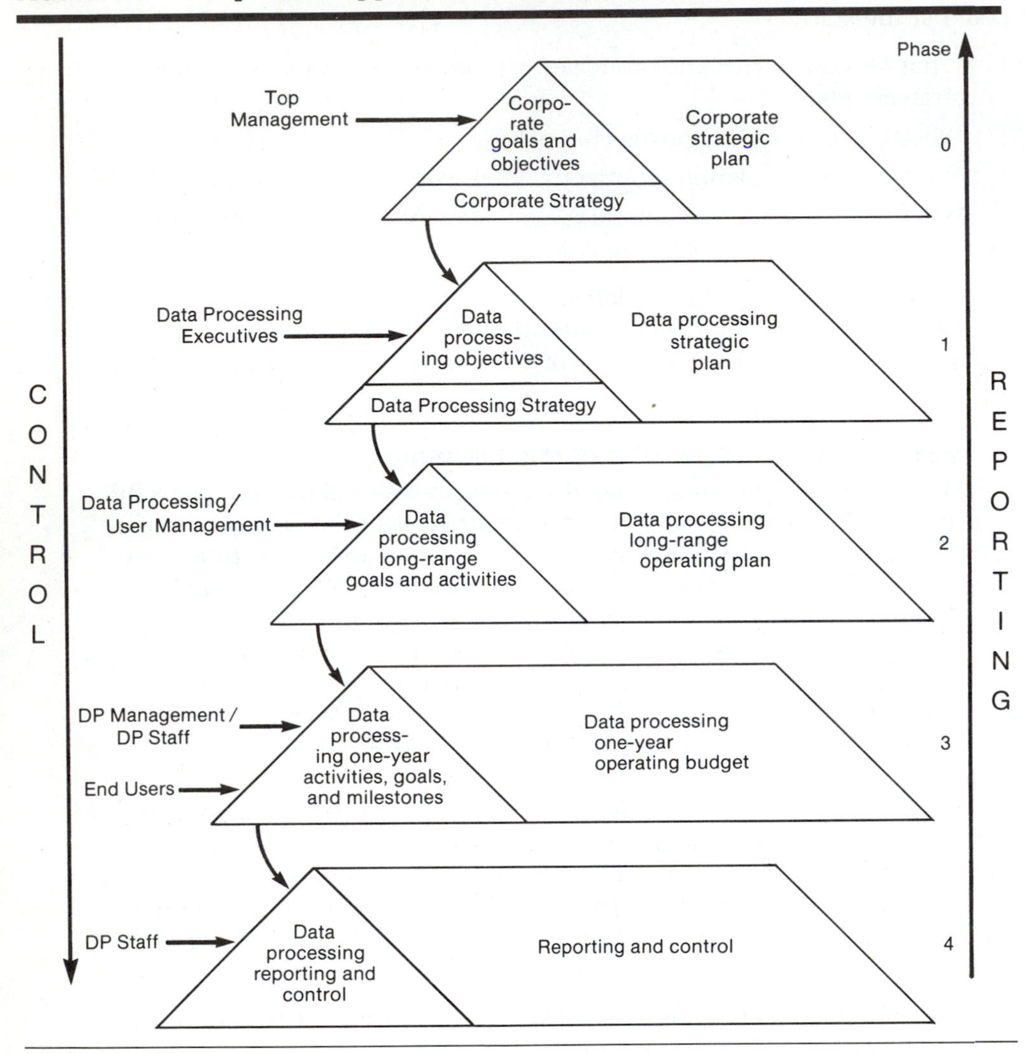

system development objectives, which dictate operating goals in the form of specific action plans. Formalizing the planning process makes it easier to reorient and gain the support of upper, middle, and operating management for candidate systems. Figure 4–3 illustrates a top-down approach to planning and the relationship between the corporate strategic plan and the goals and activities of the system development function.[5]

[5] Robert E. Lief, Robert D. Dodge, and Ralph L. Ogden, "Adapting DP Strategy to Management Style," *Computerworld (In Depth)*, December 5, 1983, pp. 25–32.

INITIAL INVESTIGATION

As mentioned in Chapter 2, the first step in the system development life cycle is the identification of a need. This is a user's request to change, improve, or enhance an existing system. Because there is likely to be a stream of such requests, standard procedures must be established to deal with them. The *initial investigation* is one way of handling this. The objective is to detemine whether the request is valid and feasible before a recommendation is reached to do nothing, improve or modify the existing system, or build a new one.

The user's request form (Figure 4–4) specifies the following:

1. User-assigned title of work requested.
2. Nature of work requested (problem definition).
3. Date request was submitted.
4. Date job should be completed.
5. Job objective(s)—purpose of job requested.
6. Expected benefits to be derived from proposed change.
7. Input/output description—quantity (number of copies or pages) and frequency (daily, weekly, etc.) of inputs and outputs of proposed change.
8. Requester's signature, title, department, and phone number.
9. Signature, title, department, and phone number of person approving the request.

The user request identifies the need for change and authorizes the initial investigation. It may undergo several modifications before it becomes a written commitment. Once the request is approved, the following activities are carried out: background investigation, fact-finding and analysis, and presentation of results—called project proposal. The proposal, when approved, initiates a detailed user-oriented specification of system performance and analysis of the feasibility of the candidate system. A feasibility study focuses on identifying and evaluating alternative candidate systems with a recommendation of the best system for the job. This chapter deals with the initial investigation. Chapter 7 discusses the feasibility study.

Needs Identification

The success of a system depends largely on how accurately a problem is defined, thoroughly investigated, and properly carried out through the choice of solution. User need identification and analysis are concerned with what the user *needs* rather than what he/she *wants*. Not until the problem has been identified, defined, and evaluated should the analyst think about solutions and whether the problem is worth solving. This step is intended to help the user and the analyst understand the real problem rather than its symptoms.

FIGURE 4-4 User's Request Form—An Example

REQUEST FORM
INFORMATION SERVICE

Job Title:	Nature of job	Request date	To be completed no later than:
Investigation of an on-line safe deposit system	X NEW __ REVISION	*12/08/84* dy mo yr.	*01 /03 /85* dy mo yr.

Job Objective(s): *To improve customer service, reduce paperwork, better billing system, and possible reduction in staffing requirements.*

Expected Benefits:
- *shorten or eliminate customer lines in lobby*
- *provide quick up-to-date information regarding box availability*
- *more accurate billing procedure*

Output Specifications	Input Specifications
Report title *customer monthly statement*	Document Title *billing notices*
Quantity *2-5* No. of pages/report *4* No. of copies/report	Quantity *100-400*
Frequency *X* daily __ weekly	Frequency *X* daily __ weekly
Remarks *letter quality printing*	Remarks *number of notices varies with season. in January, range may be 200-1,000.*
Report title ____	Document Title ____
Quantity __ No. of pages/report __ No. of copies/report	Quantity __
Frequency __ daily __ weekly __ other	Frequency __ daily __ weekly __ other
Remarks ____	Remarks ____

REQUESTER—Please Fill Out

Name of Requester (please print)	Signature:	Title:	Dept.:	Phone:
Barbara Betolatti	*[signature]*	*supervisor*	*safe deposit*	*X 5511*
Approved by (please print)	**Signature:**	**Title:**	**Dept./Div.**	**Phone:**
Curtiss Sibley	*[signature]*	*sr. vice pres.*	*operations*	*X4324*

FOR MIS DEPT. ONLY

Job No. *124*	Status: *X* accept __ return with comments __ reject		
Acceptance authorized by: *[signature]* B. Solen	Title: *Mgr./systems*	Phone: *X4430*	Remarks: *assign Bob Hays to job. Check to make sure boxes in new wing are included in study.*

Form #1A-14 Rev. 8/85

The user or the analyst may identify the need for a candidate system or for enhancements in the existing system. For example, the cashier (chief operations officer) of a bank may become concerned about the long customer lines in the lobby or about the number of tellers who are "over" or "short" when they balance their cash. Similarly, an analyst who is familiar with the operation may point out a bottleneck and suggest improvements.

TABLE 4-1 Objectives of Decision Situations

Objective Question	Typical Objective	Real Objective
What is the objective of a city library?	To impart wisdom of the ages to the uneducated	Majority of books checked out are for enjoyment rather than fact-finding
Why do most people buy furniture?	To furnish new homes, apartments; to sit on, eat on, etc.	Marketing informs us that people discard old furniture not because it is nonfunctional but because it does not look good anymore
What is the objective of a two-car garage?	To store two automobiles and protect them from the elements	Families desire to park one car only and save room to store more important equipment (or junk)
What is the objective of an inventory control system?	To ensure prompt availability of needed material	Isn't locating a key factor? Aren't costs a primary consideration?

Source: Adapted from Jack Caldwell, "The Misunderstanding of Objectives," *Journal of Systems Management,* June 1982, p. 30.

Often problems come into focus after a joint meeting between the user and the analyst. In either case, the user initiates an investigation by filling out a request form for information. The request provides for statements of objectives and expected benefits.

The objectives of the problem situation must be understood within the framework of the organization's MIS objective, as discussed earlier under system planning. If objectives are misunderstood, it is easy to solve the wrong problem. In Table 4-1, Caldwell cites examples of "typical" versus "real" objectives of decision situations. It illustrates that the successful design of a system requires a clear knowledge of what the system is intended to do.

Determining the User's Information Requirements

Shared, complete, and accurate information requirements are essential in building computer-based information systems. Unfortunately, determining the information each user needs is a particularly difficult task. In fact, it is recognized as one of the most difficult tasks in system development.[6] The Association for Computing Machinery (ACM) Curriculum Committee on Computing Education for Management recognized this by suggesting two distinct job titles for systems development: "information analyst" and "systems designer" rather than the more general term "systems analyst."[7] The

[6] George Pitzgorsky, "Analyzing, Defining Systems Needs," *MIS Week,* August 24, 1983, p. 30.

[7] J. D. Couger, "Comparative Analysis of Information Systems Curricula," *Computer Newsletter for Schools of Business,* vol. XVII, No. 2, (October 1983), p. 1.

information analyst determines the needs of the user and the information flow that will satisfy those needs. The usual approach is to ask the user what information is currently available and what other information is required. Interaction between the analyst and the user usually leads to an agreement about what information will be provided by the candidate system.

There are several reasons why it is difficult to determine user requirements:[8]

1. System requirements change and user requirements must be modified to account for these changes.
2. The articulation of requirements is difficult, except for experienced users. Functions and processes are not easily described.
3. Heavy user involvement and motivation are difficult. Reinforcement for their work is usually not realized until the implementation phase—too long to wait.
4. The pattern of interaction between users and analysts in designing information requirements is complex.

Users and analysts traditionally do not share a common orientation toward problem definition. For example, in the analyst's view, the problem definition must be translatable into a system design expressed quantitatively in terms of outputs, inputs, processes, and data structures. This is an ideal way to develop a good system when all features are known, under the best of situations, and within time constraints. In contrast, the user seems to be satisfied with a qualitative definition that specifies the system in generalities. Flexibility is a key consideration. System specifications must change with their needs, as must the system after implementation.

Based on these contrasting views, users who try to define their information requirements with the analyst's views find themselves in a predicament. According to Scharer, they defend themselves by producing strategies that will satisfy the analyst.[9]

1. In the *kitchen sink* strategy the user throws everything into the requirement definition—overstatement of needs such as an overabundance of reports, exception processing, and the like. This approach usually reflects the user's lack of experience in the area.[10]

2. The *smoking* strategy sets up a smoke screen by requesting several system features when only one or two are needed. The extra requests are used as bargaining power. This strategy usually reflects the user's experi-

---

[8] Laura Scharer, "Pinpointing Requirements," *Datamation,* April 1981, pp. 139–40.

[9] Scharer, "Pinpointing Requirements," p. 140.

[10] See Wm. J. Doll and U. Mesbah Ahmed, "Managing User Expectations," *Journal of Systems Management,* June 1983, p. 6.

ence in knowing what he/she wants. Requests have to be reduced to one that is realistic, manageable, and achievable.

3. The *same thing* strategy indicates the user's laziness, lack of knowledge, or both. "Give me the same thing but in a better format through the computer" is a typical statement. Here the analyst has little chance of succeeding because only the user can fully discover the real needs and problems.

Humans have problems specifying information requirements. "Asking" the user what is needed of a candidate system does not often yield accurate and complete requirements. According to Davis, humans have these limitations:[11]

1. *Humans as information processors.* The human brain has both high-capacity, long-term memory and limited-capacity (but fast), short-term memory. The limits of short-term memory affect the information requirements obtained, because the user who is interviewed has a limited number of requirements that he/she defines as important. This limits processing responses. He/she may have selectively emphasized a few items of information and recorded them in long-term memory as being the most important. They may be the only ones that are recalled during the interveiw.

2. *Human bias in data selection and use.* Humans are generally biased in their selection and use of data. Their behavior becomes a representation of the bias. For example, users are influenced more by recent events than by past events. Thus, an information need that was discovered recently tends to carry greater weight than a need experienced in the distant past. This is called the *recency effect.* In another bias, users tend to use only information that is available in the form in which it is displayed. This means that the requirements provided by the user are biased by currently available information.

3. *Human problem-solving behavior.* Humans have a limited capacity for rational thinking. According to Simon, they must simplify it in order to deal with it. Coined as the concept of *bounded rationality,* it means that rationality for determining information requirements is "bounded" by a simplified model (as well as by limited training, prejudice, and attitude of user) that may not reflect the real situation.[12] Bounded rationality is often reflected in the behavior of systems analysts. A successful analyst uses a general model to search for information requirements. It includes the consideration of organizational and policy issues in arriving at realistic requirements. The poorly rated analyst does not consider these issues, but focuses on the immediate (short-term) requirements facing the user.

---

[11] Gordon B. Davis, "Strategies for Information Requirements Determination," *IBM Systems Journal* 21, no. 1 (1982), p. 5.

[12] A. Newell and H. A. Simon, *Human Problem Solving* (Englewood Cliffs, N.J.: Prentice-Hall, 1972).

Strategies for Determining Information Requirements

There are three key strategies or general approaches for eliciting information regarding the user's requirements: (1) asking, (2) getting information from the existing information system, and (3) prototyping.[13]

Asking. This strategy obtains information from users by simply asking them about the requirements. It assumes a stable system where users are well informed and can overcome biases in defining their problem. There are three key asking methods:

1. *Questions* may be open-ended or closed. An open-ended question allows the respondent to formulate a response. It is used when feelings or opinions are important. For example, "How do you evaluate the latest addition to your hardware?" In contrast, a closed question requests one answer from a specific set of responses. It is used when factual responses are known. For example, "How long have you been manager of the computer center?" More on question construction is covered in Chapter 5.

2. *Brainstorming* is a technique used for generating new ideas and obtaining general information requirements.[14] This method is appropriate for eliciting nonconventional solutions to problems. A guided approach to brainstorming asks each participant to define ideal solutions and then select the best feasible one. It works well for users who have system knowledge but have difficulty accepting new ideas.[15]

3. *Group consensus* asks participants for their expectations regarding specific variables. In a *Delphi* inquiry, for example, each participant fills out a questionnaire. The results are summarized and given to participants along with a follow-up questionnaire. Participants are invited to change their responses. The results are again summarized and fed back to the participants. This debate by questionnaire continues until participants' responses have converged enough. This method has an advantage over brainstorming in that participants are not subjected to psychological pressure from others with presumed authority or influence.[16]

Getting Information from the Existing Information System. Determining information from an existing application has been called the *data analysis* approach.[17] It simply asks the user what informa-

---

[13] See Davis, "Strategies," pp. 12–19.

[14] Albert T. Lederer, "Information Requirements Analysis," *Journal for Systems Management,* December 1981, p. 17.

[15] Gil Mosard, "Problem Definition: Tasks and Techniques," *Journal of Systems Management,* June 1983, p. 16.

[16] Olaf Helmer, "Analysis of the Future: The Delphi Method," in *Technological Forecasting for Industry and Government: Methods and Applications,* ed. James R. Bright (Englewood Cliffs, N.J.: Prentice-Hall, 1968), pp. 116–122. For examples of its use, see M. A. Linstone and M. Turoff, eds., *The Delphi Method: Techniques and Applications* (Reading, Mass.: Addison-Wesley Publishing, 1975).

[17] Malcolm C. Munro, "Determining the Manager's Information Needs," *Journal of Systems Management,* June 1978, pp. 34–36.

tion is currently received and what other information is required. It relies heavily on the user to articulate information needs. The analyst examines all reports, discusses with the user each piece of information examined, and determines unfulfilled information needs by interviewing the user. The analyst is primarily involved in improving the existing flow of data to the user. In contrast to this method is *decision analysis*. This breaks down a problem into parts, which allows the user to focus separately on the critical issues. It also determines policy and organizational objectives relevant to the decision areas identified and the specific steps required to complete each major decision. Then the analyst and the user refine the decision process and the information requirements for a final statement of information requirements.

The data analysis method is ideal for making structured decisions, although it requires that users articulate their information requirements. A major drawback is a lack of established rules for obtaining and validating information needs that are not linked to organizational objectives.[18]

In the decision analysis method, information needs are clearly linked to decision and organizational objectives. It is useful for unstructured decisions and information tailored to the user's decision-making style. The major drawback, though, is that information requirements may change when the user is promoted or replaced.

Prototyping. The third strategy for determining user information requirements is used when the user cannot establish information needs accurately before the information system is built. The reason could be the lack of an existing model on which to base requirements or a difficulty in visualizing candidate systems. In this case, the user needs to anchor on real-life systems from which adjustments can be made. Therefore, the iterative discovery approach captures an initial set of information requirements and builds a system to meet these requirements. As users gain experience in its use, they request additional requirements or modifications (iterations), in the system. In essence, information requirements are discovered by using the system. Prototyping is suitable in environments where it is difficult to formulate a concrete model for defining information requirements and where the information needs of the user are evolving, such as in DSS.

Which of the three strategies is selected depends on uncertainties in the process of determining information requirements—that is, uncertainty with respect to the stability of information requirements, the user's ability to articulate information requirements, and the ability of the analyst to elicit requirements and evaluate their accuracy.[19] Thus, the asking strategy is appropriate for low-uncertainty information requirements determination,

[18] Ibid., p. 39.

[19] Marvin O. Tharp and Wm. Taggart, "Management Information Analysis: A Situational Perspective," *Management Datamatics* 5, no. 6 (1976), p. 231.

FIGURE 4–5 Strategies for Determining Information Requirements

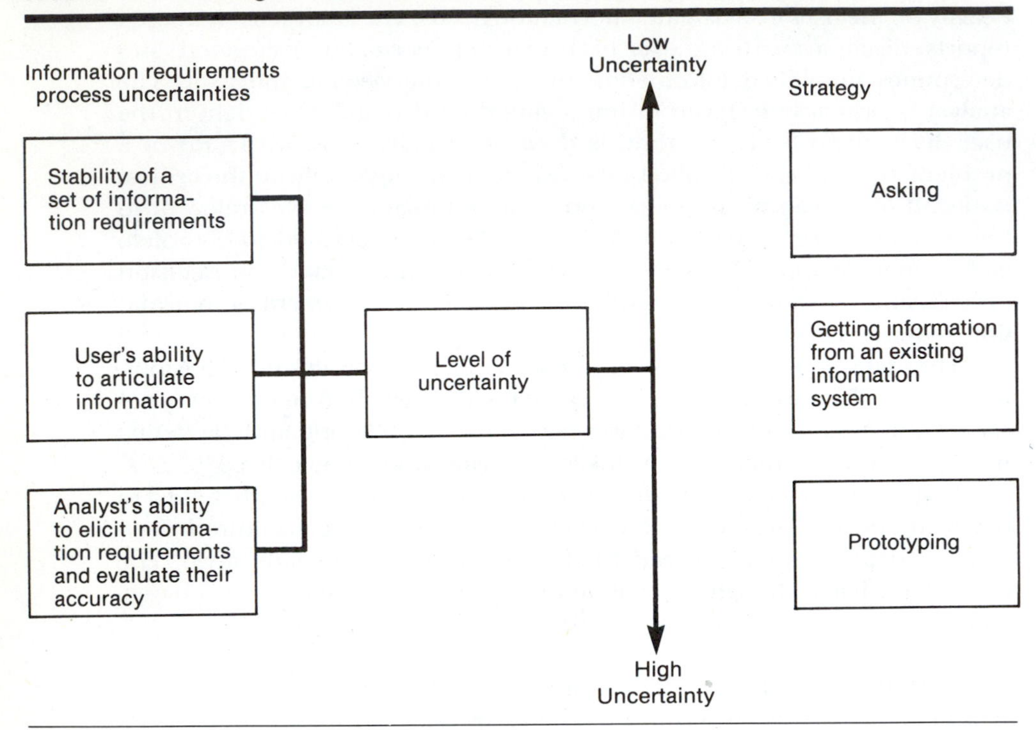

Source: Adapted from Gordon B. Davis, "Strategies for Information Requirements Determination, *IBM Systems Journal* 21, no. 1 (1982), p. 21.

whereas the prototyping strategy is appropriate for high-uncertainty information requirements determination (see Figure 4–5).

Case Scenario[20]

To apply the steps undertaken in an initial investigation, we use a case based on a real system development project that occurred in a medium-size commercial bank in a large city. For the purposes of this investigation, we shall call it the First National Bank of South Miami. The city has a population of 200,000.

In 1980, the bank had a safe deposit department that consisted of three employees and 4,000 boxes of various sizes. They are rented to bank customers, jewelers, and retailers in the neighborhood. Late in the year, an influx of illegal immigrants contributed to a rise in the number of burglaries and break-ins. This created a demand for safe deposit boxes for storing personal effects (china, jewelry, cash, coins, etc.). With increased demand,

[20] This case will be used to illustrate systems analysis and design throughout the text.

the board of directors authorized management to double the size of the facility. By the end of the year, capacity increased to 8,000 boxes. The increase required two additional employees, bringing the total staff to five.

The rental procedure begins when a customer applies for a box of a certain size and pays a year's rent in advance. The customer is issued a key that is used simultaneously with the attendant's key to open or close the box. As shown in Figure 4–6, the customer takes the box to a cubicle. When finished, he/she inserts the box in the vault, closes the door, and walks out.

Each new box rental generates a transaction for future billing. Bills are mailed to approximately 5,300 customers, some of whom have two or more boxes. Bills are processed manually on a cycle basis: every six days, beginning with the first of each month. A customer receives a renewal notice with the amount due about a month before the expiration date of the contract.

You are the analyst in the bank's MIS department. Your phone rings one morning and Sibley, the senior vice president in charge of operations, says, "Hi, Jack, we have a problem in safe deposit. Can you spare a minute? Let's get together over lunch and talk. . . ." As you interpret the conversation during lunch, Sibley has a problem. Bills are prepared by hand and mailed to all customers—a time-consuming job that keeps a clerk busy all day. You want to formalize that initial investigation, so you ask him to fill out a user request form. Barbara Betolatti, the supervisor, fills out the details. Sibley

FIGURE 4–6 Safe Deposit Department—Physical Layout

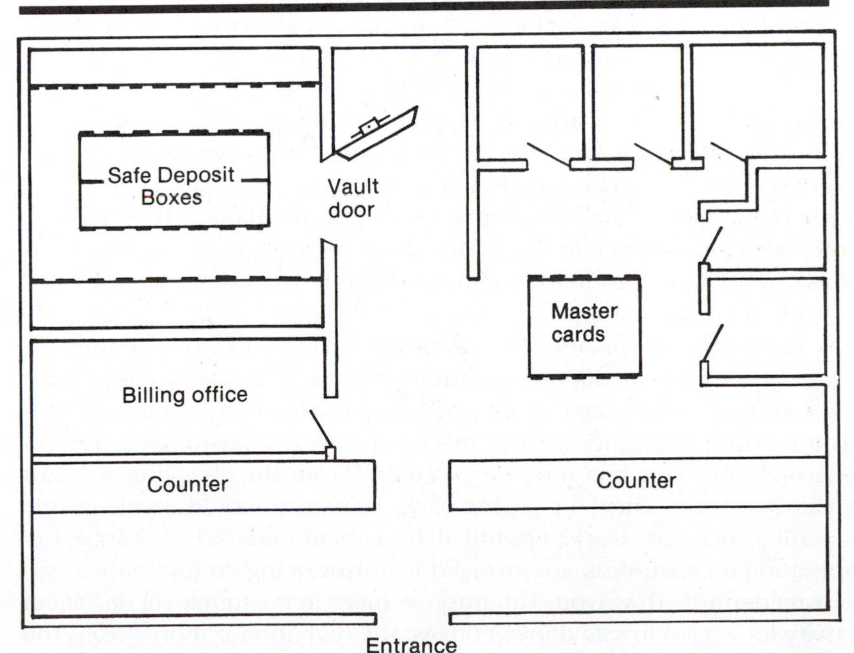

verifies them, signs the form, and hands it to you. The form is shown in Figure 4–4. Let's follow the initial investigation process using our scenario.

Problem Definition and Project Initiation

The first step in an initial investigation is to define the problem that led to the user request. The problem must be stated clearly, understood, and agreed upon by the user and the analyst. It must state the objectives the user is trying to achieve and the results the user wants to see. Emphasis should be on the logical requirements (what must be the results) of the problem rather than the physical requirements. For example, in the user request form, a job objective is improved customer service (logical objective; *how* the objective should be achieved (physical requirement) is not as important at this time.

Given user identification of need, the analyst proceeds to verify the problem by separating symptoms from causes. Long customer lines, for example, are not the problem per se, but are the symptoms of staff shortage, a manual procedure that does not retrieve customer information (master cards) for clearance, or both. The latter reason is the problem.

Related to the problem definition is the verification of user requirements—in our example, the quick and accurate availability of safe deposit information and shorter lines. As we discussed earlier in the chapter, requirements may be confirmed by eliciting information through one or more strategies: (1) asking the clerks and the supervisor what they must have to run the department (questions, brainstorming, or group consensus); (2) eliciting information from the existing manual system—how data are stored, how up to date they are, how efficient, and so on; or (3) prototyping—start improving the existing system, a step at a time, anchoring real-life change from which further adjustments can be made.

Of these strategies, asking would be the most suitable, since the safe deposit department is considered to have low uncertainty. This is attributable to the department's stable information requirements (rarely a change), Betolatti's ability to articulate the information requirements, and the analyst's ease in eliciting information about the same requirements and evaluating their accuracy.

It is important to consider the planning phase and how planning a candidate system for safe deposit fits into the larger MIS environment of the bank. For example, how congruent is designing a safe deposit billing system with the existing computer environment? Suppose a customer's box renewal is due and he wants the rent deducted from the checking account rather than writing a check or paying cash. How well would an electronic debit-credit procedure work? Should it be considered? What managerial and operational constraints are involved in introducing an alternative system for safe deposit? How would an improvement in customer service affect the activity for access to safe deposit boxes, the demand for more boxes, the image of the department (and the bank) as an all-service bank? Are satisfied safe deposit customers likely to open other accounts with or take loans from the bank? How would a computer-based safe deposit billing system affect the morale of users in the trust and personnel departments who are in need of improving their operations? These questions relating the proposed system to the larger system should be carefully evaluated.

Background Analysis

Once the project is initiated, the analyst begins to learn about the setting, the existing system, and the physical processes related to the revised system. For example, it is important to understand the structure of the bank; who runs it; who reports to whom in the safe deposit area; the relationship between safe deposit and the teller line, accounting and customer service; and the nature, frequency, and level of interaction between the safe deposit staff and these departments. The existing billing system could be the result of ill-trained staff or inefficient organization, or both. It could be that reorganization might be a solution. Therefore, the analyst should prepare an organization chart with a list of the functions and the people who perform them. In doing so, he/she would have a better feel for the work environment in which safe deposit operates, the kinds of customers involved, and the procedure employees follow in conducting business in safe deposit.

Fact-Finding

After obtaining this background knowledge, the analyst begins to collect data on the existing system's outputs, inputs, and costs. The tools used in data collection are covered in detail in Chapter 5. They are (1) review of written documents, (2) on-site observations, (3) interviews, and (4) questionnaires.

Review of Written Documents

When available, all documentation on data carriers (forms, records, reports, manuals, etc.) is organized and evaluated. Included in procedures manuals are the requirements of the system, which helps in determining to what extent they are met by the present system. Unfortunately, most manuals are not up to date or may not be readable. Day-to-day problems may have forced changes that are not reflected in the manual. Furthermore, people have a tendency to ignore procedures and find shortcuts as long as the outcome is satisfactory.

Regarding existing forms, the analyst needs to find out how they are filled out, how useful they are to the user (in our scenario, the customer), what changes need to be made, and how easy they are to read. Figure 4–7 illustrates the forms used in our safe deposit case. Forms design is discussed in Chapter 9.

FIGURE 4–7 Key Forms for Safe Deposit Accounting

SIZE 3X5	AVAILABLE YES	AVAILABLE NO	RENTAL 25.00		BOX NO. 204	
RENTER	FROM	TO	LOCK TO DIEBOLD	LOCK APPLIED	APPLIED BY	WITNESSED BY
DOROTHY GARRISON	1-5-80	Closed	6-3-81		2 Keys	Returned
JOHN SMITH	9-12-81	Closed	9-14-82		2 Keys	"
DENISE AGUIAR	12-29-83	Closed	1-2-84		1 Key	"

THIS IS A HISTORY CARD FOR BOX RENTALS. THE ONLY INFORMATION WE ARE CURRENTLY RECORDING IS WHEN THE BOX IS OPENED AND CLOSED AND THE AMOUNT OF KEYS RETURNED. THESE CARDS HAVE BEEN KEPT SINCE THE VAULT FIRST OPENED

HISTORY CARD

AGUIAR, DENISE OR SCOTT 3X5

BOX NO. Dec 204

SIGNATURE	ATTENDANT	BOOTH NO	DATE	HOUR
Denise Aguiar	[initials]	N/A		
Scott Aguiar	[initials]			
Denise Aguiar				

SIGNATURE CARDS SERVE TWO PURPOSES. ONE, IDENTIFICATION OF PATRON, TWO, ACCURATE RECORD OF ADMITTANCE TO THE VAULT. THE ATTENDING CUSTODIAN INITIALS THE CARD AFTER THE SIGNATURE IS VERIFIED, THEN TIME + DATE POSTED.

TR1

Advice of Charge

ADVICE OF CHARGE

THE FIRST NATIONAL BANK OF SOUTH MIAMI

DATE

WE CHARGE YOUR ACCOUNT AS FOLLOWS:	AMOUNT
Safe Deposit Box 204 from 12/29/83 thru 12/29/84	25 —
REFER TO / GENERAL LEDGER NO. / TOTAL	$ 25 —

26-312-5
ACCT. NO.

[initials]
APPROVED

ROD REAMS MOTIVATIONS, INC. - MIAMI, FLA. 33142

These are used to draft customer accounts per approval of same.

On-Site Observations

Another fact-finding method used by the systems analyst is on-site or direct observation. The analyst's role is that of an information seeker. One purpose of on-site observation is to get as close as possible to the "real" system being studied. As an observer, the analyst follows a set of rules. While making observations, he/she is more likely to listen than talk, and to listen with interest when information is passed on. He/she avoids giving advice, does not pass moral judgment on what is observed, does not argue with the user staff, and does not show undue friendliness toward one but not others.

On-site observation is the most difficult fact-finding technique. It requires intrusion into the user's area and can cause adverse reaction by the user's staff if not handled properly. The analyst observes the physical layout of the current system, the location and movement of people, and the work flow. He/she is alert to the behavior of the user staff and of the people with whom they come into contact. A change in behavior provides an experienced analyst with clues that can help put the behavior observed in perspective.

The following questions should precede a decision to use on-site observation:

1. What behavior can be observed that cannot be described in other ways?
2. What data can be obtained more easily or more reliably by observation than by other means?
3. What assurances can be given that the observation process is not seriously affecting the system or the behavior being observed?
4. What interpretation needs to be made on observational data to avoid being misled by the obvious?

If on-site observation is to be done properly in a complex situation, it can be time-consuming. Proper sampling procedures must be used to ascertain the stability of the behavior being observed. Without knowledge of stability, inferences drawn from small samples of behavior (small time slices) can prove inaccurate and, therefore, unreliable.

Interviews and Questionnaires

As we have discussed, on-site observation is directed toward describing and understanding events and behavior as they occur. This method, however, is less effective for learning about people's perceptions, feelings, and motivations. The alternative is the personal interview and the questionnaire. In either method, heavy reliance is placed on the interviewee's report for information about the job, the present system, or experience. The quality of the response is judged in terms of its reliability and validity. *Reliability* means that the information gathered is dependable enough to be used for making decisions about the system being studied. *Validity* means that the questions asked are so worded as to elicit the intended information. So, the reliability and validity of the data gathered depend on the design of the interview or questionnaire and the manner in which each instrument is administered.

In an interview, since the analyst (interviewer) and the person(s) interviewed meet face to face, there is an opportunity for greater flexibility in eliciting information. The interviewer is also in a natural position to observe the subjects and the situation to which they are responding. In contrast, the information obtained through a questionnaire is limited to the written responses of the subjects to predefined questions. Details on the advantages and limitations of each instrument and the stages of construction and administration are given in Chapter 5.

Fact Analysis

As data are collected, they must be organized and evaluated and conclusions drawn for preparing a report to the user for final review and approval. Many tools are used for data organization and analysis. As will be discussed in detail in Chapter 6, among the tools used are input/output analysis, decision tables, and structure charts.

Input/output analysis identifies the elements that are related to the inputs and outputs of a given system. Flowcharts and data flow diagrams are excellent tools for input/output analysis. For example, Figure 4–8 is an information-oriented flowchart of our safe deposit problem. It is a systems flowchart that displays the relationships among the forms used in the existing billing system. Figure 4–9 is an input/output analysis sheet that describes the relationships among inputs, processing functions, and outputs for the safe deposit function of the bank.

Data flow diagrams are also used to analyze the safe deposit billing system. They can be very effective in settings that pose few constraints on the development or modification of the system under study. Figure 4–10 is a data flow diagram whose content is equivalent to the information-oriented flowchart in Figure 4–8. It shows the general steps in the safe deposit billing system. As an input/output analysis technique, it enables the analyst to focus on the logic of the system and develop feasible alternatives. The circles (also called *bubbles*) represent a processing point within the system. The open rectangles represent data stores or points where data are stored. The squares are departments or people involved in the billing system. The major steps in the billing process are extracting the customer account, applying the renewal cycle, preparing the bill, processing the payment, and accounting for cash receipts.

Decision tables describe the data flow within a system. They are generally used as a supplement when complex decision logic cannot be represented clearly in a flowchart. As a documenting tool, they provide a simpler form of data analysis than the flowchart. When completed, they are an easy-to-follow communication device between technical and nontechnical personnel. They are verbally oriented to managers, easy to learn and update,

FIGURE 4–8 Information-Oriented Flowchart for a Safe Deposit Billing System

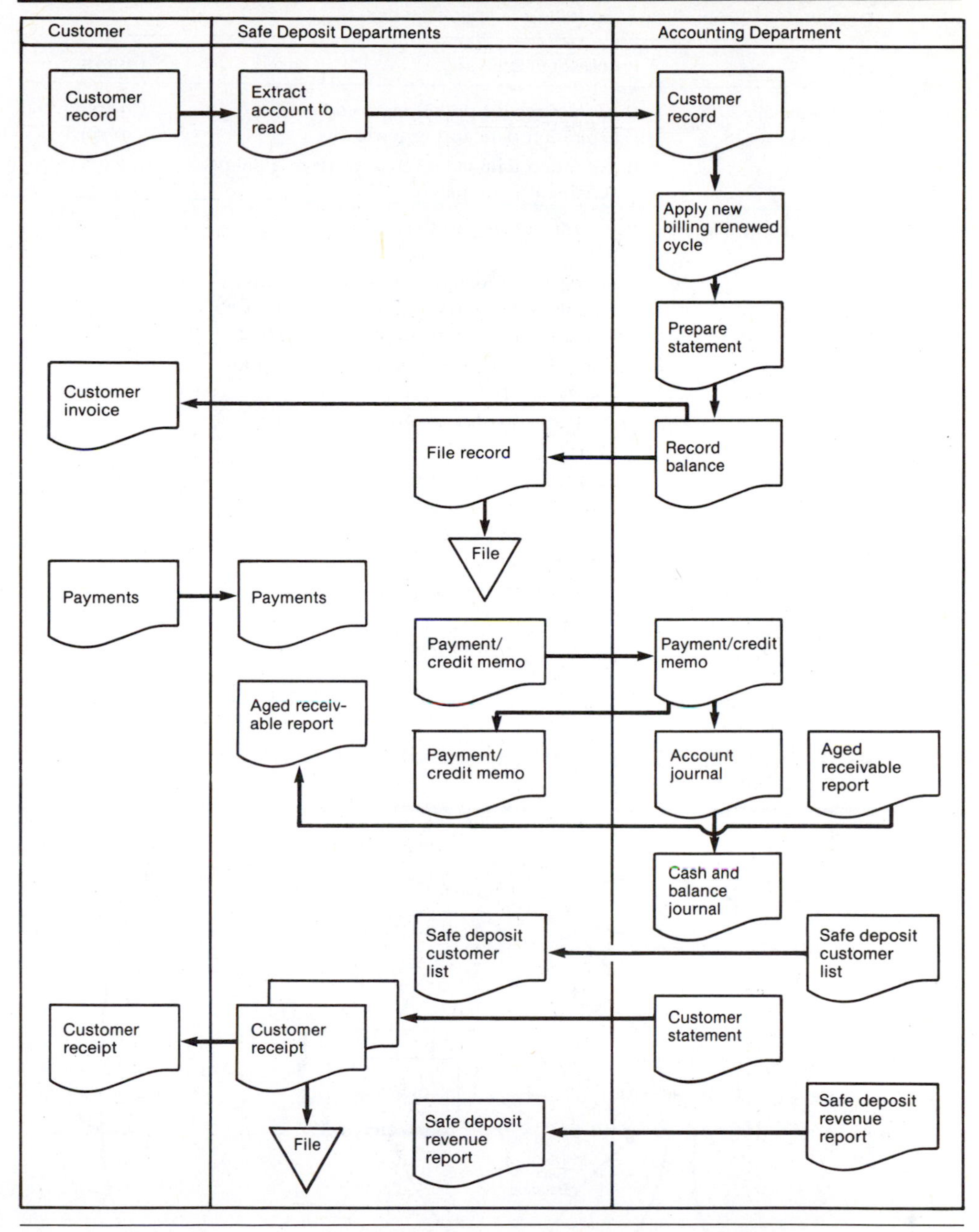

FIGURE 4–9 Input/Output Analysis Sheet

Dept: Safe Deposit **System: Billing** **Date: 1/14/85**

Input	Processing/Files	Output
Customer file record	Customer record pulled out to determine expiration date and renewal rate.	Customer file record
	If expiration date is less than 30 days, a billing statement is prepared.	
Customer payment	Payments are sent to the accounting department.	
	Payment/credit memo is sent to data processing for verification and crediting customer's account. The memo is returned for filing.	
	Memo creates a journal entry in a cash and balance journal.	
	Data processing produces aged receivable report.	
	Memo becomes a part of customer safe deposit active list.	
	Data processing produces safe deposit revenue report.	
	Copy is sent to safe deposit and another copy kept on file.	
	Customer receipt of payment mailed to customer.	

FIGURE 4–10 Simplified Data Flow Diagram of the Existing Safe Deposit Billing System

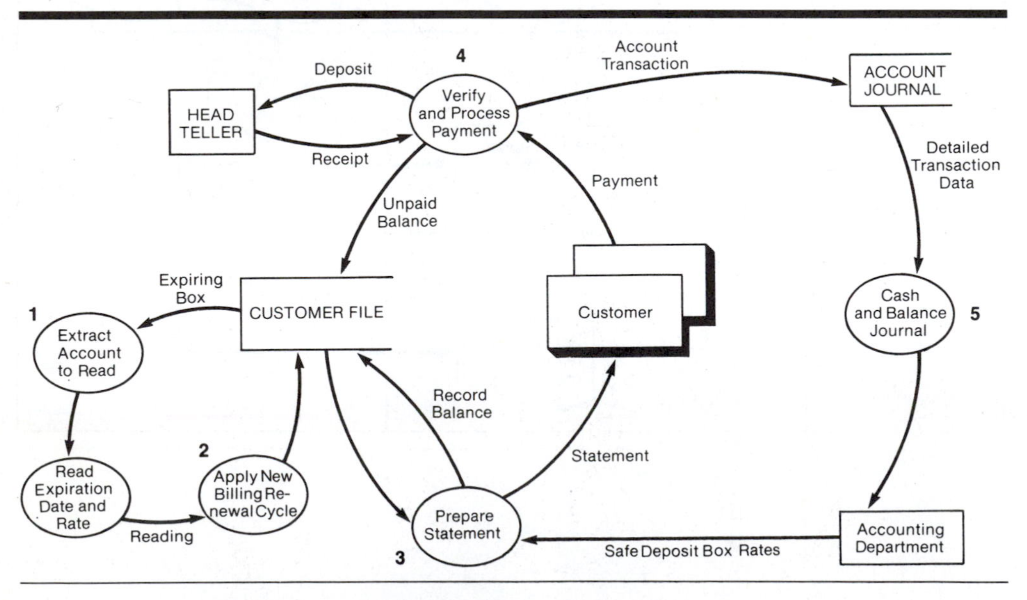

and continue to function once the logic is developed (see Figure 4–11). Details on the structure and uses of decision tables are covered in Chapter 6.

A *structure chart* is a working tool and an excellent way to keep track of the data collected for a system. There are several variations of a structure chart. Briefly, the analyst starts with a single input/processing/output (IPO) chart, locates the module associated with the IPO on the hierarchy chart, and identifies the data elements along the line linking the module to a higher level (parent).

Determination of Feasibility

After organizing and summarizing the data, the analyst has a thorough knowledge of the system. The following information should be available:

- Interview and correspondence records.
- Updated system documentation.
- Flowcharts.
- Familiarity with names, positions, and personalities of user personnel.
- Specification of the good and bad features of the current system.
- Understanding of how well actual problems facing the system are in line with the problem(s) stated in the user request form.

The outcome of the initial investigation is to determine whether an alternative system is feasible. A report summarizing the thinking of the analyst is presented to the user for review. At this time, it is important to identify the principal user. In our safe deposit department example, although the supervisor is the person in charge of the operation, it was found that the senior vice president is the principal user—the one who accepts or

FIGURE 4–11 Decision Table—Safe Deposit Billing Routine

	Billing	1	2	3	4	5
IF . . .	New rental application?	Y	N	N	N	N
	Box rental expires in 60 days?	—	Y	N	N	N
	Box rental expires in 30 days?	—	—	Y	N	N
	Box rental expired?	—	—	—	Y	N
	Box rental past due?	—	—	—	—	Y
THEN . . .	Open safe deposit account	X	—	—	—	—
	Issue two keys	X	—	—	—	—
	Mail bill	—	X	—	—	—
	Mail reminder	—	X	X	—	—
	Mail past due letter	—	—	—	X	—
	Contact customer	—	—	—	X	—
	Call security	—	—	—	—	X
	Drill box	—	—	—	—	X

rejects the candidate system. He is the one who also issued project directives and has a final say on what can and cannot be done. When the final report was turned in, he reviewed it (with the supervisor) for content, justification, and authorization to implement the online safe deposit billing system.

The final decision is the end user's response to a *project directive.* When approved, it becomes an authorization document that also reflects the results of the discussions made during the final review. More and more organizations have a computer user committee as a final approval authority for the project undertaken.

A signature on the directive by the end user and its acceptance by the MIS department make it a formal agreement to proceed with the design and implementation of the candidate system. So, the directive initiates a feasibility study, which essentially involves the description and evaluation of the candidate system and the selection of the best system that meets system performance requirements. These two steps are described in Chapter 7.

Summary

1. Planning information systems has become increasingly important because information is a vital resource and company asset, more and more funds are committed to information systems, and system development is a serious business for computers that incorporate data bases and networking.
2. Planning for information systems has a time horizon and a focus dimension. The time horizon dimension specifies the time range of the plan, whereas the focus dimension relates whether the primary concern is strategic, managerial, or operational.
3. The initial investigation has the objective of determining the validity of the user's request for a candidate system and whether a feasibility study should be conducted. The objectives of the problem posed by the user must be understood within the framework of the organization's MIS plan.
4. Determining user requirements is not easy. System requirements change, the articulation of requirements is difficult, and heavy user involvement and motivation are uncertain. Problems with the user/analyst interface add further difficulties to the procedure.
5. There are three strategies for eliciting information regarding the user's requirements: asking questions, obtaining information from the present system, and prototyping. The asking strategy assumes a stable system where the user is well informed about information requirements. In contrast, the prototyping strategy is appropriate for high-uncertainty information requirements determination.
6. Fact-finding is the first step in the initial investigation. It includes a review of written documents, on-site observations, interviews, and

questionnaires. The next step is fact analysis, which evaluates the elements related to the inputs and outputs of a given system. Data flow diagrams and other charts are prepared during this stage.

7. The data flow diagram (DFD) shows the flow of data, the processes, and the areas where they are stored. It is a commonly used structured tool for displaying the logical aspects of the system under study. Decision tables are used as a supplement when complex decision logic cannot be represented clearly in a DFD.
8. The outcome of the initial investigation is to determine whether an alternative system is feasible. The proposal details the findings of the investigation. Approval of the document initiates a feasibility study, which leads to the selection of the best candidate system.

Key Words

Bounded Rationality
Brainstorming
Data Flow Diagram
Decision Table
Delphi Method
Initial Investigation
Kitchen Sink Strategy
Management Planning
Operational Planning
Planning
Project Directive
Project Proposal
Prototyping
Recency Effect
Reliability
Strategic Planning
Structure Chart
Validity

Review Questions

1. Why is it so critical to manage system development? Explain.
2. What planning dimensions determine information system development? Elaborate.
3. What is the difference between managerial and operational MIS planning? Discuss.
4. Elaborate on the top-down approach to system planning—what it means, its uses, and its implications for system development.
5. Distinguish between the following:
   a. Brainstorming and the Delphi method.
   b. Validity and reliability.
   c. Strategic and operational planning.
   d. Decision table and structure chart.
6. What important information does the user's request form provide? Why is it so important in the initial investigation? Explain in detail.
7. Why is it difficult to determine user requirements? Illustrate.
8. According to Scharer, users use various strategies to define their information requirements. Give examples and explain four strategies.

9. Discuss and illustrate the key strategies for eliciting information about the user's requirements. Which strategy would you select? Why?
10. A question may be closed or open-ended. Illustrate the difference.
11. Review the literature on prototyping. Give a five-minute summary in class.
12. Describe the data analysis method. How does it differ from the decision analysis method? Elaborate on the pros and cons of each method.
13. Illustrate the difference between input/output analysis and systems flowcharts. When is each tool used?
14. What are data flow diagrams? How do they differ from structure charts?

Application Problems

1 JEFFERSON CREDIT CENTER

The Mid-Atlantic Credit Center for Belk-Jefferson is operated by approximately 30 employees, 20 of whom are directly involved in credit and payment processing and customer service. The credit center is concerned with the receipt and credit of payments, check encoding, and payment documentation and microfilm storage. Most of the center's operations are organized by an interactive IBM 3031 and 4341 mainframe located in Charlotte, North Carolina—the central processing unit for the entire Maryland-based Jefferson retail stores. Twelve CRTs allow customer service employees to access customer accounts and payment information when needed.

Payment processing begins in the mailroom. Two employees open incoming payments (approximately 5,000 per day) and place payment slips and checks in alternating order, grouping them in 250-payment batches. The batches are then encoded by a magnetic ink character recognition (MICR) device that separates the payment slips from the checks while encoding both with the amount of the payment. Although this encoding process is performed by the bank, Jefferson has found it more expedient to do it in house.

Next the two stacks (checks and payment slips) are taken to the data entry station, where the amount of payment is entered into the computer and credited to the proper account. Although the routine is automated, an employee monitors the operation for encoding errors. Following data entry, payment slips and checks are taken to microfilm for documentation. The reels are kept randomly in boxes with only the date of their creation printed on the outside. Payment slips are filed in boxes and indexed by date only. The actual checks are mailed to the bank for processing (see Exhibit 4–1).

EXHIBIT 4-1 Present Information Flow at Jefferson Credit Center

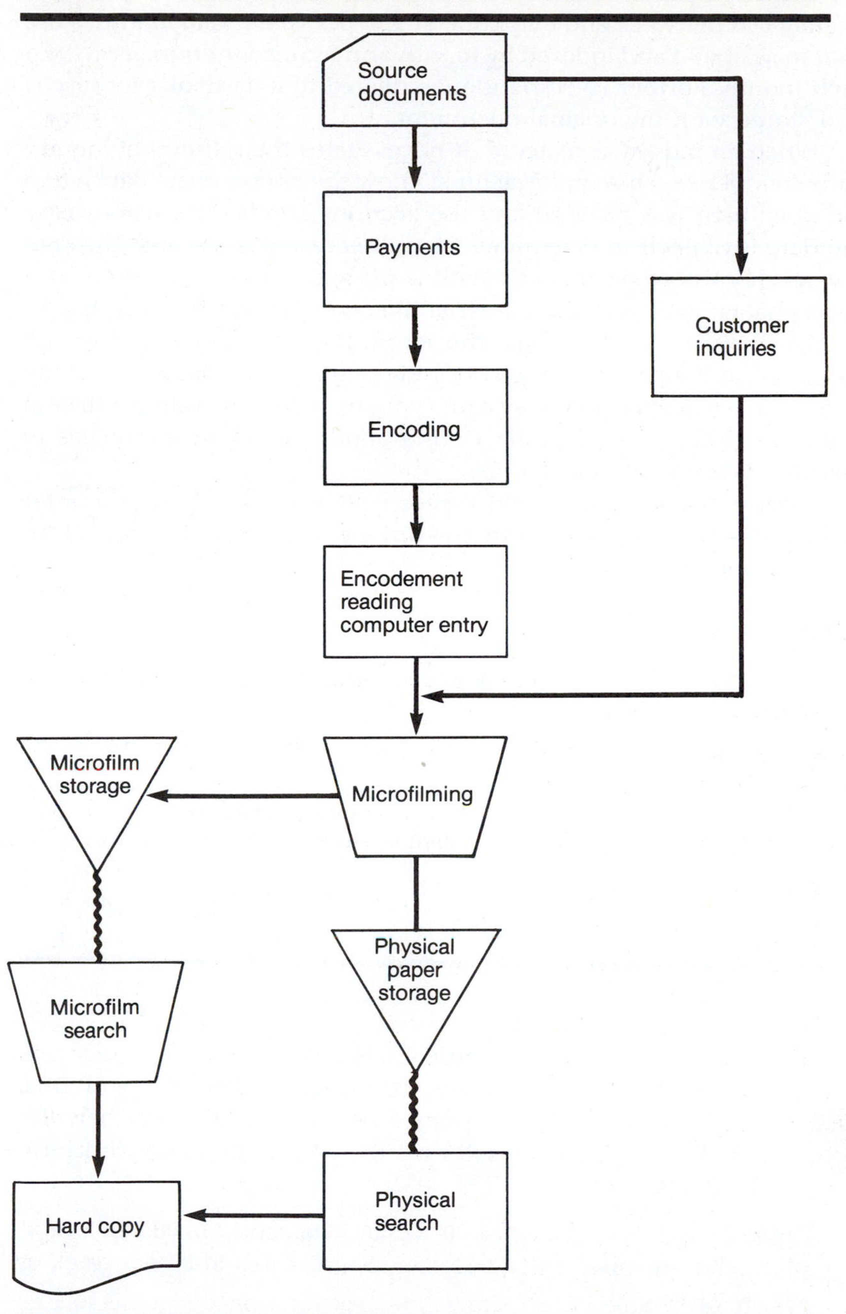

Customer inquiries are received through the mailroom at approximately 660 per week and by phone at 150 per week. Mail inquiries are filed in a cabinet and indexed by month and by account number within each month. Further correspondence related to a particular inquiry is also stored with the original document.

When an inquiry is received, it necessitates the retrieval of inquiry correspondence. The employee must know the approximate date when the document was received and the account number. In many cases, the date is difficult to determine, which necessitates a search through several months of documents. With a physical file system, there is a good chance of misfiling documents; this complicates the search.

To retrieve payment slips, the employee invariably searches the paper file, although they are also stored on microfilm. This is due to the lack of a reasonable index system, which makes microfilm retrieval almost as slow as the paper file. Consequently, search time averages 10 minutes. Often it takes half a day.

Clearly, the Jefferson credit center uses state-of-the-art computer hardware—IBM interactive system—but inefficient manual document-retrieval operations.

Assignment

a. What is the main problem facing Jefferson stores and the credit center? Be specific.

b. In doing an initial investigation, what goals and considerations would you focus on?

c. Provide the necessary information to determine the feasibility of an alternative system. Is an alternative system feasible? Justify your answer.

2

An in-house analyst of a large commercial bank received an inquiry (see the box) from Mrs. Mattes, the second vice president of operations, concerning Christmas Club coupons. After a brief talk with her, the analyst decided to see for himself how the whole operation works. He observed the following:

a. Coupons are classified as $1.00, $2.00, $5.00, and $10.00. For example, a club member with 52 $1.00 coupons saves $1.00 per week or $52 by Christmas.

b. A customer comes in with the coupon book for deposit. The teller receives the cash, detaches the coupon, stamps the stub (in the coupon book) with the date of payment and the amount, and returns the coupon book to the customer.

c. At the end of the day, the coupons are sorted by denomination, counted, and the total value balanced against the cash received. The entire process takes one person about three hours.

After observing the operation, the analyst determined the problem to be too many categories of coupons rather than a shortage of clerical help. Increases in people's incomes, interest-bearing checking accounts, and the like would make it unprofitable for the bank to maintain the $1.00, $2.00, and $5.00 coupon books. Therefore, they should be discontinued. Effective January of next year, club members should be sold $5.00 or $10.00 coupon books or encouraged to open a savings account.

PROBLEM REPORT FORM 16-A

Statement of Problem

Because of the increase in the number of Christmas Club accounts, it has become necessary to seek full-time clerical help to process the daily coupons. Furthermore, the manual handling of each coupon has made it more costly to maintain the club.

Reason(s) for Reporting Problem

The manual handling of coupons makes it difficult to attract regular help. Tellers complain that coupons interrupt their work. Immediate solution to this problem can improve the service and provide efficient operation of the club.

NAME: Dixie Mattes DEPARTMENT: Operations

TITLE: Second vice president EXT: 5421 DATE: 3/15/85

Assignment

a. Do you agree with the analyst's definition of the problem? If not, how would you define it? Why? Explain.

b. If you were to do the initial investigation, how would you handle it? Elaborate.

3 Allied Concrete, Inc., has had to revamp its approach to maintaining a computer system and converting applications. Recently management has established a steering committee to oversee and approve all applications before they are run on the mainframe. The committee consists of one member from each of the following areas: accounting, sales, production, and information systems. The committee is chaired by the vice president in charge of production. The primary charge is to review each user request and approve or disapprove it based on feasibility and priority. If a request is approved, the user department is billed for its development by a debit transaction against its budget. The amount includes computer time, analyst and programmer time, and supplies. All department heads have agreed to the new policy.

In formalizing the committee's authority and responsibilities, serious questions were raised by several user departments about whether the committee has the authority to turn down a project even if it is technically feasible. They argued that since they are paying for the project out of their budget, there is no reason for it to be rejected.

Assignment

a. Should all user projects that are operationally and technically feasible be developed as long as the user is paying the price? If so, what should be the role of the steering committee?

b. What do you think of the makeup of the steering committee? What role should the analyst, programmer, or data base specialist play in a steering committee? Elaborate.

4

The steering committee for the information system of a large savings and loan bank is evaluating a request from the mortgage loan department to provide an online system in all branches. The bank's mainframe is operating at 55 percent of capacity. It has adequate memory to handle the new application. The only equipment needed is an online terminal in each of the bank's 27 branches and a software package that can be installed in five weeks using the existing telecommunications network. The terminals are available through the vendor within the week at $2,100 each. Branch tellers could be well trained in less than four working days. The software package costs $18,000.

The existing mortgage loan applications are handled in a batch mode. At the end of the day, each branch sends the mortgage payments and documents to the computer center, located 18 miles away. When the documents are received, data entry operators enter each payment and account number directly on disk. When all transactions are entered, they are processed. All accounts are updated and the resulting report (1,400 pages long) is sent to various branches for reference. Obviously, in a batch environment, all information is based on the previous day's activities.

The bank is a leader in the industry for introducing new laborsaving and income-generating applications. In the past, when a new application was implemented, it set the tone for other banks to duplicate. The systems group is highly motivated and well paid and works closely with management on a regular basis.

In the proposal, the vice president of mortgage loans reasons that with an online mortgage loan system, tellers can answer inquiries about mortgage payments, balances, and other matters in just seconds. Within the year, she expects customers to call the branch rather than the main office for all information regarding their respective loans. This means a savings in human resources and a more efficient distribution of the work load among the branches.

Assignment

a. Based on the information provided, is this proposal feasible? Should it be pursued? Why? Elaborate.

b. What other information does the steering committee need to do a thorough investigation? What source(s) would it come from? Be specific.

Selected References

Bariff, M. L. "Information Requirements Analysis: A Methodological Review." *Working Paper* 76-08-02, the Wharton School, University of Pennsylvania, Philadelphia, 1976.

Bowman, Brent; Gordon Davis; and **James C. Wetherbe.** "Modeling for MIS." *Datamation,* July 1981, pp. 155–64.

Business Systems Planning—Information Systems Planning Guide. Application Manual, GE 20-0527-3, 3d ed. IBM Corp, July 1981. Available through IBM branch offices.

Caldwell, Jack. "The Misunderstanding of Objectives." *Journal of Systems Management,* June 1982, p. 30.

Cerullo, Michael J. "MIS: What Can Go Wrong?" *Management Accounting,* April 1979, pp. 43–49.

Cooper, Roldolph B., and **E. Burton Swanson.** "Management Information Requirements Assessment: The State of the Art." *Data Base,* Fall 1979, pp. 5–16.

Couger, J. D. "Comparative Analysis of Information Systems Curricula." *Computing Newsletter for Schools of Business,* vol. XVII, no. 2 (October 1983), p. 1.

Davis, Gordon B. "Strategies for Information Requirements Determination." *IBM Systems Journal* 21, no. 1 (1982), pp. 4–30.

Doll, Wm. J., and **Mesbah U. Ahmed,** "Managing User Expectations." *Journal of Systems Management,* June 1983, pp. 6–11.

Gore, Marvin, and **John Stubbe.** *Elements of Systems Analysis.* 3d ed. Dubuque, Iowa: Wm. C. Brown, 1983, pp. 178–207.

Haughey, Thomas F., and **Robert M. Rollason.** "Function Analysis: Refining Information Engineering." *Computerworld (In-Depth),* August 22, 1983, pp. 24–26ff.

Helmer, Olaf. "Analysis of the Future: The Delphi Method." In *Technological Forecasting for Industry and Government: Methods and Applications* ed. James R. Bright. Englewood Cliffs, N.J.: Prentice-Hall, 1968, pp. 116–22.

Lederer, Albert T. "Information Requirements Analysis." *Journal of Systems Management,* December 1981, pp. 15–19.

Leif, Robert E.; Robert D. Dodge; and **Ralph L. Ogden.** "Adapting Data Processing Strategy to Management Style." *Computerworld (In-Depth),* December 5, 1983, pp. 25–32.

Lientz, Bennet P., and **Myles Chen.** "Long Range Planning for Information Services." *Long Range Planning,* vol. 13 (February 1980), pp. 55–61.

Linstone, M. A., and **M. Turoff,** eds. *The Delphi Method: Techniques and Applications.* Reading, Mass.: Addison-Wesley Publishing, 1975.

McFarlan, Warren. "Portfolio Approach to Information Systems." *Journal of Systems Management,* January 1982, pp. 11–19.

McLean, Ephraim R., and **John V. Soden.** *Strategic Planning for MIS.* New York: John Wiley & Sons, 1977.

Miller, Wm. B. "Developing a Long Range EDP Plan." *Journal of Systems Management,* July 1979, pp. 36–39.

Mosard, Gil. "Problem Definition: Tasks and Techniques." *Journal of Systems Management,* June 1983, pp. 16–21.

Munro, Malcolm C. "Determining the Manager's Information Needs." *Journal of Systems Management,* June 1978, pp. 34–39.

Newell, A., and **H. A. Simon.** *Human Problem Solving.* Englewood Cliffs, N.J.: Prentice-Hall, 1972.

Nolan, Richard L. "Managing Information Systems by Committee." *Harvard Business Review,* July–August 1982, pp. 72–79.

Pitagorsky, George. "Analyzing, Defining Systems Needs." *Management Information Systems Week,* August 24, 1983, p. 30.

Poppel, Harvey. *Strategic Impact of Information Technology.* New York: Deltak Corp., 1982, pp. 5–9.

Powers, Michael; David Adams; and **Harlan D. Mills.** *Computer Information System Development: Analysis & Design.* Cincinnati: South-Western Publishing, 1984, pp. 60–82.

Scharer, Laura. "Pinpointing Requirements." *Datamation,* April 1981, pp. 139–40.

Steiner, George A. "Formal Strategic Planning in the United States Today." *Long Range Planning* 16, no. 3 (1983), pp. 13–17.

Tharp, Marvin O., and **Wm. Taggart.** "Management Information Analysis: A Situation Perspective." *Management Datamatics* 5, no. 6, (1976), pp. 231–39.

Thierauf, Robert J., and **George W. Reynolds.** *Effective Information Systems Management.* Columbus, Ohio: Charles E. Merrill Publishing, 1982.

Wetherbe, James C. *Systems Analysis and Design: Traditional, Structured, and Advanced Concepts and Techniques.* Minneapolis, Minn.: West Publishing, 1984, pp. 319–59.

Chapter 5

Information Gathering

At a Glance

A key part of feasibility analysis is gathering information about the present system. The analyst must know what information to gather, where to find it, how to collect it, and what to make of it. The proper use of tools for gathering information is the key to successful analysis. The tools are the traditional interview, questionnaire, and on-site observation. We need to know, for example, how to structure an interview, what makes up a questionnaire, and what to look for in on-site observations. These tools, when learned, help analysts assess the effectiveness of the present system and provide the groundwork for recommending a candidate system.

By the end of this chapter, you should know:

1. What categories of information are available for systems analysis.
2. The sources of information.
3. How to arrange an interview.
4. The types of interviews and questionnaires.
5. How to construct a questionnaire.

INTRODUCTION

Chapters 5 and 7 describe the early phase of system development. Whether the thrust of the activities is the initial investigation or a feasibility study, the aim is primarily to develop an understanding of the problem facing the user and the nature of the operation. Understanding how each activity operates requires access to information.

Information gathering is an art and a science. The approach and manner in which information is gathered require persons with sensitivity, common sense, and a knowledge of what and when to gather and what channels to use in securing information. Additionally, the methodology and tools for information gathering require training and experience that the analyst is expected to have. This means that information gathering is neither easy nor routine. Much preparation, experience, and training are required.

This chapter addresses the categories and sources of information and the functions, uses, and relevance of key information-gathering tools during the phases of system analysis. The phases are:

1. Familiarity with the present system through available documentation, such as procedures manuals, documents and their flow, interviews of the user staff, and on-site observation.
2. Definition of the decision making associated with managing the system. This is important for determining what information is required of the system. Conducting interviews clarifies the decision points and how decisions are made in the user area.
3. Once decision points are identified, a series of interviews may be conducted to define the information requirements of the user. The information gathered is analyzed and documented. Discrepancies between the decision system and the information generated from the information system are identified. This concludes the analysis and sets the stage for system design.[1]

WHAT KINDS OF INFORMATION DO WE NEED?

Before one determines where to go for information or what tools to use, the first requirement is to figure out what information to gather. Much of the information we need to analyze relates to the organization in general, the user staff, and the work flow (see Figure 5–1).

[1] For details on the application of system analysis activities, refer to James Wetherbe, *Systems Analysis and Design: Traditional, Structured, and Advanced Concepts and Techniques* (St. Paul, Minn.: West Publishing, 1984), pp. 127–54.

FIGURE 5-1 Categories of Information

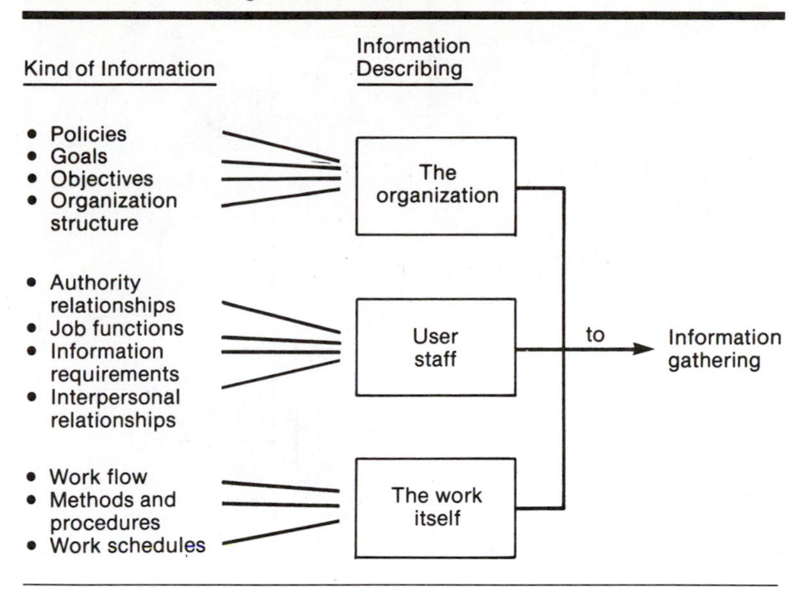

Information about the Firm

Information about the organization's policies, goals, objectives, and structure explains the kind of environment that promotes (or hinders) the introduction of computer-based systems. Company policies are guidelines that determine the conduct of business. Policies are translated into rules and procedures for achieving goals. A statement of goals describes management's commitment to objectives and the direction system development will follow. Objectives are milestones of accomplishments toward achieving goals. Information from employee manuals, orientation pamphlets, annual company reports, and the like helps an analyst form opinions about the goals of the organization.

After policies and goals are set, a firm is organized to meet these goals. The organization structure, via the organization chart, indicates management directions and orientation (see Figure 5-2). For example, a family-owned firm often has a rigid, centralized structure and a conservative approach to implementing change. This suggests that when it comes to computerizing applications, the analyst has a challenge to sell the project before a final approval is secured.

The organization chart represents an achievement-oriented structure. It helps us understand the general climate in which candidate systems will be considered. In gathering information about the firm, the analyst should watch for the correspondence between what the organization claims to achieve (goals) and actual operations. Policies, goals, objectives, and struc-

FIGURE 5–2 Organization Chart

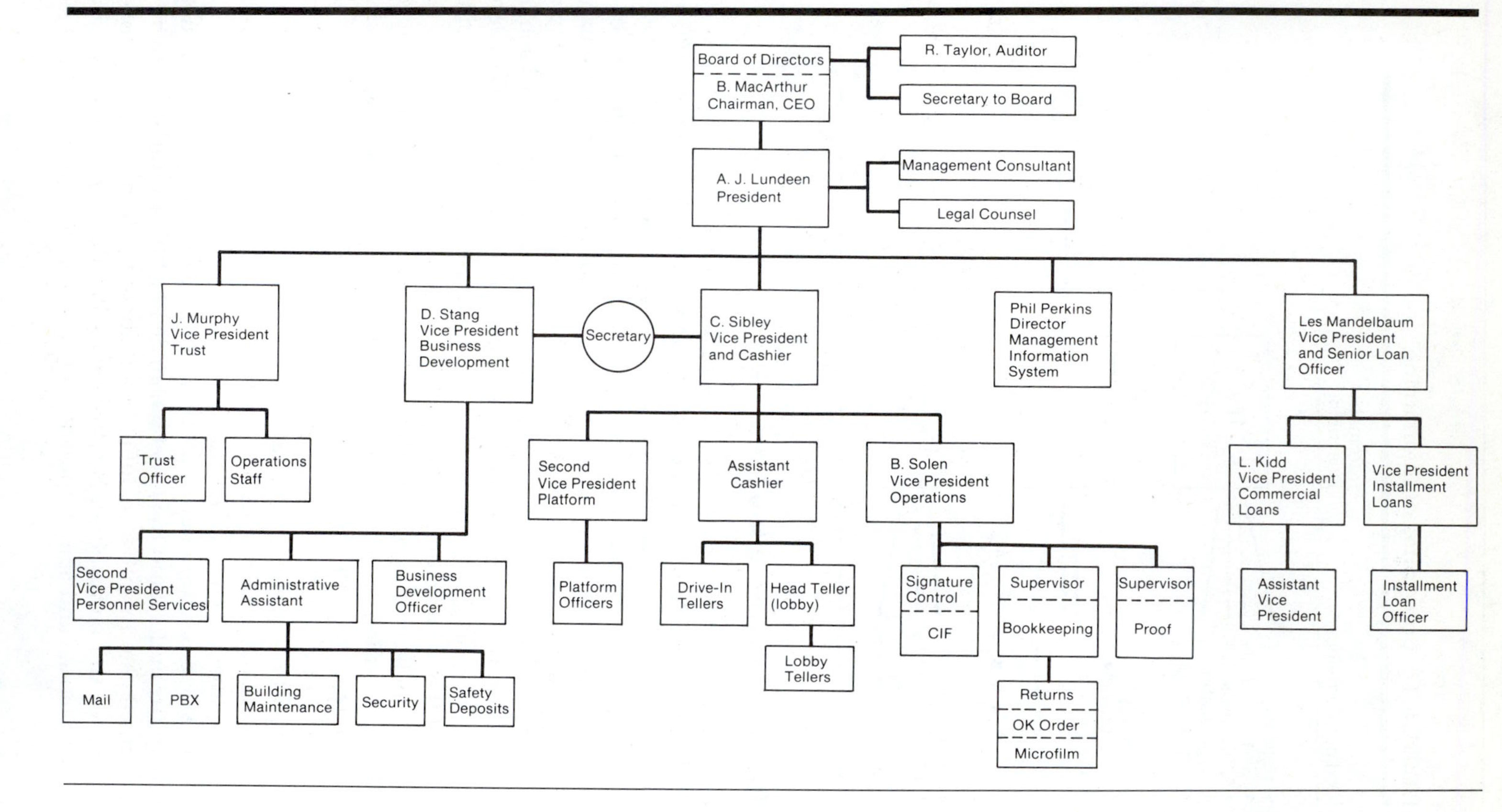

ture are important elements for analysis. Requests for computer service must be evaluated in the light of these elements.

Information about User Staff

Another kind of information for analysis is knowledge about the people who run the present system—their job functions and information requirements, the relationships of their jobs to the existing system, and the interpersonal network that holds the user group together. We are actually focusing on people's roles, authority relationships, job status and functions, information requirements, and interpersonal relationships. Information of this kind highlights the organization chart and establishes a basis for determining the importance of the existing system for the organization.

In summary, the major focus is to find out what people the analyst is going to be dealing with and what each person expects to get out of a candidate system before it goes through design and final implementation. Once such information is secured, the next step is to show how various jobs hang together within work schedules and procedures.

Information about Work Flow

Work flow focuses on what happens to the data through various points in a system. This can be shown by a data flow diagram or a system flowchart. A data flow diagram represents the information generated at each processing point in the system and the direction it takes from source to destination (see Figure 5–3). In contrast, a system flowchart describes the physical system (see Figure 5–4). The information available from such charts explains the procedures used for performing tasks and work schedules. Details on charts are covered in Chapter 6.

WHERE DOES INFORMATION ORIGINATE?

Information is gathered from two principal sources: personnel or written documents from within the organization and from the organization's environment. The primary *external* sources are:

1. Vendors.
2. Government documents.
3. Newspapers and professional journals.

The primary *internal* sources are:

1. Financial reports.
2. Personnel staff.
3. Professional staff (legal counsel, EDP [electronic data processing] auditor, etc.).

FIGURE 5-3 Data Flow Diagram of a Payroll System

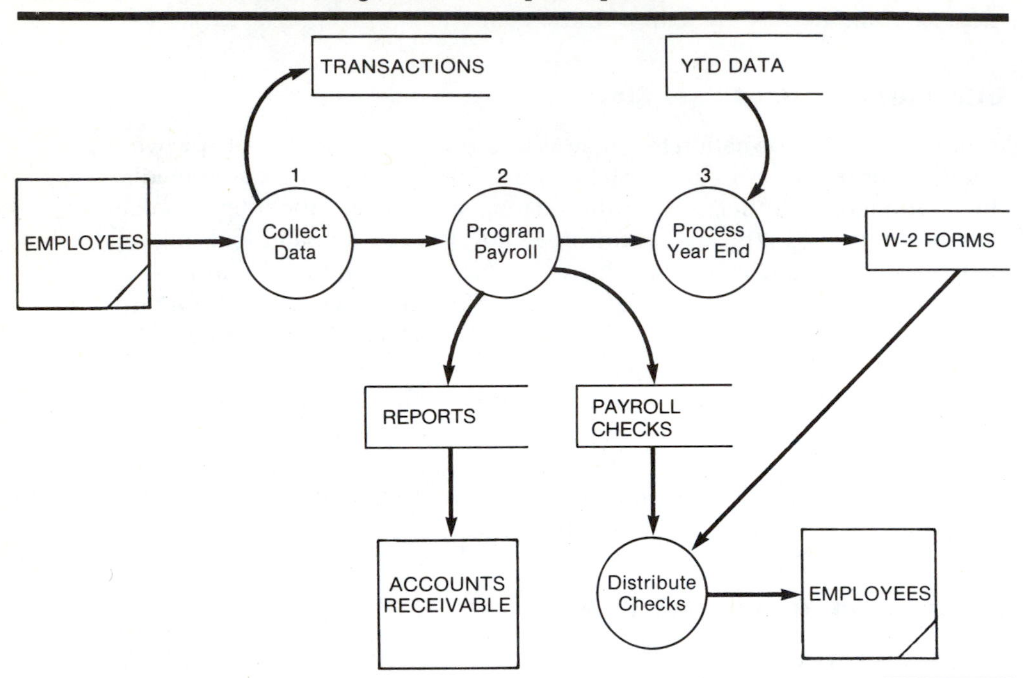

4. System documentation or manuals.
5. The user or user staff.
6. Reports and transaction documents.

Hardware vendors are traditional sources of information about systems and software. Other equipment manufacturers provide information about competitive systems. A third source that has experienced tremendous growth during the past decade is the software house. There are thousands of software packages on the market to suit virtually every problem area with reasonable modifications. Independent listings of software packages and their vendors are available through associations such as Computerworld and DATAPRO, or other organizations with experience in the application under consideration.

Other external sources of information are government documents, technical newspapers, and professional journals. *Computerworld,* for example, provides weekly information about new hardware, hardware installations, software developments, and trends in the field. Articles are also published in system development, documentation, and EDP journals, such as *Communications of the ACM* and *Journal for System Management*. They provide invaluable updates in the systems area.

Internal sources of information are limited to the user staff, company personnel, and various reports. User personnel are the front-line contacts

FIGURE 5-4 System Flowchart of a Payroll System

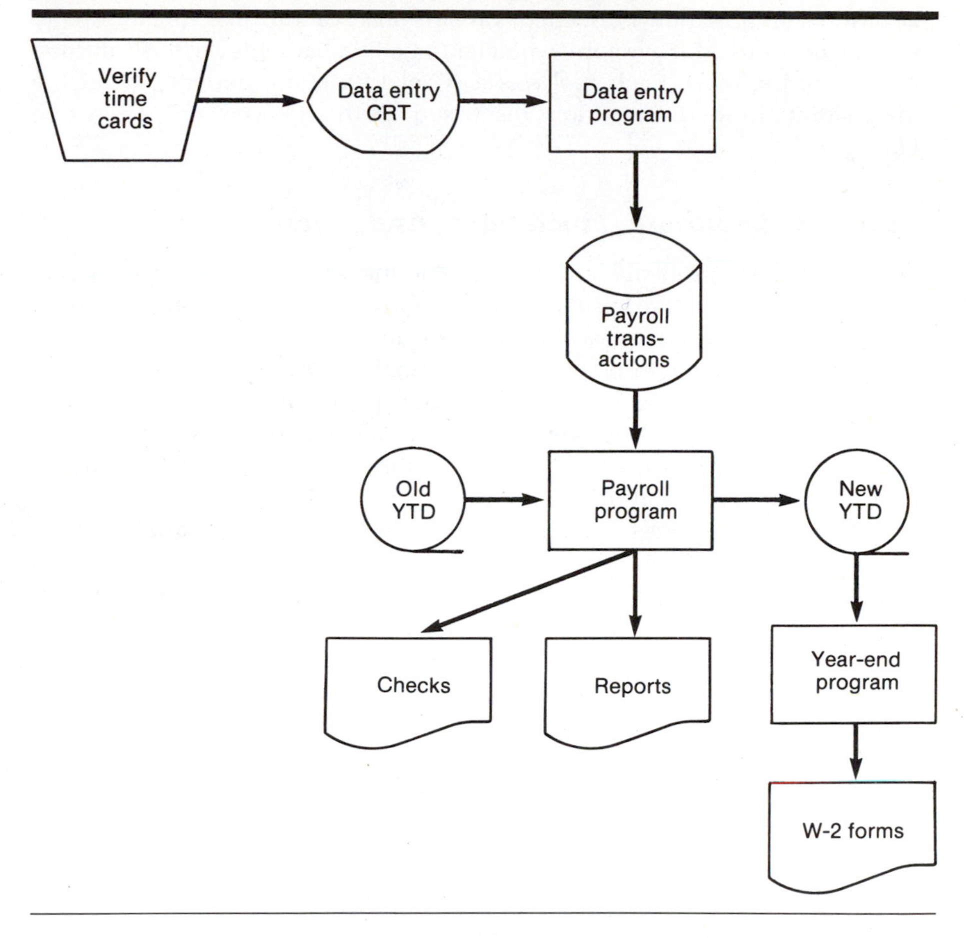

for acquiring and validating information about a system. An important source of information is the key employee who has been in the user area for years and is familiar with present activities and applications. As we shall see later, historical and sensitive information is often acquired from informants. In some cases, that is the only source available to the analyst.

INFORMATION-GATHERING TOOLS

No two projects are ever the same. This means that the analyst must decide on the information gathering tool and how it must be used. Although there are no standard rules for specifying their use, an important rule is that information must be acquired accurately, methodically, under the right conditions, and with minimum interruption to user personnel. For example, if the analyst needs only information available in existing manuals, then interviewing is unnecessary except where the manual is not up to date. If additional information is needed, on-site observation or a questionnaire may be considered. Therefore, we need to be familiar with various information-gathering tools. Each tool has a special function, depending on the information needed. The tools discussed in this chapter are shown in Figure 5-5.

Review of Literature, Procedures, and Forms

Very few system problems are unique. The increasing number of software packages suggests that problem solutions are becoming standardized. Therefore, as a first step, a search of the literature through professional references and procedures manuals, textbooks, company studies, government publications, or consultant studies may prove invaluable. The primary drawback of this search is time. Often it is difficult to get certain reports, publications may be expensive, and the information may be outdated due to a time lag in publication.

Procedures manuals and forms are useful sources for the analyst. They describe the format and functions of the present system. Included in most manuals are system requirements that help determine how well various objectives are met. Up-to-date manuals save hours of information-gathering time. Unfortunately, in many cases, manuals do not exist or are seriously out of date.

FIGURE 5-5 Information-Gathering Methods

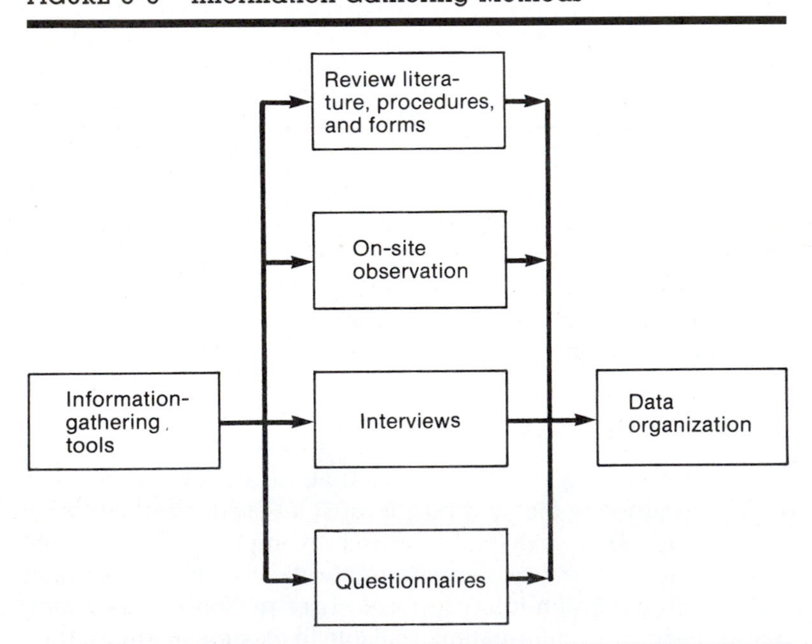

Included in the study of procedures and manuals is a close look at existing forms. Printed forms are widely used for capturing and providing information. Figure 5–6 illustrates the flow of a purchase order in a production system. The objective is to understand how forms are used. The following questions may be useful:

1. Who uses the form(s)? How important are they to the user?
2. Do the forms include all the necessary information? What items should be added or deleted?
3. How many departments receive the existing form(s)? Why? In Figure 5–6, each department has a reason for receiving a copy of the purchase order. It would make little sense, for instance, if the manager of the production department required copies of each purchase order even though puchase requisitions were initiated by the department.
4. How readable and easy to follow is the form?
5. How does the information in the form help other users make better decisions? What other uses does the form offer the user area?

On-Site Observation

Another information-gathering tool used in system studies is on-site observation. It is the process of recognizing and noting people, objects, and

FIGURE 5–6 Distribution Flow of a Purchase Order in a Production System

PURCHASE ORDER
A. B. Dick & Son, Inc.
4117 Waukegan Road,
Deerfield, Ill. 60015

To: E. B. Electronics Co.
Lake Geneva, Ill.
60100

Date Aug. 30, 1985
Deliver Before Sept. 30
Ship via Best way
f.o.b.
Arrive 2/10; 3/30

Quantity	Description	Unit Price	Total price
200	Transformers 392K	3.00	600.00
300	Switches 410A	2.25	675.00
130	Capacitors 17C	4.10	410.00
400	2 ft. wires 15	.35	140.00
60	Reactors 072	10.00	600.00

A.B. Dick & Son, Inc.
Reg. No. M110
By Eleanor Page
Purchasing Agent

Source: Purchasing department
Destination: Receiving department; Accounts payable department; Vendor; Computer center (processing)
1 → 1 → 1 Inventory control
2 → 2 → 2 Billing routine
3 → 3 → 3 Purchase analysis
4 (in file) (copies of purchase order)

occurrences to obtain information.[2] The analyst's role is that of an information seeker who is expected to be detached (therefore unbiased) from the system being observed. This role permits participation with the user staff openly and freely.

The major objective of on-site observation is to get as close as possible to the "real" system being studied. For this reason it is important that the analyst is knowledgeable about the general makeup and activities of the system. For example, if the focus of the analysis is communication, one needs to know as much as possible about the modes of communication available through the organization structure and the aspects of the physical layout that might adversely affect communication. The following questions can serve as a guide for on-site observations:

1. What kind of system is it? What does it do?
2. Who runs the system? Who are the important people in it?
3. What is the history of the system? How did it get to its present stage of development?
4. Apart from its formal function, what kind of system is it in comparison with other systems in the organization? Is it a primary or a secondary contributor to the organization? Is it fast paced or is it a leisurely system that responds slowly to external crises?

As an observer, the analyst follows a set of rules. While making observations, he/she is more likely to listen than talk and to listen with a sympathetic and genuine interest when information is conveyed. The emphasis is not on giving advice or passing moral judgment on what is observed. Furthermore, care is taken not to argue with the persons being observed or to show hostility toward one person and undue friendliness toward another.

When human observers are used, four alternative observation methods are considered:

1. *Natural or contrived.* A natural observation occurs in a setting such as the employee's place of work; a contrived observation is set up by the observer in a place like a laboratory.
2. *Obtrusive or unobtrusive.* An obtrusive observation takes place when the respondent knows he/she is being observed; an unobtrusive observation takes place in a contrived way such as behind a one-way mirror.
3. *Direct or indirect.* A direct observation takes place when the analyst actually observes the subject or the system at work. In an indirect observation, the analyst uses mechanical devices such as cameras and videotapes to capture information.

---

[2] Harper Boyd; Ralph Westfall; and Stanley Stasch, *Marketing Research: Text and Cases*, 5th ed. (Homewood, Ill.: Richard D. Irwin, 1981), p. 125.

4. *Structured or unstructured.* In a structured observation, the observer looks for and records a specific action such as the number of soup cans a shopper picks up before choosing one. Unstructured methods place the observer in a situation to observe whatever might be pertinent at the time.

Any of these methods may be used in information gathering. Natural, direct, obtrusive, and unstructured observations are frequently used to get an overview of an operation. The degree of structure is increased when observations have a specific purpose. An example is tracing the route of a sales invoice through a system. The degree of obtrusiveness may decrease when one wants to observe the tasks that make up a given job. For example, the analyst may want to create a list of the activities of a production supervisor by observing him/her from a remote location. Indirect observations could be used in a similar manner. For instance, the daily routine of a bank teller may be observed indirectly via a video camera. Finally, contrived situations are used to test or debug a candidate system. They are also used in training programs to help evaluate the progress of trainees.

Electronic observation and monitoring methods are becoming widely used information-gathering tools because of their speed, efficiency, and low cost. For example, some truck fleets use an electronic recorder system that records, analyzes, and reports information (online) about the hours and minutes a vehicle was driven faster than 60 miles per hour, the number of hours an engine was idle in a day, and how much out-of-service time a vehicle had.[3] These and other electronic methods expedite the information-gathering process in systems analysis.

On-site observations are not without problems:

1. Intruding into the user's area often results in adverse reactions by the staff. Therefore, adequate preparation and training are important.
2. Attitudes and motivations of subjects cannot be readily observed—only the actions that result from them.
3. Observations are subject to error due to the observer's misinterpretation and subjective selection of what to observe, as well as the subjects' altered work pattern during observation.
4. Unproductive, long hours are often spent in an attempt to observe specific, one-time activities or events.

In deciding to use an on-site observation, several questions are considered:

1. What behavior can be observed that cannot be described in other ways?
2. What data can be obtained more easily or more reliably by observation than by other means?

[3] "Electronic: Data for Fleet Management," *Fleetowner*, 76, no. 6 (June 1981), pp. 76–78.

3. What assurances can be given that the observation process is not seriously affecting the system or the behavior being observed?
4. What interpretation needs to be made about observational data to avoid being misled by the obvious?
5. How much skill is required and available for the actual observation?

For on-site observation to be done properly in a complex situation it can be very time-consuming. Proper sampling procedures must be used to ascertain the stability of the behavior being observed. Without a knowledge of stability, inferences drawn from small samples of behavior (small time slices) can be inaccurate.

Interviews and Questionnaires

As we have seen, on-site observation is directed primarily toward describing and understanding events as they occur. It has limitations when we need to learn about people's perceptions, feelings, or motivations, however. Therefore, other information-gathering tools are also used for analysis.

Information-gathering tools can be categorized by their degree of directness. If we wish to know about something, we simply ask someone about it directly, but we may not get an answer. Most of the information-gathering tools used in systems analysis are relatively direct. This is a strength because much of the information needed can be acquired by direct questions. There is information of a more difficult nature that user staff may be reluctant to give directly, however—for example, information on company politics or satisfaction with the supervisor. When asked by direct questions, the respondent may yield information that is invalid; yet properly handled, information can be successfully obtained with interviews or questionnaires.

Interviews

The interview is a face-to-face interpersonal role situation in which a person called the interviewer asks a person being interviewed questions designed to gather information about a problem area.[4] The interview is the oldest and most often used device for gathering information in systems work. It has qualities that behavioral and on-site observations do not possess. It can be used for two main purposes: (1) as an exploratory device to identify relations or verify information, and (2) to capture information as it exists.

Validity is no small problem. Special pains are taken to eliminate interview bias. We assume that information is more valid, the more freely it is given. Such an assumption stresses the voluntary character of the interview as a relationship freely and willingly entered into by the respondent. If the

[4] Fred N. Kerlinger, *Fundamentals of Behavioral Research*, 2d ed. (New York: Holt, Rinehart & Winston, 1973), p. 481.

interview is considered a requirement, the interviewer might gain the respondent's time and attention, but cannot be certain of the accuracy of the information gathered during the interview.

In an interview, since the analyst (interviewer) and the person interviewed meet face to face, there is an opportunity for flexibility in eliciting information. The analyst is also in a position to observe the subject. In contrast, the information obtained through a questionnaire is limited to the subject's written responses to predefined questions.

There are four primary advantages of the interview:

1. Its flexibility makes the interview a superior technique for exploring areas where not much is known about what questions to ask or how to formulate questions.
2. It offers a better opportunity than the questionnaire to evaluate the validity of the information gathered. The interviewer can observe not only what subjects say but also how they say it.
3. It is an effective technique for eliciting information about complex subjects and for probing the sentiments underlying expressed opinions.
4. Many people enjoy being interviewed, regardless of the subject. They usually cooperate in a study when all they have to do is talk. In contrast, the percentage of returns to a questionnaire is relatively low: often less than 20 percent. Attractively designed questionnaires that are simple to return, easy to follow, and presented in a context that inspires cooperation improve the return rate.

The major drawback of the interview is the long preparation time. Interviews also take a lot of time to conduct, which means time and money. So whenever a more economical alternative captures the same information, the interview is generally not used.

The Art of Interviewing. Interviewing is an art. Few analysts learn it in school, but most of them develop expertise through experience. The interviewer's art consists of creating a permissive situation in which the answers offered are reliable. Respondents' opinions are offered with no fear of being criticized by others. Primary requirements for a successful interview are to create a friendly atmosphere and to put the respondent at ease. Then the interview proceeds with asking questions properly, obtaining reliable responses, and recording them accurately and completely.

Arranging the Interview. The interview should be arranged so that the physical location, time of the interview, and order of interviewing assure privacy and minimal interruption. Usually a neutral location that is nonthreatening to the respondent is preferred. Appointments should be made well in advance and a fixed time period adhered to as closely as possible. Interview schedules generally begin at the top of the organization structure and work down so as not to offend anyone.

Guides to a Successful Interview. Interviewing should be approached as logically as programming. In an interview, the following steps should be taken:

1. Set the stage for the interview.
2. Establish rapport; put the interviewee at ease.
3. Phrase questions clearly and succinctly.
4. Be a good listener; avoid arguments.
5. Evaluate the outcome of the interview.

1. *Stage setting*. This is an "ice breaking," relaxed, informal phase where the analyst opens the interview by focusing on (a) the purpose of the interview, (b) why the subject was selected, and (c) the confidential nature of the interview.

After a favorable introduction, the analyst asks the first question and the respondent answers it and goes right through the interview. The job of the analyst should be that of a reporter rather than a debater. The direction of the interview is controlled by discouraging distracting conversation.

During stage setting the interviewer evaluates the cooperation of the interviewee. Both the content and tone of the responses are evaluated. How well the interview goes depends on whether the interviewee is the *friendly* type, the *timid* type who needs to be coaxed to talk, or the *resident expert*, who bombards the analyst with opinions disguised as facts. In any case, the analyst adjusts his/her own image to counter that of the interviewee.

2. *Establishing rapport*. In one respect, data collection is an imposition on user staff time and an intrusion into their privacy. Even though the procedure is authorized by management in advance, many staff members are reluctant to participate. There is seldom a direct advantage in supplying information to outsiders, regardless of their credentials. There is a strong perception that it may do them harm. This factor makes it important to gain and maintain rapport with the user staff. The investigation is an art. Although there are no ground rules to follow, there are pitfalls to avoid.

a. Do not deliberately mislead the user staff about the purpose of the study. A careful, well-thought-out briefing of participants should not provide any more detail than is necessary. Too much technical detail may tend to confuse people. The briefing should be consistent for all participants to avoid rumors.

b. Assure interviewees confidentiality that no information they offer will be released to unauthorized personnel. The promise of anonymity is very important.

c. Avoid personal involvement in the affairs of the user's department or identification with one faction at the cost of another. This may be difficult when several groups are involved in the study.

d. Avoid showing off your knowledge or sharing information received from other sources.

e. Avoid acting like an expert consultant or confidant. This can reduce the objectivity of the approach and discourage people from freely giving information.

f. Respect the time schedules and preoccupations of your subjects. Do not make an extended social event out of the meeting. If the subject does not complain, subordinates might, especially if they are waiting to see the subject (boss).

g. Do not promise anything you cannot or should not deliver, such as advice or feedback.

h. Dress and behave appropriately for the setting and the circumstances of the user contact.

i. Do not interrupt the interviewee. Let him/her finish talking.

3. *Asking the questions*. Except in unstructured interviews, it is important that each question is asked exactly as it is worded. Rewording or impromptu explanation may provoke a different answer or bias the response. The questions must also be asked in the same order as they appear on the interview schedule. Reversing the sequence could destroy the comparability of the interviews. Finally, each question must be asked unless the respondent, in answering a previous question, has already answered the next one.

4. *Obtaining and recording the response*. Interviewers must be prepared to coax respondents to elicit further information when necessary. The "probing" technique enables the interviewer to act as a catalyst, for example:

a. *Interviewer:* I see what you mean. Could you elaborate further on that?

b. *Analyst (interviewer):* How do you feel about separating the present loan division into commercial and loan departments?

Financial vice president (respondent): Well, I'm not sure. Sometimes I think that we have to take this route eventually.

Analyst: I see. Can you tell me more about that?

These statements indicate that the analyst is listening, is interested, understands what the respondent is trying to say, and is making an effort to gain more information. The information received during the interview is recorded for later analysis.

5. *Data recording and the notebook*. Many system studies fail because of poor data recording. Care must be taken to record the data, their source, and the time of collection. If there is no record of a conversation, the analyst runs the risk of not remembering enough details, attributing them to the wrong source, or otherwise distorting the data.

The form of the notebook varies according to the type of study, the amount of data, the number of analysts, and their individual preferences. The "notebook" may be a card file, a set of carefully coded file folders, or a looseleaf binder. It should be bound and the pages numbered. The information shown in Figure 5–7 should be included in the notebook.

FIGURE 5-7 Data Capture and the Notebook

1. Originals or duplicate copies of all notes taken during investigation are documented. They are the chief sources of interview and observational data, as well as background information on the system. Each page of notes should be numbered serially, and a running chronological record of them should be kept. The name of the analyst, the date the notes were taken, and surrounding circumstances are all important. Since handwritten notes often are not intelligible to others, it is good to have them transcribed or typed soon after they are taken.
2. Copies of all information-gathering tools—questionnaires, interview schedules, observation guides—are placed in the notebook for future reference.
3. Copies of all data—originals or duplicates—are included. Loss of key data, even temporarily, can be costly.
4. Minutes of all meetings as well as a record of discussions, decisions, and changes in design all become part of the notebook.

The organization of the notebook is also important. In some cases, a purely chronological arrangement will suffice. In others, a system of categories with cross-classification would be appropriate. Proper indexing makes it easier to retrieve information when needed.

Questionnaires

In contrast to the interview is the *questionnaire*, which is a term used for almost any tool that has questions to which individuals respond. It is usually associated with self-administered tools with items of the closed or fixed alternative type. By its nature, a questionnaire offers the following advantages:

1. It is economical and requires less skill to administer than the interview.
2. Unlike the interview, which generally questions one subject at at time, a questionnaire can be administered to large numbers of individuals simultaneously.
3. The standardized wording and order of the questions and the standardized instructions for reporting responses ensure uniformity of questions. In contrast, the interview situation is rarely uniform from one interview to the next.
4. The respondents feel greater confidence in the anonymity of a questionnaire than in that of an interview. In an interview, the analyst usually knows the user staff by name, job function, or other identification. With a questionnaire, respondents give opinions without fear that the answer will be connected to their names.
5. The questionnaire places less pressure on subjects for immediate responses. Respondents have time to think the questions over and do calculations to provide more accurate data.

The advantages of the self-administered questionnaire outweigh disadvantages, especially when cost is a consideration. The principal disadvantage is a low percentage of returns. Another disadvantage is that many people have difficulty expressing themselves in writing, especially when responding to essay (open) questions. Many dislike writing. Because of these disadvantages, the interview is probably superior to the questionnaire.

Types of Interviews and Questionnaires

Interviews and questionnaires vary widely in form and structure. Interviews range from the highly unstructured, where neither the questions nor the respective responses are specified in advance, to the highly structured alternative in which the questions and responses are fixed. Some variation within this range is possible.

The Unstructured Alternative

The unstructured interview is a relatively nondirective information-gathering technique. It allows respondents to answer questions freely in their own words. The responses are spontaneous rather than forced. They are self-revealing and personal rather than general and superficial. The role of the analyst as an interviewer is to encourage the respondent to talk freely and serve as a catalyst to the expression of feelings and opinions. This is best achieved in a permissive atmosphere in which the subjects have no feeling of disapproval.

The Structured Alternative

In the structured approach, the questions are presented with exactly the same wording and in the same order to all subjects. If the analyst asks a subject, "Would you like to see a computerized approach to solving your accounts receivable problem?" and asks another subject, "How do you feel about computers handling accounts receivable?" the response may not be the same even though the subjects both have the same opinion. Standardized questions improve the reliability of the responses by ensuring that all subjects are responding to the same questions.

Structured interviews and questionnaires may differ in the amount of structuring of the questions. Questions may be either closed or open-ended. An *open-ended* question requires no response direction or specific response (see Figure 5–8). In a questionnaire, it is written with space

FIGURE 5–8 Examples of Open-Ended Questions

- Now that you have had the new installation for six months, how would you evaluate the benefits?
- What is your opinion regarding the "no smoking" policy in the DP center?
- If you had a choice, how would you have designed the present information center?

FIGURE 5-9 Examples of Fill-in-the Blank Questions

- What is the name of the MIS director of your firm?

  ________________.

- How many analysts handle the accounts receivable conversion?

  ____

- What is the average number of calls you receive from clients?

  ____

provided for the response. Such questions are more often used in interviews than in questionnaires because scoring takes time.

Closed questions are those in which the responses are presented as a set of alternatives. There are five major varieties of closed questions:

1. *Fill-in-the-blanks* questions request specific information (Figure 5-9). These responses can then be statistically analyzed.

2. *Dichotomous (yes/no type)* questions that offer two answers (Figure 5-10) have advantages similar to those of the multiple-choice type (explained later). The problem is making certain that a reliable response can be answered by yes or no; otherwise, an additional choice (e.g., yes, no, I don't know) should be included. The question sequence and content are also important.

3. *Ranking scales* questions ask the respondent to rank a list of items in order of importance or preference. In Figure 5-11, the first question asks the respondent to rank five statements on the basis of how they describe his/her present job.

4. *Multiple-choice* questions offer respondents specific answer choices (Figure 5-12). This offers the advantage of faster tabulation and less analyst bias due to the order in which the answers are given. Respondents have a favorable bias toward the first alternative item. Alternating the order in which answer choices are listed may reduce bias but at the expense of additional time to respond to the questionnaire. In any case, it is important

FIGURE 5-10 Examples of Dichotomous Questions

- Are you personally using a microcomputer in your business? (please circle one)

  yes no

- If not, do you plan to be using one in the next 12 months? (please circle one)

  yes no

- In the performance of your work, are you personally involved in computer hardware/software purchase decisions? (please circle one)

  yes no

FIGURE 5–11 An Example of a Ranking Scales Question

Please rank the five statements in each group on the basis of how well they describe the job mentioned on the front page. Write a "1" by the statement that best describes the job; write a "2" by the statement that provides the next best description, and continue ranking all five statements, using a "5" for the statement that describes the job least well.

Workers on this job . . .

_____ are busy all the time.

_____ have work where they do things for other people.

_____ try out their own ideas.

_____ are paid well in comparison with other workers.

_____ have opportunities for advancement.

---

to be aware of these types of bias when constructing multiple-choice questions.

5. *Rating scales* questions are an extension of the multiple-choice design. The respondent is offered a range of responses along a single dimension. In Figure 5–13, the respondent is asked to rate various aspects of his/her job on a scale of 1–5.

Open-ended and closed questions have advantages and limitations. Open-ended questions are ideal in exploratory situations where new ideas and relationships are sought. The main drawback is the difficulty of inter-

FIGURE 5–12 Examples of Multiple-Choice Questions

- What is the average salary of an entry-level analyst? (please check one)

  _____ Under $15,000

  _____ $15,000–$19,999

  _____ $20,000–$24,999

  _____ Over $25,000

- Please check one category that best describes the business of the firm where you are employed.

  _____ Savings bank

  _____ College, school, library, association

  _____ Computer service

  _____ Industrial company

  _____ Outside computer consulting

  _____ Other (please describe) ____________________

FIGURE 5–13 An Example of a Rating Scale Question

- How satisfied are you with the following aspects of your present job? (please circle one for each question)

	Very Dissatisfied	Dissatisfied	No Opinion	Satisfied	Very Satisfied
1. The way my job provides for steady employment	1	2	3	4	5
2. The chance to be responsible for the work of others	1	2	3	4	5
3. The pleasantness of the working conditions	1	2	3	4	5
4. The chance to make use of my best abilities	1	2	3	4	5

preting the subjective answers and the tedious responses to open-ended questions. Other drawbacks include potential analyst bias in interpreting the data and time-consuming tabulation. Closed questions are quick to analyze, but typically most costly to prepare. They are more appropriate for securing factual information (for example, about age, education, sex, and salary). They have the additional advantage of ensuring that answers are given in a frame of reference consistent with the line of inquiry. A summary of structured and unstructured interview techniques is given in Figure 5–14.

FIGURE 5–14 Structured and Unstructured Interview Techniques—A Summary

Interview Type	Advantages	Drawbacks
Structured	1. Easy to administer and evaluate due to standardization 2. Requires limited training 3. Easy to train new staff	1. High initial preparation cost 2. Standardization of questions tends to reduce spontaneity 3. Mechanizes interviewing, which makes it impractical for all interview settings
Unstructured	1. Provides for greater creativity and spontaneity in interviewing 2. Facilitates deeper understanding of the feelings and standing of the interviewee 3. Offers greater flexibility in conducting an overall interview	1. More information of questionable use is gathered 2. Takes more time to conduct—therefore, costly 3. Requires extensive training and experience for effective results

Procedure for Questionnaire Construction

The procedure for constructing a questionnaire consists of six steps:

1. Decide what data should be collected; that is, define the problem to be investigated.
2. Decide what type of questionnaire (closed or open-ended) should be used.
3. Outline the topics for the questionnaire and then write the questions.
4. Edit the questionnaire for technical defects or biases that reflect personal values.
5. Pretest (try out) the questionnaire to see how well it works.
6. Do a final editing to ensure that the questionnaire is ready for administration. This includes a close look at the content, form, and sequence of questions as well as the appearance and clarity of the procedure for using the questionnaire.

A critical aspect of questionnaire construction is the formulation of reliable and valid questions. To do a satisfactory job, the analyst must focus on question content, wording, and format. The following is a checklist of what to consider:

1. Question content.
   a. Is the question necessary? Is it a part of other questions?
   b. Does the question adequately cover the area intended?
   c. Does the subject(s) have proper information to answer the question?
   d. Is the question biased in a given direction?
   e. Is the question likely to generate emotional feelings that might lead to untrue responses?
2. Question wording.
   a. Is the question worded for the subject's background and experience?
   b. Can the question be misinterpreted? What else could it mean to a respondent?
   c. Is the frame of reference uniform for all respondents?
   d. Is the wording biased toward a particular answer?
   e. How clear and direct is the question?
3. Question format.
   a. Can the question best be asked in the form of check answer (answered by a word or two or by a number) or with a follow-up free answer?
   b. Is the response form easy to use or adequate for the job?
   c. Is the answer to the question likely to be influenced by the preceding question? That is, is there any *contamination effect*?

Reliability of Data from Respondents. The data collected from the user staff are presumed to accurately correspond with the actual way in which events occur. If such reports are the only source of data, there may be several uncontrolled sources of error:

1. The respondent's perceptual slant. Perceptual ability is known to vary. Reports of a given event from several staff members who have no training in careful observation often have little resemblance to one another.
2. The respondents failure to remember just what did happen. Assuming that he or she received a fairly reliable impression of an event at the time that it happened, it generally becomes more difficult with the passage of time to describe the details of an event.
3. Reluctance of persons being interviewed to report their "true" impressions of what occurred. A subject often distorts descriptions of events for fear of retaliation, a desire not to upset others, or a general reluctance to verbalize a particular type of situation.
4. Inability of subjects to communicate their reports or inability of the analyst to get from subjects the information that they are qualified to provide.

The Reliability-Validity Issue

An information-gathering instrument faces two major tests: reliability and validity. Before administering the instrument, the analyst must ask and answer the questions: What is the reliability of the measuring instrument? What is its validity? The term *reliability* is synonymous with dependability, consistency, and accuracy. Concern for reliability comes from the necessity for dependability in measurement. Using the questionnaire as an example, reliability may be approached in three ways:

1. If we administer the same questionnaire to the same subjects, will we get the same or similar results? This question implies a definition of reliabilty as *stability, dependability,* and *predictability*.
2. Does the questionnaire measure the true variables it is designed to measure? This question focuses on the *accuracy* aspect of reliability.
3. How much error of measurement is there in the proposed questionnaire? Errors of measurement are random errors stemming from fatigue or fortuitous conditions at a given time, or fluctuations in mood that temporarily affect the subjects answering the questionnaire. To the extent that errors of measurement are present in a questionnaire, it is unreliable. Thus, reliability is viewed as the relative absence of errors of measurement in a measuring instrument.[5]

[5] Ibid., p 442.

To illustrate, suppose we administered a questionnaire to measure the attitude of the user staff toward a new computer installation. The "true" scores of the five staff members were 92 (excellent attitude), 81, 70, 59, and 40. Suppose further that the same questionnaire was administered again to the same group within the same time period and the scores were 96, 82, 69, 61, and 55. Although not a single case hit the "true" score again, the second test showed the same rank order. The reliability in this example is extremely high.

Now suppose that the last set of scores had been 72, 89, 51, 74, and 67. They are the same five scores, but they have a different rank order. In this case, the test is unreliable. Figure 5–15 shows the three sets of scores. The rank orders of the first two sets of scores covary exactly. Even though the test scores in the two columns are not the same, they are in the same rank order. To this extent, the test is reliable. The opposite case is shown in columns (1) and (3). The rank order changed, making the test unreliable.

It can be seen, then, that for an information-gathering instrument to be interpretative, it must be reliable. Unreliable measurement is overloaded with error. Although high reliability is no guarantee of good questionnaire results, there can be no good results without reliability. It is a necessary, but not sufficient, condition for the value of questionnaire results and their interpretation.

The most common question that defines *validity* is: Does the instrument measure what we think it is measuring? It refers to the notion that the questions asked are worded to produce the information sought. In contrast, reliability means that the information gathered is dependable enough to be used for decision making. In validity, the emphasis is on what is being measured. For example, an analyst administers a questionnaire to a user group to measure their understanding of a billing procedure and has included in the questionnaire only factual items that identify the parts of the billing system. The questionnaire is not valid because, whereas it may measure employees' factual knowledge of the billing system, it does not measure their understanding of it. In other words, it does not measure what

FIGURE 5–15 Reliable and Unreliable Test Scores

(1) "True" Scores	Rank	(2) Scores from Reliable Questionnaire	Rank	(3) Scores from Unreliable Questionnaire	Rank
92	1	96	1	72	3
81	2	82	2	89	1
70	3	69	3	51	5
59	4	61	4	74	2
40	5	55	5	67	4

the analyst intended to measure. For this reason, it is important to pretest a questionnaire for validity as well as for reliability.

It can be concluded, then, that the adequacy of an information-gathering tool is judged by the criteria of validity and reliability. Both depend on the design of the instrument as well as the way it is administered.

Summary

1. Much of the information that we need to analyze a system relates to the organization, the user staff, and the work flow. Organization-based information deals with policies, objectives, goals, and structure. User-based information focuses on job functions, information requirements, and interpersonal relationships. Work-based information addresses the work flow, methods and procedures, and work schedules. We are interested in what happens to the data through various points in the system.
2. Information is gathered from sources within the organization and from the organization's environment. External sources include vendors, government documents, and professional journals. The primary internal sources are financial reports, personnel, system documentation, and users.
3. The primary information-gathering tools are documentation, on-site observation, interviews, and questionnaires. The most commonly used tool is the interview.
4. The major objective of on-site observation is to get close to the "real" system being studied. The methods used may be natural or contrived, obtrusive or unobtrusive, direct or indirect, and structured or unstructured. The main limitation of observation is the difficulty of observing attitudes and motivation and the many unproductive hours that often are spent in observing one-time activities.
5. The interview is a face-to-face interpersonal meeting designed to identify relations and capture information as it exists. It is a flexible tool, offering a better opportunity than the questionnaire to evaluate the validity of the information gathered. The major drawback is preparation time. Interviewing is an art that requires experience in arranging the interview, setting the stage, establishing rapport, phrasing questions clearly, avoiding arguments, and evaluating the outcome.
6. The questionnaire is a self-administered tool that is more economical and requires less skill to administer than the interview. It examines a large number of respondents at the same time, provides standardized wording and instructions, and places less pressure on subjects for immediate response. The main drawback is the low percentage of returns.
7. An interview or a questionnaire may be structured or unstructured. The unstructured approach allows respondents to answer questions freely

in their own words, wheras the structured approach requires a specific response to open-ended or closed questions.

8. There are five major varieties of closed questions:
   a. Fill-in-the-blanks questions request specific information.
   b. Dichotomous questions offer a two-answer choice.
   c. Ranking scales questions ask the respondent to rank a list of items in order of importance or preference.
   d. Multiple-choice questions ask for specific answer choices.
   e. Rating scales questions ask the respondent to rank various items along a single dimension (scale).
9. In constructing a questionnaire, the analyst must focus on question content, wording, and format. These are considered with validity and reliability in mind. There are uncontrolled sources of error, however, that stem from the respondent's perceptual slant, failure to remember specific details, reluctance to report the "true" impressions of what occurred, or inability to communicate information.
10. An information-gathering instrument faces the tests of reliability and validity. Reliability is synonymous with dependability, consistency, and accuracy, whereas validity emphasizes what is being measured. It should measure what the analyst intended to measure.

Key Words

Closed Question
Contrived Observation
Dichotomous Question
Direct Observation
Fill-in-the-Blanks Question
Indirect Observation
Informant
Interview
Multiple-Choice Question
Natural Observation
Obtrusive Observation
On-Site Observation
Open-Ended Question
Questionnaire
Ranking Scales Question
Rating Scales Question
Reliability
Structured Interview
Structured Observation
Unobtrusive Observation
Unstructured Interview
Unstructured Observation
Validity

Review Questions

1. What categories of information are available for analysis? How would one decide on the category for a given project?
2. Why is it important that the analyst learns about an organization's policies and objectives?
3. Information is available from internal and external sources. Under what circumstances would the analyst depend more heavily on external than internal information? Why?

4. How is the informant useful in systems analysis? Explain.
5. What traditional information-gathering tools are available for the analyst? Explain each tool briefly.
6. Visit the computer center of a local firm. Review a user manual and report your findings.
7. What is considered in evaluating forms? Explain.
8. How would one conduct an on-site observation? Lay out a plan and specify the pros and cons of this tool.
9. If you were asked to observe a computer operator at work, what observation method would you select? Why?
10. Summarize the advantages and limitations of interviews and questionnaires.
11. Under what circumstances or for what purposes would one use an interview rather than other data-collection methods? Explain.
12. Explain the difference between (a) structured and unstructured interviewing and (b) open-ended and closed questions. Give an example of each.
13. List and explain the primary steps in interviewing.
14. Explain briefly the procedure used to construct questionnaires.
15. If you were to interview a user to obtain biographical information (age, education, years of experience on the job, and so forth) about the staff of 10 employees and you have only one hour to acquire the information, which of the following methods would you use and why?
    a. Structured interviews using open-ended questions.
    b. Unstructured interviews of five minutes each.
    c. Self-administered questionnaires.
    d. Structured interviews using closed questions.
16. In what respect is interviewing an art? Explain.
17. Suppose you have completed the first draft of a questionnaire, how would you pilot test it?
18. What sources of error affect the reliability of data from respondents? Elaborate.
19. What is rapport? As an analyst, how do you gain and maintain rapport with the user's staff? Give an example.
20. What kinds of data should be recorded? Why?
21. Distinguish between validity and reliability. How are they related?
22. Explain and give an example of each variety of closed questions:
    a. Fill-in-the-blanks questions.
    b. Dichotomous questions.
    c. Ranking scales questions.
    d. Multiple-choice questions.
    e. Rating scales questions.

Application Problems

1 The systems analyst of a radio assembly plant contacts the manager of the production department. She briefs him on the survey she is taking and asks the manager to help her get answers to some questions. The manager is cordial, and he invites her to come over. The following interview takes place.

Analyst: What is the main function of your department?

Manager: We assemble radios from components and ship them to order.

Analyst: How many people work here?

Manager: Why do you want to know?

Analyst: It could be that you have too many people on your payroll.

Manager: Maybe I should be the judge of that.

Analyst [*ignores answer*]: What's that girl doing in the room across the hall? She hasn't done a thing since I walked in here.

Manager: She verifies shipping orders. It could be that she is waiting for more orders from purchasing.

Analyst: Why do you need to check these orders when they have already been cleared for production?

Manager: We've had occasions when the units ordered belong to more than one person or to another address.

Analyst: I want to talk to her.

The manager reluctantly agrees. The manager walks with the analyst to the clerk's desk. She is idle. He introduces the analyst to her.

Analyst: What work do you do, Miss Meyer?

Meyer: I verify the goods against shipping orders.

Analyst: How do you know that the shipping orders are correct?

Meyer: I guess I don't, but I verify the type, number of units ordered, and shipping address against the units produced before they are loaded on the truck.

Analyst: Aren't you wasting your time doing this?

Meyer: You'll have to ask Mr. Kehoe (the manager) that.

The manager, standing by, begins to get irritated. The analyst now talks to the manager while in Meyer's area.

Analyst: That's all I wanted to find out from this area. What are those other girls doing there?

Manager: They're preparing bills of lading, taping the firms's logo on the cartons, and making sure that the bill accompanies each order.

Analyst: I'd like to walk over there and talk with one of them. It won't take a minute.

Manager: They're pretty busy right now. Jane over on the right is breaking in a new girl we just hired. If you are after the procedure, we have it all documented. I'd be glad to give you a copy in my office.

Analyst: I'm not sure how up to date your documentation is. I'd rather hear it from them.

The manager leads the analyst to the west corner of the warehouse where four girls are typing. He introduces the analyst to the senior clerk.

Analyst: How many bills of lading does your average typist prepare per day?

Senior clerk: Around 60; maybe 70.

Analyst: You have five typists here, including yourself, and your total output yesterday was only 200. What happended?

Senior clerk: First, as you can see, we're training a new person here. The girls also file, call customers to tell them that the order is on its way, and the like. We stay busy.

Analyst: This is fine, but what else do they do?

Senior clerk: Well, we take the bills of lading to the drivers at the dock and have them sign for the shipment.

Analyst: Don't you think that this running around is a waste of time?

Senior clerk: [no answer]

Assignment

a. How do you rate the interview? Explain.

b. What type of interview was conducted by the analyst?

c. What questions were open-ended? Closed?

d. Should the analyst have asked the questions in the same sequence to all respondents? Why?

e. Critique the analyst's questions in terms of their content, wording, and format.

f. If you were the analyst, illustrate how you would have conducted the interview.

2 The dean of students of a major university requested the development of a nonacademic transcript (NAT). The NAT has the objective of maintaining a record of each student's extracurricular activities while in school. Three benefits are listed:

a. Faculty may gain better insight into student activities when making recommendations to employers, graduate schools, or awards committees, without much effort.

b. Faculty can work with students to coordinate career or educational goals with extracurricular commitments.

c. Students may submit their resumes and produce a separate record of extracurricular activities for a prospective employer. The employer, in turn, may access such records, with the student's permission.

To determine the feasibility, the college of business at the university was used as a prototype. It was hoped that the results could then be applied across the university. To determine the various organizations on campus, interviews were conducted with the staff of the dean's office, the career planning director of the business school, and the registrar. Extracurricular activities were classified into four groups:

a. Greek organizations, including fraternities, sororities, and interfraternity and intersorority councils.

b. Organizations recognized by the school council.

c. University-affiliated organizations.

d. Athletics (men's and women's) as well as varsity and junior varsity.

The questionnaire shown here was used to collect data from fourth-year MIS students at the college of business. After the data were tabulated, the specific format of the NAT was developed and information was entered into the data base. Since an IBM PC lab was readily available, a dBASE II package was used to implement the prototype.

Assignment

a. What type of questionnaire was used?

b. Critique the questionnaire in terms of its length, completeness, organization, and sequence. What changes would you make?

c. Since a prototype was selected for determining feasibility, should all university activities have been included rather than only those that were unique to business students? Why?

NONACADEMIC TRANSCRIPT
DATA COLLECTION FORM

Instructions: After filling out the top portion of this form, please check each activity on the list below that you were involved in during your enrollment at the university.

Name: ______________________________ SSN ____________
first m.i. last

Current Address: ______________________________

____

____

Phone: ______________

School: ______________

Year: ______________

Major: ______________________________

Please check each activity or activities in which you have been or are currently involved. Indicate your level of involvement and each year of participation (e.g., 2 = second year).

Example:

Activity	Level of Involvement	Years Participated
✓ Alpha Epsilon Pi	Vice president, rush chairman	(1)(2)(3)(4)

Greek Organizations

Activity	Level of Involvement	Years Participated
____ Alpha Epsilon Pi		1 2 3 4
____ Alpha Phi Alpha		1 2 3 4
____ Alpha Tau Omega		1 2 3 4
____ Beta Theta Pi		1 2 3 4
____ Chi Phi		1 2 3 4
____ Chi Psi		1 2 3 4
____ Delta Kappa Epsilon		1 2 3 4
____ Delta Sigma Phi		1 2 3 4

	Activity	Level of Involvement	Years Participated
____	Delta Tau Delta		1 2 3 4
____	Delta Upsilon		1 2 3 4
____	Kappa Alpha		1 2 3 4
____	Kappa Alpha Psi		1 2 3 4
____	Kappa Sigma		1 2 3 4
____	Omega Psi Phi		1 2 3 4
____	Phi Beta Sigma		1 2 3 4
____	Phi Delta Theta		1 2 3 4
____	Phi Epsilon Pi of ZBT		1 2 3 4
____	Sigma Phi Epsilon		1 2 3 4
____	Sigma Pi		1 2 3 4
____	Tau Kappa Epsilon		1 2 3 4
____	Theta Chi		1 2 3 4
____	Theta Delta Chi		1 2 3 4
____	Zeta Psi		1 2 3 4
____	Inter-Fraternity Council		1 2 3 4

Organizations Recognized by Student Council

	Activity	Level of Involvement	Years Participated
____	Aikido Club		1 2 3 4
____	AFROTC Cadet Group		1 2 3 4
____	Akindelas Fraternity, Inc.		1 2 3 4
____	Alpha Phi Omega Theta Chapter		1 2 3 4
____	American Advertising Federation		1 2 3 4
____	Amnesty International, USA		1 2 3 4
____	AROTC Cadet Assoc.		1 2 3 4
____	Asian Studies Club		1 2 3 4
____	Assoc. for Arab American Understanding		1 2 3 4
____	Assoc. for Computing Machinery (ACM)		1 2 3 4
____	Baptist Student Union		1 2 3 4
____	Beta Alpha Psi		1 2 3 4
____	Black Engineering Society		1 2 3 4
____	Black Student Alliance		1 2 3 4
____	Black Voices		1 2 3 4
____	Blue Ridge Mountain Rescue Group		1 2 3 4
____	Bowling Club		1 2 3 4
____	B'nai Brith Hillel		1 2 3 4

	Activity	Level of Involvement	Years Participated
_____	Campus Girl Scouts		1 2 3 4
_____	Catholic Fellowship		1 2 3 4
_____	Cavalier Club		1 2 3 4
_____	Chess Club		1 2 3 4
_____	Chinese Student Assoc.		1 2 3 4
_____	Circle K		1 2 3 4
_____	Corks and Curls		1 2 3 4
_____	Country Dance and Song Society		1 2 3 4
_____	Creator Magazine		1 2 3 4
_____	Croquet Club		1 2 3 4
_____	Cycling Team		1 2 3 4
_____	Dance Club		1 2 3 4
_____	The Declaration		1 2 3 4
_____	Deliverance Crusade for Christ		1 2 3 4
_____	German Honor Society		1 2 3 4
_____	Ecology Club		1 2 3 4
_____	Economics Club		1 2 3 4
_____	El Limited		1 2 3 4
_____	Engineering Council		1 2 3 4
_____	English Club		1 2 3 4
_____	Environmental Awareness Assoc.		1 2 3 4
_____	Environmental Forum and Research		1 2 3 4
_____	Episcopal Assoc./Canterbury Student Fellowship		1 2 3 4
_____	Fencing Club		1 2 3 4
_____	Frisbee Club		1 2 3 4
_____	Gamma Nu Psi Fraternity		1 2 3 4
_____	Gay Student Union		1 2 3 4
_____	German Club		1 2 3 4
_____	Glee Club		1 2 3 4
_____	Gymnastic Club		1 2 3 4
_____	Indian Student Assoc.		1 2 3 4
_____	Intercollegiate Policy Analysis Council		1 2 3 4
_____	International Business Society		1 2 3 4
_____	International Club		1 2 3 4
_____	International Relations Club		1 2 3 4
_____	Irish Cultural Society		1 2 3 4
_____	Israel Interest Group		1 2 3 4

	Activity	Level of Involvement	Years Participated
____	Italian Club		1 2 3 4
____	Jazz Ensemble		1 2 3 4
____	Karate and Self-Defense Club		1 2 3 4
____	Korean Club		1 2 3 4
____	Latin American Studies Group		1 2 3 4
____	Libertarian Student Assoc.		1 2 3 4
____	Loki Science Magazine		1 2 3 4
____	McIntire Marketing Assoc.		1 2 3 4
____	Madison House		1 2 3 4
____	Marantha Christian Fellowship		1 2 3 4
____	John B. Minor Pre-Legal Society		1 2 3 4
____	Modulus		1 2 3 4
____	Musique		1 2 3 4
____	NAACP		1 2 3 4
____	NROTC Honor Guard		1 2 3 4
____	Navigators		1 2 3 4
____	Okinawam Kempo Karate Club		1 2 3 4
____	Omega Rho		1 2 3 4
____	Opportunity Consultants		1 2 3 4
____	Othermind Magazine		1 2 3 4
____	Peer Alcohol Educators		1 2 3 4
____	Peer Sexuality Educators		1 2 3 4
____	Polo Club		1 2 3 4
____	Racquetball Club		1 2 3 4
____	Republican Club		1 2 3 4
____	Riding Club		1 2 3 4
____	Rifle and Pistol Club		1 2 3 4
____	Right to Life Committee		1 2 3 4
____	Rowing Assoc.		1 2 3 4
____	Rugby Club, UVA Men's		1 2 3 4
____	Rugby Club, UVA Women's		1 2 3 4
____	Running Club		1 2 3 4
____	Sailing Assoc.		1 2 3 4
____	Skiing, Alpine Assoc.		1 2 3 4
____	Slavic Society		1 2 3 4
____	Society for Creative Anachronism		1 2 3 4
____	Society of Women Engineers		1 2 3 4
____	Spanish Club		1 2 3 4
____	Students for Handgun Control		1 2 3 4
____	Students' International Meditation Society		1 2 3 4

	Activity	Level of Involvement	Years Participated
____	Students Together Against Racial Separation		1 2 3 4
____	Symphonic Band		1 2 3 4
____	Tennis Club		1 2 3 4
____	United Student for America		1 2 3 4
____	The University Journal		1 2 3 4
____	University Singers		1 2 3 4
____	Volleyball Club		1 2 3 4
____	Volunteer Income Tax Assistance Program		1 2 3 4
____	Volunteers for Youth		1 2 3 4
____	Washington Literary Society and Debating Union		1 2 3 4
____	Water Polo Club		1 2 3 4
____	Weightlifting Club		1 2 3 4
____	Women's Chorus		1 2 3 4
____	Young Americans for Freedom		1 2 3 4
____	Young Democrats		1 2 3 4

University-Affiliated Organizations

	Activity	Level of Involvement	Years Participated
____	Engineering Council		1 2 3 4
____	Student Nurses Assoc.		1 2 3 4
____	Assoc. of Education Students		1 2 3 4

Men's Varsity and Junior Varsity Athletics

	Activity	Level of Involvement	Years Participated
____	Baseball		1 2 3 4
____	Basketball		1 2 3 4
____	J. V. Basketball		1 2 3 4
____	Cross Country		1 2 3 4
____	Track		1 2 3 4
____	Football		1 2 3 4
____	J. V. Football		1 2 3 4
____	Golf		1 2 3 4
____	Lacrosse		1 2 3 4
____	Soccer		1 2 3 4
____	Swimming		1 2 3 4
____	Tennis		1 2 3 4
____	Wrestling		1 2 3 4

Women's Varsity and Junior Varsity Athletics

	Activity	Level of Involvement	Years Participated
____	Basketball		1 2 3 4
____	Cross Country		1 2 3 4
____	Track		1 2 3 4
____	Field Hockey		1 2 3 4
____	Lacrosse		1 2 3 4
____	J. V. Lacrosse		1 2 3 4
____	Softball		1 2 3 4
____	Swimming		1 2 3 4
____	Tennis		1 2 3 4
____	Volleyball		1 2 3 4

Please respond to the following questions:

1. Do you feel that a nonacademic transcript is needed at the University?
   ____ Yes ____ No ____ Undecided

2. Who do you feel would benefit most from a nonacademic transcript?
   ____ Students
   ____ Recruiters
   ____ Faculty
   ____ Administration
   ____ Other, please specify ______________________

3. In your opinion, for which use would a nonacademic transcript be most effective?
   ____ Resume supplement (employment purposes)
   ____ Part of a developmental program (advisory program coordinating activities toward specific educational or career goals)
   ____ Other, please specify ______________________

4. In your opinion, how could a nonacademic transcript be most effectively updated?
   ____ At registration
   ____ Through faculty advisors
   ____ Distributed questionnaires
   ____ Other, please specify ______________________

5. Please list any additional activities you feel should be included on a nonacademic transcript (e.g., student government, intramural sports, employment, etc.):

3 A large wholesale liquor distributor has been having difficulty keeping inventory up to date because incoming shipments are not processed quickly enough. The sales force can never be certain which brands and quantities are available for sale. The vice president of sales asks an outside analyst to investigate the problem.

The analyst arrived at the computer center Monday at 8:00 A.M. He asked to see the manager. The receptionist told him that Mr. Sibley came around 9:00 A.M. Not wanting to waste an hour, the analyst decided to interview the programmers to learn about inventory control.

Mr. Sibley arrived at 8:45. He was furious to find that the analyst (an outsider) had taken the liberty to talk to his staff without his consent. He promptly told the analyst that Monday is a bad day. The programmers could not be interrupted before 3:30 P.M. on Tuesday. The analyst agreed to come then. Meanwhile, he went to the stockroom to observe stockkeeping activities.

The next day, the analyst interviewed more employees in the stockroom and the clerical staff of the warehouse. It was 4:30 P.M. when he remembered his appointment with the manager of the computer center. He abruptly terminated his work in the warehouse and rushed back to the computer center for the interview. The manager had been waiting for an hour and was in an irritable mood. To make things worse, the firm's employees quit work at 5:00 P.M. The manager decided to go ahead with the interview.

The analyst inquired about data capture, stock activities, data flow, processing routines, and stock status reports. After a brief rundown on the procedures used and the reports generated, the analyst was curtly dismissed. It was 5:00 P.M.

The analyst had more questions to ask but had to stop. At the same time, he was wondering why the manager was so irritable throughout the interview.

Assignment

a. How do you assess the analyst's performance on the job? Explain.

b. Evaluate the procedure the analyst used in meeting the manager of the computer center.

c. How adequately prepared was the analyst for the first interview?

d. If you were the systems analyst, how would you have handled this project? Elaborate.

Selected References

Athey, Thomas H. "Information Gathering Techniques." *Journal of Systems Management*, January 1980, pp. 11–14.

Boyd, Harper; Ralph Westfall; and Stanley Stasch. *Marketing Research: Text & Cases*. 5th ed. Homewood, Ill.: Richard D. Irwin, 1981.

"Electronics: Data for Fleet Management." *Fleetowner* 76, no. 6 (June 1981), pp. 76–78.

Kerlinger, Fred N. *Fundamentals of Behavioral Research*. 2nd ed. New York: Holt, Rinehart & Winston, 1973.

Powers, Michael; Davis Adams; and Harlan D. Mills. *Computer Information System Development: Analysis & Design*. Cincinnati: South-Western Publishing, 1984, pp. 83–119.

Renney, Mark, and Everett C. Hughes. "Of Sociology and the Interview: Editorial Preface." *American Journal of Sociology*, vol. LXII (September 1956), pp. 137–42.

Schwartz, Morris S., and Charlotte G. Schwartz. "Problems in Participant Observation." *American Journal of Sociology*, vol. LXIX (1955), pp. 343–53.

Senn, James A. *Analysis and Design of Information Systems*. New York: McGraw-Hill, 1984, pp. 73–86.

Wetherbe, James C. *Systems Analysis and Design: Traditional, Structured, and Advanced Concepts and Techniques*. St. Paul, Minn.: West Publishing, 1984, pp. 127–54.

Chapter 6

The Tools of Structured Analysis

At a Glance

Chapter 5 discussed the traditional tools used in data gathering. These tools have drawbacks. An English narrative description of a system is often too vague. English is inherently difficult to use where precision is needed. Furthermore, system flowcharting based on the data gathered commit to a physical implementation of the candidate system before one has a complete understanding of the logical requirements of the system. Finally, system specifications are often redundant. To find information about one part of the system, one has to search through the entire document.

Because of these drawbacks, the analyst needs to focus on functions rather than physical implementation. Therefore, structured tools such as the data flow diagram, data dictionary, and structured English provide alternative ways of designing a candidate system. In real-life applications, a combination of the traditional and structured tools is used.

By the end of this chapter, you should know:

1. The meaning of structured analysis.
2. What tools are used in structured analysis.
3. How to construct a data flow diagram.
4. The advantages and uses of a data dictionary and structured English.
5. The elements and construction of decision trees and decision tables.

INTRODUCTION

In the preceding chapters, we discussed the procedures used in building computer-based systems and the role of the analyst in the system development life cycle. The goal of system development is to deliver systems in line with the user's requirements. Analysis is the heart of the process. It is the key component of the first two phases of the cycle. In phase one, we focused on problem definition and the initial investigation, where analysis helps us understand the present system. Phase two, the feasibility study, goes into detail studying the present system and determining potential solutions. The outcome is system specifications that initiate system design. The feasibility study is covered in Chapter 7.

In analyzing the present system, the analyst collects a great deal of relatively unstructured data through interviews, questionnaires, on-site observations, procedures manuals, and the like. The traditional approach is to organize and convert the data through system flowcharts, which support future developments of the system and simplify communication with the user. But the system flowchart represents a physical rather than a logical system. It makes it difficult to distinguish between *what* happens and *how* it happens in the system.

There are other problems with the traditional approach:

1. The system life cycle provides very little quality control to ensure accurate communication from user to analyst. They have no language in common.
2. The analyst is quickly overwhelmed with the business and technical details of the system. Much of the time is spent gathering information. The details are needed and must be available, but the analyst does not have the tools to structure and control the details.
3. Present analytical tools have limitations.
   a. English narrative descriptions of a system are often too vague and make it difficult for the user to grasp how the parts fit together. Furthermore, English is inherently difficult to use where precision is needed.
   b. System and program flowcharts commit to a physical implementation of the system before one has complete understanding of its logical requirements.
4. Problems also relate to system specifications:
   a. System specifications are difficult to maintain or modify. A simple change in the user's requirements necessitates changes in several parts of the document.
   b. They describe user requirements in terms of physical hardware that will implement the system rather than *what* the user wants the system to do.
   c. They are monolithic and redundant; that is, to find out information about a particular part of the system, the user has to search the

entire document. Furthermore, the same information is found in numerous locations with no cross-reference.

Because of these drawbacks, the analyst needs something analogous to the architect's blueprint as a starting point for system design. It is a way to focus on *functions* rather than *physical* implementation. One such tool is the data flow diagram (DFD). There are other tools as well. This chapter discusses and illustrates the use of several tools in structured analysis, including the following:

1. Data flow diagram (DFD).
2. Data dictionary.
3. Structured English.
4. Decision trees.
5. Decision tables.

WHAT IS STRUCTURED ANALYSIS?

Structured analysis is a set of techniques and graphical tools that allow the analyst to develp a new kind of system specifications that are easily understandable to the user. Analysts work primarily with their wits, pencil, and paper. Most of them have no tools. The traditional approach focuses on cost/benefit and feasibility analyses, project management, hardware and software selection, and personnel considerations. In contrast, structured analysis considers new goals and structured tools for analysis. The *new goals* specify the following:

1. Use graphics wherever possible to help communicate better with the user.
2. Differentiate between logical and physical systems.
3. Build a logical system model to familiarize the user with system characteristics and interrelationships before implementation.

The *structured tools* focus on the tools listed earlier—essentially the data flow diagram, data dictionary, structured English, decision trees, and decision tables. The objective is to build a new document, called *system specifications*. This document provides the basis for design and implementation. The system development life cycle with structured analysis is shown in Figure 6–1. The primary steps are:

Process 2.1: *Study affected user areas, resulting in a physical DFD.* The logical equivalent of the present system results in a logical DFD.

Process 2.2: *Remove the physical checkpoints and replace them with a logical equivalent, resulting in the logical DFD.* To illustrate, consider the two DFDs shown in Figure 6–2. Figure 6–2(a) is a physical DFD. It shows how the opening of a new safe deposit box flows through the current department. Figure 6–2(b) is the logical equivalent.

FIGURE 6–1 System Development Life Cycle Using Structured Analysis

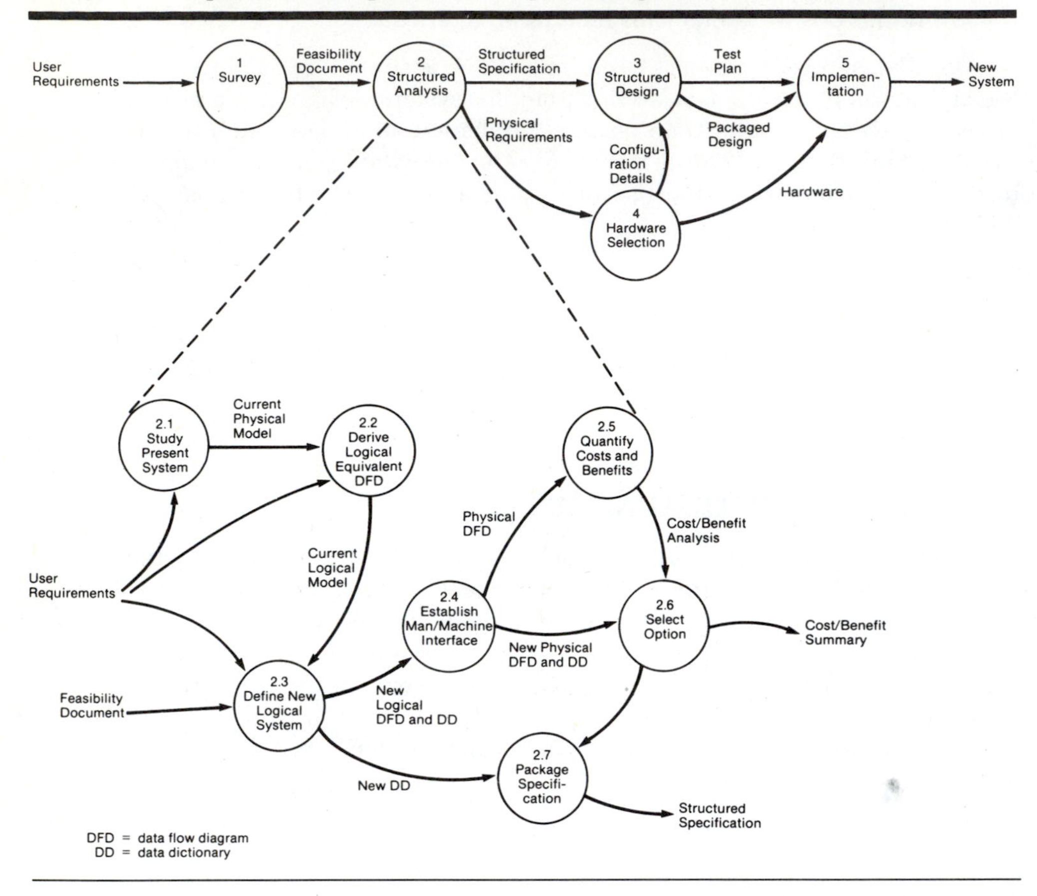

Source: Adapted from Tom De Marco, *Structured Analysis and System Specifications* (New York: Yourdon Press, 1979), p. 26.

Proces 2.3: *Model new logical system.* So far no consideration is given to modifying methods called for in the feasibility report. This step incorporates the changes and begins to describe the candidate system. It is essentially a paper model system to be installed.

Process 2.4: *Establish man/machine interface.* This process modifies the logical DFD for the candidate system and considers the hardware needed to implement the system. The combination results in the physical DFD of the candidate system.

Processes 2.5 and 2.6: *Quantify costs and benefits and select hardware.* The purpose of this step is to cost-justify the system, leading to the selection of hardware for the candidate system. All that is left after this step is writing the structured specification.

The structured specification consists of the DFDs that show the major decomposition of system functions and their interfaces, the data dictionary

FIGURE 6-2 *(a)* A Physical DFD, and *(b)* Its Logical Equivalent.

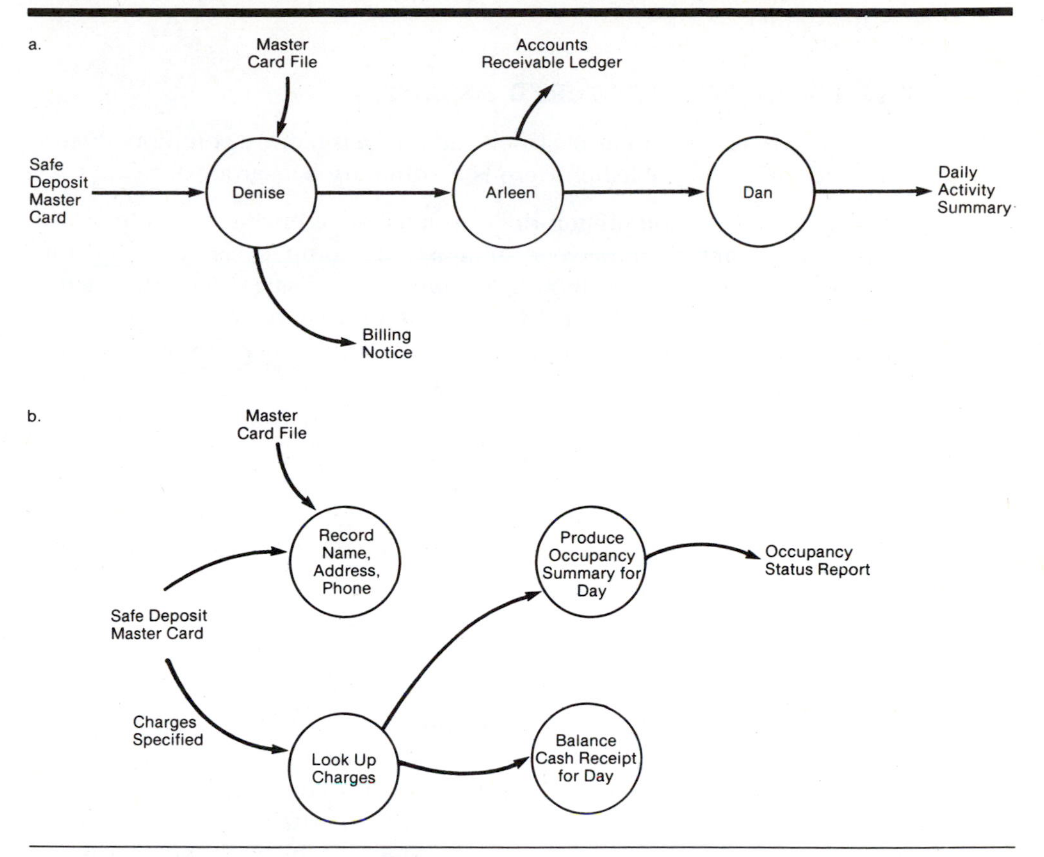

documenting all interface flows and data stores on the DFDs, and documentation of the intervals of DFDs in a rigorous manner through structured English, decision trees, and decision tables.

In summary, structured analysis has the following attributes:

1. It is graphic. The DFD, for example, presents a picture of what is being specified and is a conceptually easy-to-understand presentation of the application.
2. The process is partitioned so that we have a clear picture of the progression from general to specific in the system flow.
3. It is logical rather than physical. The elements of system do not depend on vendor or hardware. They specify in a precise, concise, and highly readable manner the workings of the system and how it hangs together.
4. It calls for a rigorous study of the user area, a commitment that is often taken lightly in the traditional approach to systems analysis.
5 Certain tasks that are normally carried out late in the system development life cycle are moved to the analysis phase. For example, user

procedures are documented during analysis rather than later in implementation.

THE TOOLS OF STRUCTURED ANALYSIS

Let's take a look at the tools of structured analysis using a common illustration—the textbook publisher. Here is a summary background:

1. ABC, Inc., is a multimillion-dollar publisher of business and technical textbooks, located in Homewood, Illinois. The company is organized into divisions such as trades, textbooks, accounting, and sales. The organization structure consists of a president, two senior vice presidents, a general manager of each division, shipping and receiving supervisors, and 45 sales representatives.
2. The college book division receives orders from bookstores for books at a discount which depends on the size of the order. The clerk in charge verifies the order and authorizes shipment through the warehouse. An invoice follows the shipment. Accounts receivable are processed through the accounting department from forms filled out by an accounting clerk.
3. Business is highly seasonal; it peaks about a month before the beginning of each school term. There is an average of 80 invoices per week, each with an average of 8 book titles and an average value of $5,000.
4. Recently, management decided to improve the availability of textbooks by holding stocks of new computer and other high-demand texts and making it possible for all bookstores to order by calling a toll-free number as well as by the present mail method. This means that an improved inventory control system must be devised along with a catalog of texts to verify authors and titles and determine the availability of the books being ordered.
5. The new system of receiving orders is expected to increase the sales volume by 80 percent within the year. Although fewer average texts per order are expected due to the use of the toll-free number, books are now shipped more quickly than before and delivered in time for the start of the semester.

An analyst has been asked to investigate the new system and build a logical model of the candidate system without abruptly jumping to conclusions about what will be automated and what will remain manual.

The Data Flow Diagram (DFD)

The first step is to draw a data flow diagram (DFD). The DFD was first developed by Larry Constantine as a way of expressing system requirements in a graphical form; this led to a modular design.[1]

---

[1] Chris Gane and Trish Sarson, *Structured Systems Analysis: Tools and Techniques* (Englewood Cliffs, N.J.: Prentice-Hall, 1979).

A DFD, also known as a "bubble chart," has the purpose of clarifying system requirements and identifying major transformations that will become programs in system design. So it is the starting point of the design phase that functionally decomposes the requirements specifications down to the lowest level of detail. A DFD consists of a series of bubbles joined by lines. The bubbles represent data transformations and the lines represent data flows in the system. A basic DFD format is shown in Figure 6–1. A general model of our publisher's ordering system is illustrated in Figure 6–3. The system takes orders from the customer (bookstore, library, etc.), checks them against an index (file) listing the books available, verifies customer credit through a credit information file, and authorizes shipment with an invoice.

DFD Symbols

In the DFD, there are four symbols, as shown in Figure 6–4:

1. A *square* defines a *source* (originator) or *destination* of system data.
2. An *arrow* identifies *data flow*—data in motion. It is a pipeline through which information flows.
3. A *circle* or a "bubble" (some people use an oval bubble) represents a *process* that transforms incoming data flow(s) into outgoing data flow(s).
4. An *open rectangle* is a data store—data at rest, or a temporary repository of data.

Note that a DFD describes *what* data flow (logical) rather than *how* they are processed, so it does not depend on hardware, software, data structure, or file organization. The key question that we are trying to answer is: What major transformations must occur for input to be correctly transformed into output?

Our example in Figure 6–3 is too general. Let's expand process orders to

FIGURE 6–3 General Model of Publisher's Present Ordering System

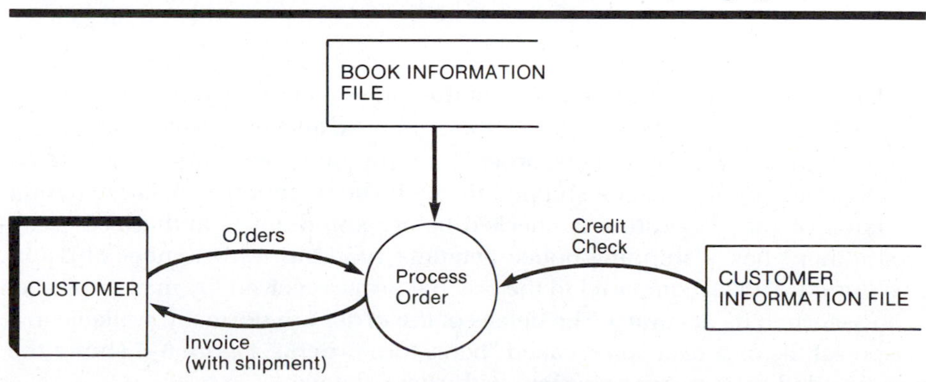

FIGURE 6-4 Data Flow Diagram: *(a)* Basic Symbols and *(b)* General Format

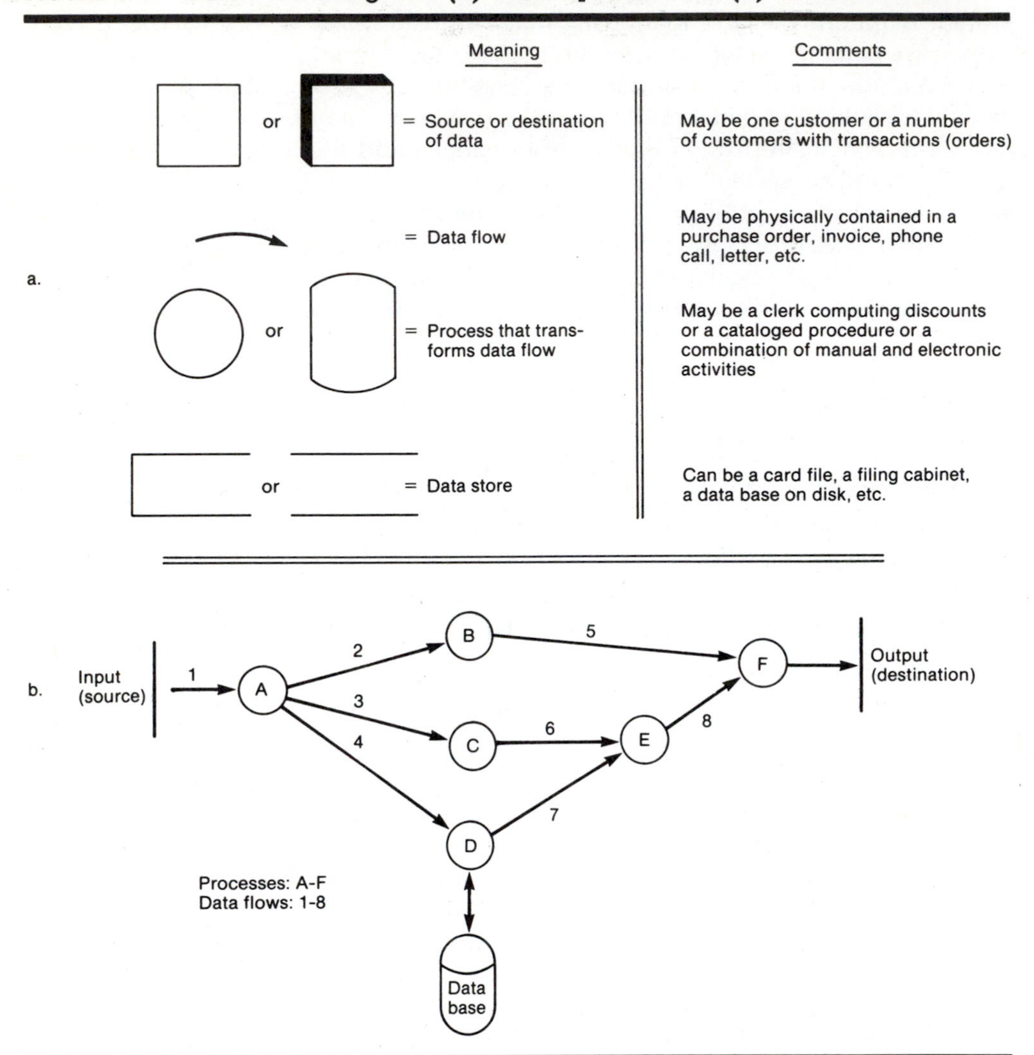

elaborate on the logical functions of the system. First, incoming orders are checked for correct book titles, authors' names, and other information and then batched with other book orders from the same bookstore to determine how many copies can be shipped through the warehouse. Also, the credit status of each bookstore is checked before shipment is authorized. Each shipment has a shipping notice detailing the kind and number of books shipped. This is compared to the original order received (by mail or phone) to ascertain its accuracy. The details of the order are normally available in a special file or a *data store,* called "bookstore orders." Figure 6-5 shows the expanded version of the system with these details.

FIGURE 6-5 Expanded DFD, Showing Order Verification and Credit Check

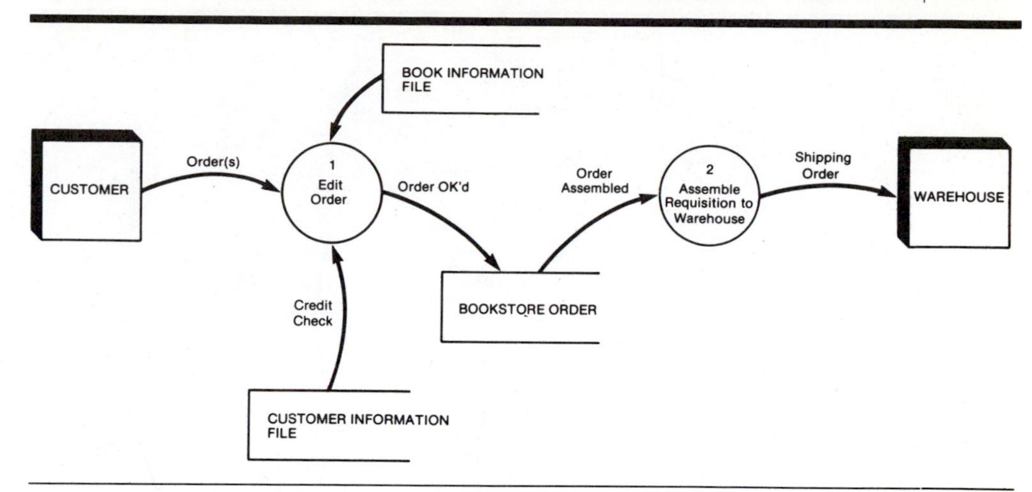

Following order verification and credit check, a clerk batches the order by assembling all the book titles ordered by the bookstore. The batched order is sent to the warehouse with authorization to pack and ship the books to the customer. Figure 6-6 shows the steps taken to finalize the process.

Further expansion of the DFD focuses on the steps taken in billing the bookstore. Figure 6-7 shows additional functions related to accounts receivable. As you can tell by now, each process summarizes a lot of information and can be exploded into several lower-level, detailed DFDs. This is often necessary to make sure that a complete documentation of the data flow is available for future reference.

Constructing a DFD

Several rules of thumb are used in drawing DFDs:

1. Processes should be named and numbered for easy reference (see Figure 6-7). Each name should be representative of the process.

2. The direction of flow is from top to bottom and from left to right. Data traditionally flow from the source (upper left corner) to the destination (lower right corner), although they may flow back to a source. One way to indicate this is to draw a long flow line back to the source. An alternative way is to repeat the source symbol as a destination. Since it is used more than once in the DFD, it is marked with a short diagonal in the lower right corner (see Figure 6-8).

3. When a process is exploded into lower-level details, they are numbered. For example, in Figure 6-7, process 5 (assemble customer order) is exploded into two subprocesses: create invoice and verify invoice. Since they are sublevels, they are numbered 5.1 and 5.2, respectively.

FIGURE 6-6 Expanded DFD, Elaborating on Order Processing and Shipping

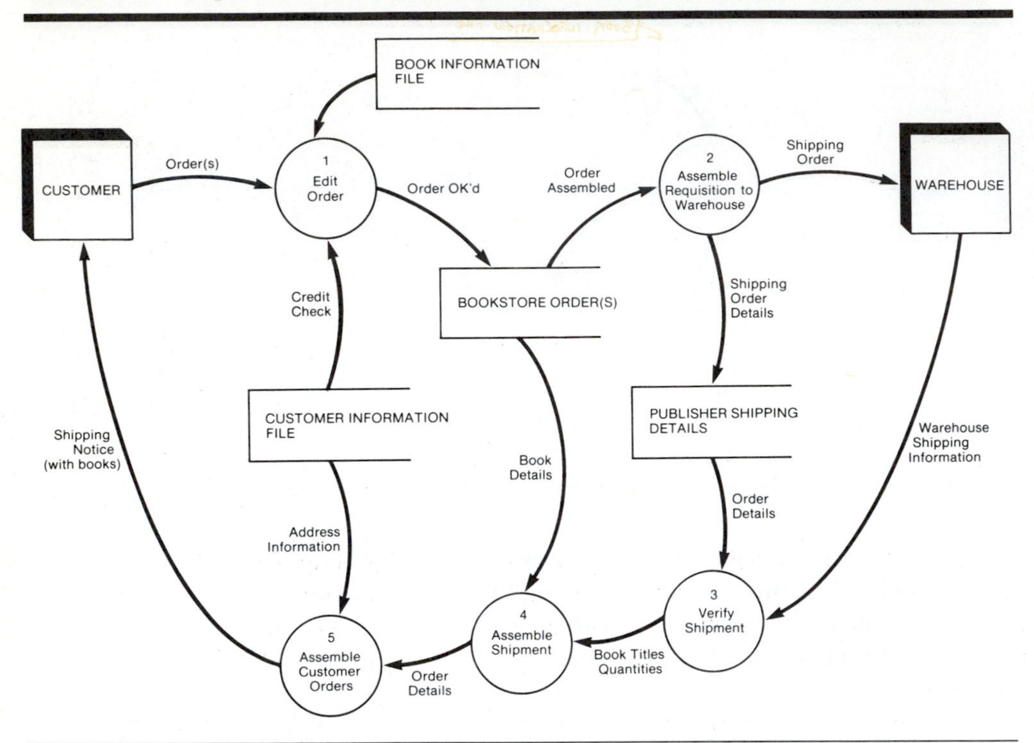

4. The names of data stores, sources, and destinations are written in capital letters. Process and data flow names have the first letter of each word capitalized.

How detailed should a DFD be? As mentioned earlier, the DFD is designed to aid communication. If it contains dozens of processes and data stores, it gets too unwieldy. The rule of thumb is to explode the DFD to a functional level, so that the next sublevel does not exceed 10 processes. Beyond that, it is best to take each function separately and expand it to show the explosion of the single process. If a user wants to know what happens within a given process, then the detailed explosion of that process may be shown.

A DFD typically shows the minimum contents of data stores. Each data store should contain all the data elements that flow in and out. Questionnaires can be used to provide information for a first cut. All discrepancies, missing interfaces, redundancies, and the like are then accounted for—often through interviews.

The DFD methodology is quite effective, especially when the required design is unclear and the user and the analyst need a notational language for communication. The DFD is easy to understand after a brief orientation.

FIGURE 6-7 Completed DFD, Showing Accounts Receivable Routine

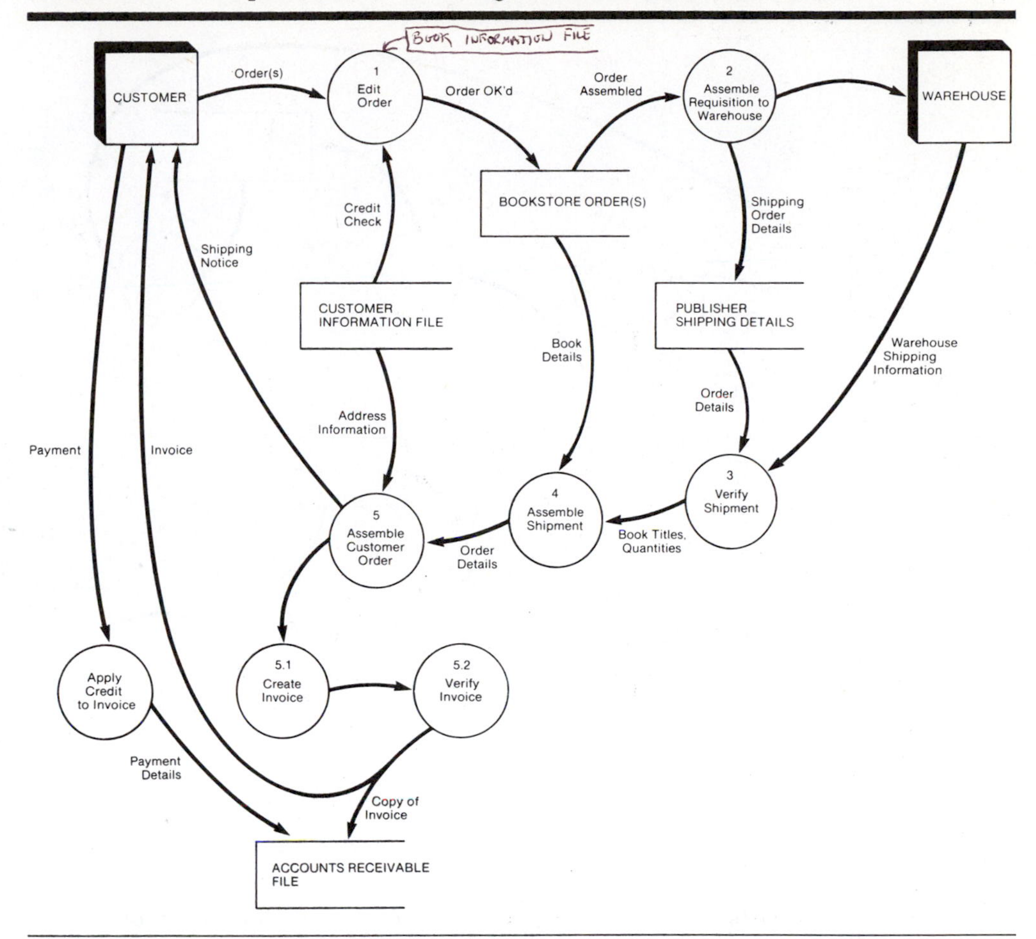

The main problem, however, is the large number of iterations that often are required to arrive at the most accurate and complete solution.

Data Dictionary

In our data flow diagrams, we give names to data flows, processes, and data stores. Although the names are descriptive of the data, they do not give details. So following the DFD, our interest is to build some structured place to keep details of the contents of data flows, processes, and data store. A *data dictionary* is a structured repository of data about data.[2] It is a set of

[2] James Martin, *Principles of Data-Base Management* (Englewood Cliffs, N.J.: Prentice-Hall, 1976), p. 4.

FIGURE 6–8 Alternative Use of Source/Destination Symbols

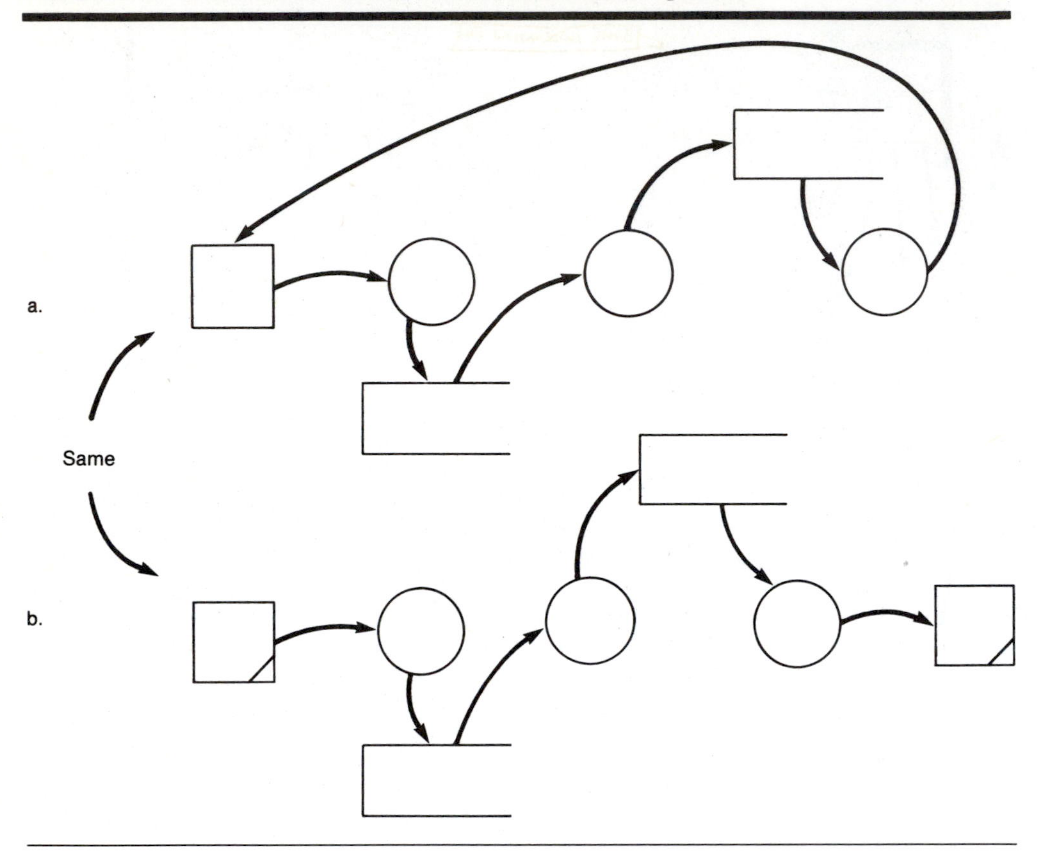

rigorous definitions of all DFD data elements and data structures (See Figure 6–9).

A data dictionary has many advantages. The most obvious is documentation; it is a valuable reference in any organization. Another advantage is improving analyst/user communication by establishing consistent definitions of various elements, terms, and procedures. During implementation, it serves as a common base against which programmers who are working on the system compare their data descriptions. Also control information maintained for each data element is cross-referenced in the data dictionary. For example, programs that use a given data element are cross-referenced in a data dictionary, which makes it easy to identify them and make any necessary changes. Finally, a data dictionary is an important step in building a data base. Most data base management systems have a data dictionary as a standard feature.

Data have been described in different ways. For example, in tape and disk processing, IBM called a file a *data set.* In data base technology, the term *file* took on a different meaning. IBM's Information Management

FIGURE 6-9 Project Data Element Form—A Sample

PROJECT DATA ELEMENT SHEET

PROJECT NAME ______________________ DATE ____________

DATA ELEMENT DESCRIPTION	DATA ELEMENT ABBREVIATION	ELEMENT PICTURE	ELEMENT LOCATION	ELEMENT SOURCE

System's (IMS) manual defines data as *fields* divided into *segments*, which, in turn, are combined into *data bases*.[3] The Conference on Data System Languages (CODASYL) defines data as *data items* combined into *aggregates*, which, in turn, are combined into *records*.[4] A group of related records is referred to as a *set*. A summary of these data definitions is given in Figure 6–10.

---

[3] *IMS/VS General Information Manual GH20-1260* (White Plains, N.Y.: IBM Corporation, 1974).

[4] *National Bureau of Standards Handbook 113*, CODASYL Data Description Language Journal of Development (Washington, D.C.: U.S. Government Printing Office, 1974).

FIGURE 6–10 Sample Data Definitions

Description	IMS	CODASYL
Smallest unit of data	Field	Data item
Groups of smallest data items	Segment	Data aggregate
Entity processed at a time	Logical record	Record
Largest grouping	Data base	Set

If we choose words that represent the general thinking of common vocabulary, there are three classes of items to be defined:

1. *Data element*: The smallest unit of data that provides for no further decomposition. For example, "date" consists of day, month, and year. They hang together for all practical purposes.
2. *Data structure*: A group of data elements handled as a unit. For example, "phone" is a data structure consisting of four data elements: Area code-exchange-number-extension—for example, 804-924-3423-236. "BOOK DETAILS" is a data structure consisting of the data elements author name, title, ISBN (International Standard Book Number), LOCN (Library of Congress Number), publisher's name, and quantity.
3. *Data flows and data stores*:[5] As defined earlier, data flows are data structures in motion, whereas data stores are data structures at rest. A data store is a location where data structures are temporarily located. The three levels that make up the hierarchy of data are shown in Figure 6–11.

FIGURE 6–11 Logical Data Description Hierarchy

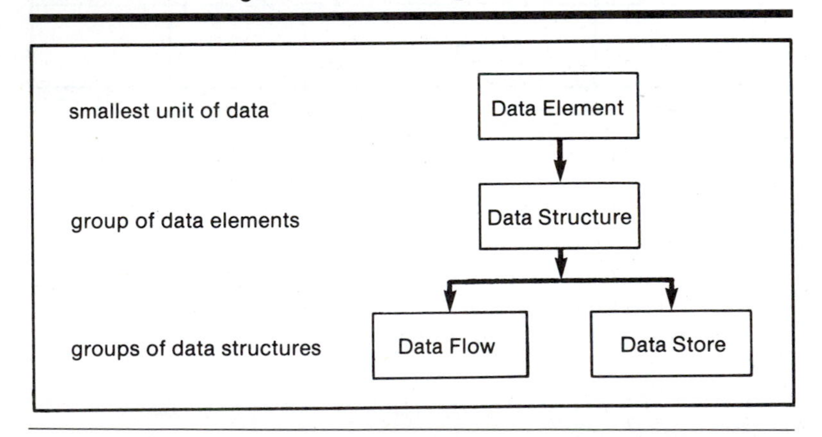

[5] For details on the data dictionary, refer to Gane and Sarson, *Structured Systems Analysis*, pp. 48–75.

Describing Data Elements

The description of a data element should include the *name, description,* and an *alias* (synonym). For example:

AUTHOR-NAME—first	WHISKEY—name
—middle	—distiller
—last	—vintage
—alias	

The description should be a summary of the data element. It may include an example. We may also want to include whether or not the data element(s) has:

1. *A different name*. For example, a PURCHASE ORDER may exist as PUR.ORDER, PURCHASE ORD., or P.O. We want to record all these in the data dictionary and include them under the PURCHASE ORDER definition and separately with entries of their own. One example is "P.O. alias of (or see also) PURCHASE ORDER." Then we look up PURCHASE ORDER to find the details. It is an index.
2. *Usage characteristics*, such as a range of values or the frequency of use or both. A value is a code that represents a meaning. Here we have two types of data elements:
   a. Those that take a value within a range: for example, a payroll check amount between $1 and $10,000 is called a *continuous* value.
   b. Those that have a specific value; for example, departments in a firm may be coded 100 (accounting), 110 (personnel), etc. In a data dictionary, it is described as follows:

   100 means "Accounting Department"
   101 means "Accounts Receivable Section"
   102 means "Accounts Payable Section"
   .
   .
   .
   108 means "General Ledger Section"

   In either type, values are codes that represent a meaning.
3. *Control information* such as the source, date of origin, users, or access authorization.
4. *Physical location* in terms of a record, file, or data base.

Describing Data Structures

We describe any data structure by specifying the name of each data structure and the elements it represents, provided they are defined elsewhere in the data dictionary. Some elements are mandatory, whereas others are optional. To illustrate, let us take "BOOK-DETAILS" from Figure 6–7. The data elements of this data structure are as follows:

		Mandatory	Optional
Data structure	BOOK-DETAILS		
Data elements	AUTHOR-NAME	X	
	TITLE OF BOOK	X	
	EDITION	X	
	ISBN (International Standard Book Number)		X
	LOCN (Library of Congress Number)		X
	PUBLISHER-NAME	X	
	QUANTITY ORDERED	X	

The data structure BOOK-DETAILS is made up of five mandatory data elements and two optional ones.

Describing Data Flows and Data Stores

The contents of a data flow may be described by the name(s) of the data structure(s) that passes along it. In our earlier example, BOOK-DETAILS expresses the content of the data flow that leads to process 4 (see Figure 6–7). Additionally, we may specify the source of the data flow, the destination, and the volume (if any). Using the BOOK-ORDER example, data flows may be described as follows:

Data Flow	Comments
BOOK-DETAILS	From Newcomb Hall Bookstore (source)
AUTHOR-NAME	
TITLE OF BOOK	
EDITION	Recent edition required
.	
.	
.	
QUANTITY	Minimum 40 copies

A data store is described by the data structures found in it and the data flows that feed it or are extracted from it. For example, the data store BOOKSTORE-ORDER is described by the following contents:

	Comments
ORDER	
ORDER-NUMBER	Data flow/data structure feeding data store
CUSTOMER-DETAILS	Content of data store
BOOK-DETAIL	Data flow/data structure extracted from data store

Describing Processes

This step is the logical description. We want to specify the inputs and outputs for the process and summarize the logic of the system. In Figure 6–7, process 1, EDIT-ORDER, can be described as shown in Figure 6–12.

In constructing a data dictionary, the analyst should consider the following points:

1. Each unique data flow in the DFD must have one data dictionary entry. There is also a data dictionary entry for each data store and process.
2. Definitions must be readily accessible by name.
3. There should be no redundancy or unnecessary definitions in the data definition. It must also be simple to make updates.
4. The procedure for writing definitions should be straightforward but specific. There should be only one way of defining words.

In summary, a data dictionary is an integral component of the structured specification. Without a data dictionary, the DFD lacks rigor, and without the DFD, the data dictionary is of no use. Therefore, the correlation between the two is important.

Decision Tree and Structured English

Once the data elements are defined in the data dictionary, we begin to focus on the processes. For example, in Figure 6–7, we need to know what goes into APPLY CREDIT TO INVOICE. Since bookstores get discounts on books acquired from publishers, we can expect one lower-level process to verify the discount. This is done through a publisher discount policy known to

FIGURE 6–12 General Description of Process—An Example

1. Name of Process: EDIT-ORDER-IS-O.K.
2. Short Description: Verify and decide whether customer credit is OK for authorizing shipment, or whether it must be sent COD.

3. Inputs	Logic	Output
ORDER	Look up customer payment record If new bookstore, clear credit rating	Credit OK, no balance reminders
Customer payment file Data account opened Balance on order Credit limit	If established customer (over one year), OK order unless value exceeds $10,000 or balance overdue is more than 90 days old. If under one year, OK order, unless there is balance overdue	

the bookstore in advance. For illustration, let us assume the following discount policy:

> Bookstores get a trade discount of 25%; for orders from libraries and individuals, 5% allowed on orders of 6–19 copies per book title; 10% on orders for 20–49 copies per book title; 15% on orders for 50 copies or more per book title.

A policy statement like this can be time-consuming to describe and confusing to implement. The analyst needs to use tools to portray the logic of the policy. The first such tool is the *decision tree*. As shown in Figure 6–13, a decision tree has as many branches as there are logical alternatives. It simply sketches the logical structure based on the stated policy. In this respect, it is an excellent tool: It is easy to construct, easy to read, and easy to update. It shows only the skeletal aspects of the policy, however, in the sense that it does not lend itself to calculations or show logic as a set of instructions for action. The alternative, then, is the use of a second tool called structured English.

Structured English borrows heavily from structured programming; it uses logical construction and imperative sentences designed to carry out instructions for action. Decisions are made through IF, THEN, ELSE, and SO statements. The structured English for our publisher's discount policy is shown in Figure 6–14. Note the correlation between the decision tree and structured English.

We can actually make structured English more compact by using terms defined in the data dictionary. For example, the process ORDER may have the data element ORDER-SIZE, which defines four values:

MINIMUM: 5 or fewer copies per book title
SMALL: 6 to 19 copies

FIGURE 6–13 Decision Tree—An Example

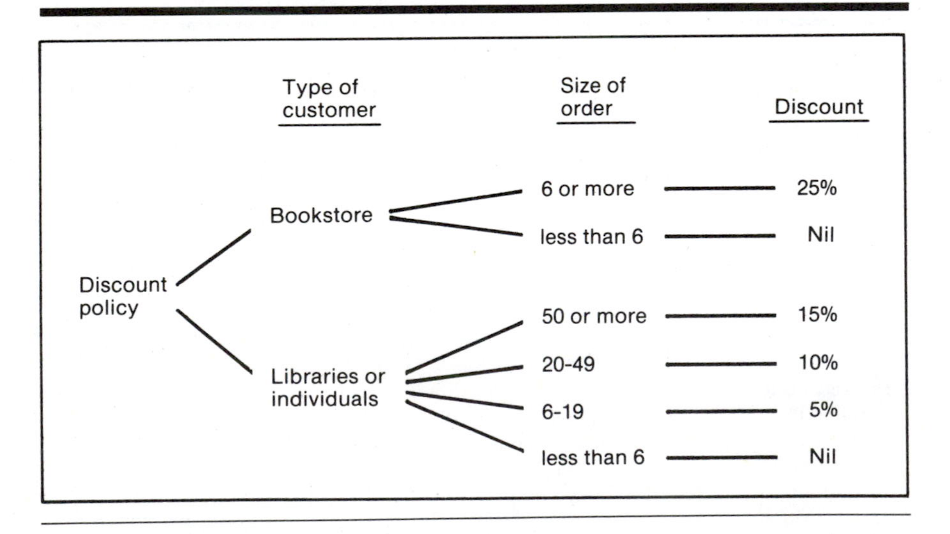

FIGURE 6–14 Structured English—An Example

```
    COMPUTE-DISCOUNT
    Add up the number of copies per book title
IF order is from bookstore
    and-IF    order is for 6 copies or more per book title
       THEN:  Discount is 25%
    ELSE      (order is for fewer than 6 copies per book title)
         SO:  no discount is allowed
ELSE (order is from libraries or individual customers)
    so-IF     order is for 50 copies or more per book title
              discount is 15%
      ELSE IF order is for 20 to 49 copies per book title
              discount is 10%
      ELSE IF order is for 6 to 19 copies per book title
              discount is 5%
      ELSE    (order is for less than 6 copies per book order)
        SO:   no discount is allowed
```

MEDIUM: 20 to 49 copies
LARGE: 50 or more copies

Using these values, the structured English in Figure 6–14 would read as shown in Figure 6–15.

From these examples we see that when logic is written out in English sentences using capitalization and multilevel indentation, it is structured English. In this tool, the logic of processes of the system is expressed by using the capitalized key words IF, THEN, ELSE, and SO. Structures are indented to reflect the logical hierarchy. Sentences should also be clear, concise, and precise in wording and meaning.

Decision Tables

A major drawback of a decision tree is the lack of information in its format to tell us what other combinations of conditions to test. This is where the decision table is useful. A *decision table* is a table of contingencies for defining a problem and the actions to be taken. It is a single representation of the relationships between conditions and actions.[6] Figure 6–16 shows a decision table that represents our discount policy (see Figure 6–13 and 6–14).

A decision table consists of two parts: *stub* and *entry*. The stub part is

[6] M. A. Colter, "Comparative Examination of Systems Analysis Techniques," *MIS Quarterly*, March 1984, p. 59.

FIGURE 6–15 Structured English—Using Data Dictionary Values

COMPUTE-DISCOUNT
Add up the number of copies per book title
IF order is from bookstore
  and-IF ORDER-SIZE is SMALL
    THEN: Discount is 25%
  ELSE (ORDER-SIZE is MINIMUM)
    SO: no discount is allowed
ELSE (order is from libraries or individual customers)
  so-IF ORDER-SIZE is LARGE
    discount is 15%
  ELSE IF ORDER-SIZE is MEDIUM
    discount is 10%
  ELSE IF ORDER-SIZE is SMALL
    discount is 5%
  ELSE (ORDER-SIZE is MINIMUM)
    SO: no discount is allowed

divided into an upper quadrant called the *condition stub* and a lower quadrant called the *action stub*. The entry part is also divided into an upper quadrant, called the *condition entry* and a lower quadrant called the *action entry*. The four elements and their definitions are summarized in Figure 6–17.

FIGURE 6–16 Decision Table—Discount Policy

	Condition Stub	Condition Entry					
		1	2	3	4	5	6
IF (condition)	Customer is bookstore?	Y	Y	N	N	N	N
	Order-size 6 copies or more?	Y	N	N	N	N	N
	Customer librarian or individual?			Y	Y	Y	Y
	Order-size 50 copies or more?			Y	N	N	N
	Order-size 20–49 copies?				Y	N	N
	Order-size 6–19 copies?					Y	N
THEN (action)	Allow 25% discount	X					
	Allow 15% discount			X			
	Allow 10% discount				X		
	Allow 5% discount					X	
	No discount allowed		X				X
	Action Stub	Action Entry					

FIGURE 6–17 Elements and Definitions in a Decision Table

Elements	Location	Definition
Condition stub	Upper left quadrant	Sets forth in question form the condition that may exist
Action stub	Lower left quadrant	Outlines in narrative form the action to be taken to meet each condition
Condition entry	Upper right quadrant	Provides answers to questions asked in the condition stub quadrant
Action entry	Lower right quadrant	Indicates the appropriate action resulting from the answers to the conditions in the condition entry quadrant

Note in Figure 6–16 that the answers are represented by a Y to signify yes, an N to signify no, or a blank to show that the condition involved has not been tested. In the action entry quadrant, an X (or a check mark will do) indicates the response to the answer(s) entered in the condition entry quadrant. Furthermore, each column represents a decision or a rule. For example, rule 1 states:

IF customer is a bookstore and order size is 6 copies or more,
THEN allow 25% discount

So, according to the decision table, we have six decisions and therefore six rules. A look at the table provides each decision (answer) immediately The following rules should be followed in constructing decision tables:

1. A decision should be given a name, shown in the top left of the table.
2. The logic of the decision table is independent of the sequence in which the condition rules are written, but the action takes place in the order in which the events occur.
3. Standardized language must be used consistently.
4. Duplication of terms or meanings should be eliminated, where possible.

Pros and Cons of Each Tool

Which tool is the best depends on a number of factors: the nature and complexity of the problem, the number of actions resulting from the decisions, and the ease of use. In reviewing the benefits and limitations of each tool, we come to the following conclusions:

1. The primary strength of the DFD is its ability to represent data flows. It may be used at high or low levels of analysis and provides good system documentation. However, the tool only weakly shows input and output detail.[7] The user often finds it confusing initially.

[7] Ibid., p. 62.

2. The data dictionary helps the analyst simplify the structure for meeting the data requirements of the system. It may be used at high or low levels of analysis, but it does not provide functional details, and it is not acceptable to many nontechnical users.
3. Structured English is best used when the problem requires sequences of actions with decisions.
4. Decision trees are used to verify logic and in problems that involve a few complex decisions resulting in a limited number of actions.
5. Decision trees and decision tables are best suited for dealing with complex branching routines such as calculating discounts or sales commissions or inventory control procedures.

Given the pros and cons of structured tools, the analyst should be trained in the use of various tools for analysis and design. He/she should use decision tables and structured English to get to the heart of complex problems. A decision table is perhaps the most useful tool for communicating problem details to the user.

The major contribution of structured analysis to the system development life cycle is producing a definable and measurable document—the structured specification. Other benefits include increased user involvement, improved communication between user and designer, reduction of total personnel time, and fewer "kinks" during detailed design and implementation. The only drawback is increased analyst and user time in the process. Overall the benefits outweigh the drawbacks, which make structured analysis tools viable alternatives in system development.

Summary

1. Traditional tools have limitations. An English narrative description is often vague and difficult for the user to grasp. System flowcharts focus more on physical than on logical implementation of the candidate system. Because of these drawbacks, structured tools were introduced for analysis. They include data flow diagrams, a data dictionary, structured English, decision trees, and decision tables.
2. The traditional approach to analysis focuses on cost/benefit and feasibility analyses, project management, hardware and software selection, and personnel considerations. In contrast, structured analysis considers new goals and structured tools for analysis. Specifically, it uses graphics wherever possible, differentiates between logical and physical systems, and builds a logical system to accentuate system characteristics and interrelationships before implementation.
3. The system development life cycle with structured analysis covers six steps:
   a. Study affected user areas, resulting in a physical DFD.
   b. Remove the physical checkpoints and replace them with the logical equivalent, resulting in the logical DFD.

   c. Model new logical system.
   d. Establish man/machine interface.
   e. Quantify costs and benefits.
   f. Select hardware/software.

4. The DFD clarifies system requirements and identifies major transformations that will become programs in system design. It is the starting point in system design that decomposes the requirements specifications down to the lowest level of detail.

5. In constructing DFDs, the analyst should name and number processes for easy reference. The direction of flow is from top to bottom and from left to right. When a process is exploded into lower-level details, they are numbered. The names of data stores, sources, and destinations are written in capital letters.

6. A data dictionary is a structured repository of data about data. It offers the primary advantages of documentation and improving analyst/user communication by establishing consistent definitions of various elements, terms, and procedures. The three classes to be defined are data elements, data structures, and data flows and data stores:
   a. *Data element* is the smallest unit of data that provides for no further decompositon.
   b. *Data structure* is a group of data elements handled as a unit.
   c. *Data flows and data stores* are data structures in motion and data structures at rest, respectively.

7. In constructing a data dictionary, the analyst considers several points:
   a. Each data flow in the DFD has one data dictionary entry.
   b. Definitions must be readily accessible by name.
   c. There should be no redundancy in the data definition.
   d. The procedure for writing definitions should be precise.

8. A decision tree sketches the logical structure based on some criteria. It is easy to construct, read, and update. It shows only the skeletal aspects of the picture however, and does not lend itself to calculations. The alternative is structured English.

9. Structured English uses logical constructs and imperative sentences designed to carry out instructions for action. Decisions are made through IF, THEN, ELSE, and SO statements. This tool is highly correlated with the decision tree.

10. A decision table is a table of contingencies for defining a problem and the actions to be taken. It is a single representation of the relationships between conditions and actions. The parts of the table are the stub and entry. The stub is divided into condition and action stubs; the entry is also divided into a condition and an action entry. The rules to follow in constructing decision tables are:
   a. A decision should be given a name.
   b. The logic of the table is independent of the sequence in which

condition rules are written, but the action takes place in the order in which the events occur.

c. Standardized language must be used consistently.

d. Duplication of terms should be eliminated.

11. In comparing the tools covered in the chapter, we find that:
    a. The primary strength of the DFD is its ability to represent data flows, but it only weakly shows input and output details.
    b. The data dictionary simplifies the structure for meeting the data requirements of the system but does not provide functional details.
    c. Structured English is best used when the problem requires using sequences of actions with decisions.
    d. Decision trees are used for logic verifications and in problems involving few complex decisions.
    e. Decision trees and decision tables are best suited for dealing with complex branching routines.

Key Words

Action Entry
Action Stub
Aggregate
Alias
Bubble Chart
Condition Entry
Condition Stub
Data Dictionary
Data Element
Data Flow
Data Flow Diagram (DFD)
Data Item
Data Set
Data Store
Data Structure
Decision Table
Decision Tree
Entry
Partitioning
Process
Record
Segment
Set
Structured Analysis
Structured English
System Flowchart
System Specification

Review Questions

1. Discuss the pros and cons of the traditional approach to systems analysis.
2. What is structured analysis? Briefly review the tools used. How does it differ from the traditional approach?
3. What steps make up the system development life cycle with structured analysis? Describe each step briefly.
4. If you were to summarize the attributes of structured analysis in four short sentences, what would you say?
5. Describe the concept and procedure used in constructing DFDs. Use an example of your own to illustrate.

6. What basic rules are relevant to constructing a DFD?
7. Using an example of your own, how detailed should a DFD be? Explain.
8. "A data dictionary is a structured repository of data about data." Discuss.
9. What advantages does a data dictionary offer in the area of documentation?
10. Define the following terms:
    a. Data set.
    b. Aggregates.
    c. Segments.
    d. Data structure.
11. Illustrate how data elements and processes are described.
12. What points should be considered in constructing a data dictionary? Be specific.
13. In what way(s) is a decision tree and a data flow diagram related? What about a decision tree and structured English?
14. List and illustrate the primary uses and elements of a decision table.

Application Problems

1 Draw an overall data flow diagram for the following applications:

a. Making auto loans.

b. A travel agency making round-trip reservations for two to Hawaii.

c. A systems analyst selling professional time by the hour and paying staff salaries.

d. Ordering supplies for a barber shop.

2 Write the following data dictionary entries in English.

a. PURCHASE-ORDER = Line-Item

b. Line-Item = Catalog-Number + Quantity + Description (+Size)(+Color) + Unit-Price + Total-Price

3 For the safe deposit case described in Chapter 4 and 7, the data flow diagram in Exhibit 6–1 A Customer Data Flow Diagram sets up a new customer account for a safe deposit box.

EXHIBIT 6–1 A Customer Data Flow Diagram

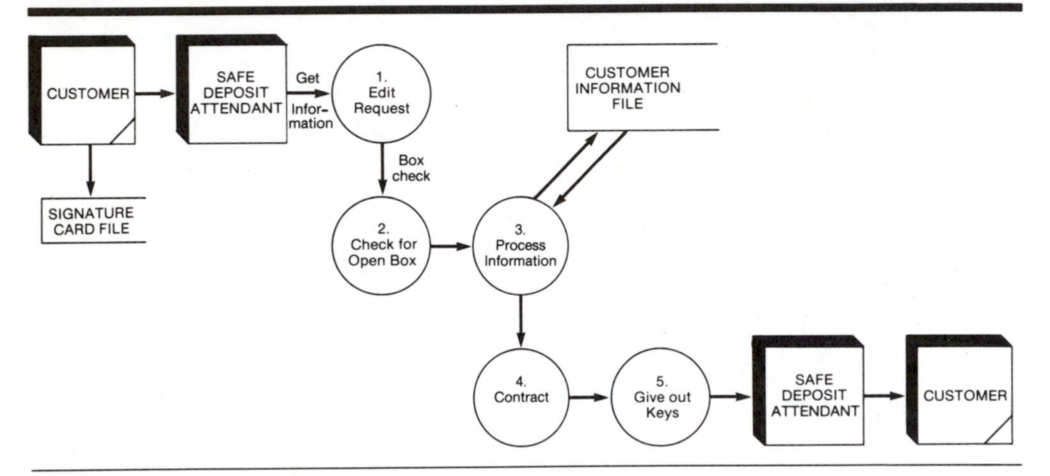

Assignment

a. Are there flaws in the data flow diagram? If so, what are they? List the corrections.

b. Draw an expanded data flow diagram of bubble 2.

c. Using the information from the safe deposit case and the data flow diagram, draw a data flow diagram to close a customer's account.

4 An international airline initiated a frequent traveler program designed to encourage passengers to fly regularly and earn awards based on miles flown. The airline policy is specified as follows:

Passengers who fly more than 100,000 miles per calendar year and, *in addition*, pay cash for tickets or have been flying the airline regularly for more than five years are to receive a free round-trip ticket around the world. Passengers who fly less than 100,000 miles per calendar year and have been flying the airline regularly for more than five years also get a free round-trip ticket around the world.

Assignment

a. Draw a decision tree based on the statement.

b. Develop a decision table for passenger free ticket.

5 A Virginia-based mail-order house specializes in microcomputers and supplies for various microcomputer makes. It offers discounts based on the number of units ordered. When an order is received, an invoice is generated that includes the quantity, unit cost, discount, and shipping and handling charges. The invoice is decomposed into the following hierarchy:

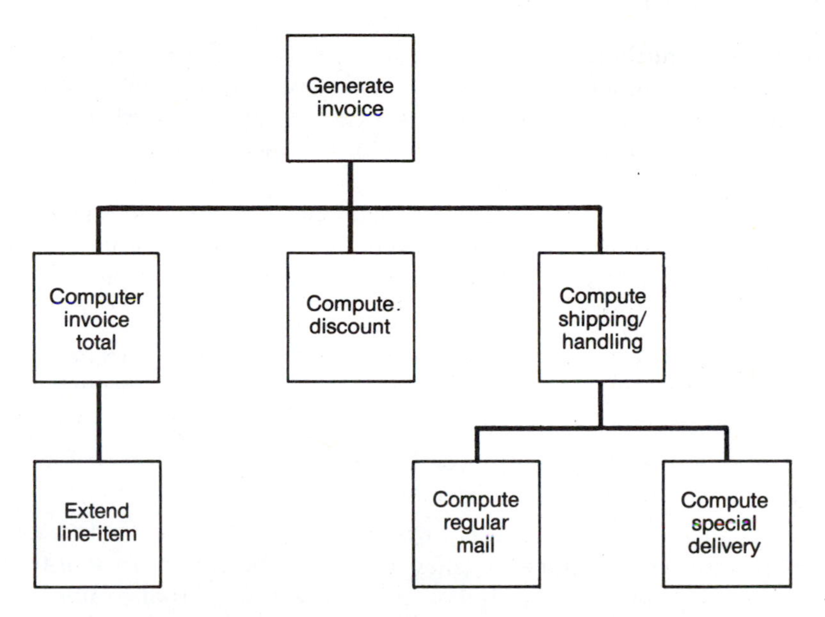

Invoice-Total is computed on each line-item. All line-item totals are then added to get invoice total. A line-item is extended by multiplying quantity by unit cost. The product is the line-item total.

Discount is computed as follows:

If Invoice-Total is $3,000 or over, discount is 20 percent.
If Invoice-Total is between $2,000 and $2,999, discount is 10 percent.
If Invoice-Total is between $1,000 and $1,999, discount is 5 percent.
If Invoice-Total is under $1,000, no discount allowed.

Shipping and handling charges depend on whether the computer is to go by UPS or air freight. If the order specifies UPS, then compute as follows:

If weight is under 35 pounds, shipping/handling rate is $30.

If weight is 35 to 55 pounds, shipping/handling rate is $45.

If weight is over 55 pounds, multiply excess (over 55) by $2 and add $45 to get shipping/handling rate.

If the order specifies air freight, then multiply each pound of weight by $1.85 to get rate.

Assignment

a. Develop a decision tree.

b. Represent the information in structured English.

---

6 A major publisher hired an analyst to study the existing system of producing textbooks. The analyst spent several weeks collecting data regarding the production process and then prepared a system flowchart. The process is initiated with the receipt of the typewritten manuscript by the book editior. The manuscript consists of the text narrative and art in the form of tables, charts, schematic charts, and figures of machines, buildings, and the like. The editor sends it to the production editor, who officially launches the production process. The steps are as follows:

a. The manuscript is assigned to a copyeditor who corrects grammatical errors, inconsistencies, and the like. The job normally takes seven to nine weeks.

b. The copyedited manuscript is sent back to the author for final review. The publisher allows two weeks for this step.

c. When the edited manuscript is returned, it is released to a compositor who produces a galley. Copies of the galley are sent to the author and the production department for proofreading. Two weeks are allowed for this step.

d. The proofread galley is returned to the compositor to incorporate final changes or corrections. Meanwhile, art is released to the art department for preparing the finished work. Five weeks are allowed to complete the job.

e. When the corrected galley and art are received, the production department "dummies" the book; that is, the galley is divided into pages that include space for illustrations, photographs, etc. This step normally takes three weeks.

f. The dummied manuscript is page-numbered and the illustrations are actually shown in the space provided. Further proofreading and checking on the exact location of the art are finalized. This step takes a week.

g. At this stage, the index is prepared, which takes a week. Front matter (preface, acknowledgment page, etc.) is also set in type.

h. The dummied galley, including index and front matter, is sent to the compositor for final production and binding. This major step takes around seven weeks. When the work is completed, the bindery ships the copies to the publisher's warehouse.

Assignment

a. Draw a system flowchart to represent the production process.

b. Use the information provided in the case to construct a Gantt chart.

7

Sanford & Sons, Inc., a small manufacturer of electronic calculators, stocks "ready to ship" units in four warehouses (A, B, C, D). Warehouse A handles all orders received from foreign customers. Warehouse B handles orders received from customers out of the state of Illinois. Warehouse C fills orders received from nongovernment customers that operate within the state of Illinois. Warehouse D handles orders received from state or federal government agencies.

Assignment

Prepare a decision table to indicate the warehouse from which customers can be accommodated.

8

A major manufacturer of floppy disks wants to have a party for employees who have been with the company between one and three months. In addition to a master list of these employees, the public relations department wishes to have a tally of those who are single, married, divorced, and "other."

Assignment

Assuming that the master personnel file of company employees will be used, develop a decision table to illustrate the job.

9

The local chapter of the Data Processing Programming Association is planning an annual picnic in July to be held in the city's main park. To encourage sustained membership, the recreation committee has decided to serve each member according to the number of years he/she has been a member, as follows:

> If a member has been with the association for one year or less, he/she is entitled to two hamburgers, one bottle of beer, and one ice cream bar. Those with two years of membership are allowed one hamburger and three bottles of beer. Those with more than three years of membership get one doubleburger and no limit on drinks.

Assignment

Construct a decision table to help the one in charge handle the orders.

---

10 A major brokerage firm wishes to have (1) a list (call it "odd lot") of all individuals who hold odd lots (less than 100 shares) of Hammermill Ball Bearing stocks, and (2) lists of each of the five types of stockholders who own 100 or more shares of the same issue, as follows:

a. List A to include individual stockholders.

b. List B to include stocks held by a trust fund.

c. List C to include stocks held by an educational institution.

d. List D to include stocks held by a financial institution.

e. List E to include stocks held by a Canadian stockholder.

Assignment

Present the request in a decision table.

Selected References

Colter, M. A. "Comparative Examination of Systems Analysis Techniques." *MIS Quarterly*, March 1984, pp 59–67.

Davis, Wm. S. *Systems Analysis and Design: A Structured Approach*. Reading, Mass.: Addison-Wesley Publishing, 1983, pp. 281–301.

DeMarco, Tom. *Structured Analysis and System Specification*. New York: Yourdon Press, 1979.

Dickinson, Brian. *Developing Structured Systems*. New York: Yourdon Press, 1980.

Gane, Chris, and Trish Sarson. *Structured Systems Analysis: Tools and Techniques*. Englewood Cliffs, N.J.: Prentice-Hall, 1979.

IMS/VS General Information Manual GH20-1260. White Plains, N.Y.: IBM Corp., 1974.

Martin, James. *Principles of Data Base Management*. Englewood Cliffs, N.J.: Prentice-Hall, 1976.

National Bureau of Standards Handbook 113, *CODASYL Data Description Language*,

Journal of Development. Washington, D.C.: U.S. Government Printing Office, 1974.

Page-Jones, M. *The Practical Guide to Structured System Design*. New York: Yourdon Press, 1980.

Senn, James A. *Analysis and Design of Information Systems*. New York: McGraw-Hill, 1984, pp. 109–134.

Shelvin, Jeffrey L. "Evaluating Alternative Methods of Systems Analysis." *Data Management*, April 1983, pp. 22–25.

Yourdon, Edward. *Managing the Structured Techniques. 2d ed.* Englewood Cliffs, N.J.: Prentice-Hall, 1979, pp. 36–57.

Chapter 7

Feasibility Study

At a Glance

An initial investigation culminates in a proposal that determines whether an alternative system is feasible. A proposal summarizing the thinking of the analyst is presented to the user for review. When approved, the proposal initiates a feasibility study that describes and evaluates candidate systems and provides for the selection of the best system that meets system performance requirements.

To do a feasibility study, we need to consider the economic, technical, and behavioral factors in system development. First a project team is formed. The team develops system flowcharts that identify the characteristics of candidate systems, evaluate the performance of each system, weigh system performance and cost data, and select the best candidate system for the job. The study culminates in a final report to management.

By the end of this chapter, you should know:

1. The steps in defining system performance.
2. What key considerations are involved in feasibility analysis.
3. How to conduct a feasibility study.

INTRODUCTION

Chapter 4 discussed the steps that make up the initial investigation. Several activities have been completed:

1. A user has recognized a need.
2. User requirements are determined and the problem has been defined.
3. An initial investigation is launched to study the existing system and verify the problem.
4. The analyst (with user participation) has verified the objectives, constraints, and required outputs, resulting in a project directive that the user has approved.

The next step is to determine exactly what the candidate system is to do by defining its expected performance. Thus, a feasibility study is carried out to select the best system that meets performance requirements. This entails identification, description, and evaluation of candidate systems and selection of the best system for the job. This chapter, then, addresses the system performance definition and expounds on the feasibility study as a second major step in the system development life cycle.

SYSTEM PERFORMANCE DEFINITION

A system's required performance is defined by describing its outputs in a user-acceptable format and at a higher level of detail than what was described in the initial investigation. This involves three steps:

1. Statement of constraints.
2. Identification of specific system objectives.
3. Description of outputs.

This phase builds on the previous phase in that much of the work may already have been done.

Statement of Constraints

Constraints are factors that limit the solution of the problem. Some constraints are identified during the initial investigation and are discussed with the user. There are general constraints that might have a bearing on the required performance of a candidate system. Let's consider our safe deposit billing system to illustrate these points. We described the current billing system and how the department handles billing and customer payments. The result of the fact-finding phase of the initial investigation revealed the following general constraints:

1. The president views safe deposit billing as a low priority. He has a false impression that computers are not needed as long as customers can have access to their boxes.

2. The senior vice president is worried that a billing system might require the transfer of safe deposit staff to other departments. Considering Florida's level of unemployment and the cost of retraining, a candidate system has to do more than produce reports.
3. The accounting department has been pushing for installing a computer-based general ledger application for months. The vice president of operations, bogged down with auditing and operations problems, continued to shelve the request.
4. Management has a limited knowledge of computers, although it has several applications on the computer: checking and savings, installment loans, commercial loans, and trusts. The president, in his early sixties and interested in "the bottom line" of the financial statement, is traditionally reluctant to spend money on computers.
5. Safe deposit, while doing better than breaking even, is not projected to grow as fast as it did in the early 1980s. The community's recent success in controlling burglaries had an adverse impact on the demand for box rentals in general.
6. If an online system is to be installed, it must interface with the existing checking/savings application to allow for the automatic payment of box rentals.
7. A proposed design must be compatible with the bank's Burroughs computer system.

Identification of Specific System Objectives

Once the constraints are spelled out, the analyst proceeds to identify the system's specific performance objectives. They are derived from the general objectives specified in the project directive at the end of the initial investigation. The steps are to state the system's benefits and then translate them into measurable objectives. In our scenario, the candidate system's anticipated benefits are as follows:

1. Improved collection schedule.
2. Cost reduction.
3. Physical space reduction.
4. Improved customer service.

Each benefit is analyzed and translated into measurable objectives.

1. Collection is improved by billing 30 days in advance of the box renewal date, and one more notice is sent within two weeks. It also improves the accounts receivables payment "float."
2. Cost reduction is realized by reducing the payroll by two employees. The new online billing system requires less than two hours of labor per day, compared with six hours under the current system.

3. Physical space requirements are reduced by placing the microcomputer in the place of one of the four existing desks. The remaining desks are removed, allowing an extra cubicle for customer use.
4. Customer service is improved by placing master cards and box rental information online, thus reducing the waiting time of entry from 3 minutes to 30 seconds.

These objectives are effective in comparing the performance of the candidate system with that of the current system. The information-oriented flowchart, input/output analysis sheet, and data flow diagram (see Chapter 4) produced in the initial investigation lead to the conclusions that (1) the current system is inefficient and (2) a new online, terminal-oriented system would be the solution. This conclusion was reflected in the general project directive submitted to the user for approval. This information is used as a basis for preparing specific objectives for the candidate system:

1. To establish a billing system with six five-day cycles per month.
2. To mail customers no later than the close of the billing cycle and no later than 25 days prior to the box renewal date.
3. To mail customers a reminder two weeks after the initial statement for box renewal.
4. To speed collections and reduce the "float" by 40 percent.
5. To examine the availability of boxes by size, rental fees, and location.
6. To evaluate the ratio of rented to available boxes at all times.
7. To produce periodic reports to management on the performance of the safe deposit department.

Description of Outputs

A final step in system performance definition is describing the outputs required by the user. An actual sketch of the format and contents of the reports (layout) as well as a specification of the media used, their frequency, and the size and number of copies required are prepared at this point. Specifying exactly what the output will look like leads to an estimate of the computer storage requirements that form the basis for the file design to be undertaken in the design phase of the life cycle. The analyst is now ready to evaluate the feasibility of candidate systems to produce these outputs.

FEASIBILITY STUDY

Many feasibility studies are disillusioning for both users and analysts. First, the study often presupposes that when the feasibility document is being prepared, the analyst is in a position to evaluate solutions. Second, most studies tend to overlook the confusion inherent in system development—the constraints and the assumed attitudes. If the feasibility study is to serve as a decision document, it must answer three key questions:

1. Is there a new and better way to do the job that will benefit the user?
2. What are the costs and savings of the alternative(s)?
3. What is recommended?

The most successful system projects are not necessarily the biggest or most visible in a business but rather those that truly meet user expectations. More projects fail because of inflated expectations than for any other reason.[1]

Feasibility Considerations

Three key considerations are involved in the feasibility analysis: economic, technical, behavioral. Let's briefly review each consideration and how it relates to the systems effort.

Economic Feasibility

Economic analysis is the most frequently used method for evaluating the effectiveness of a candidate system. More commonly known as cost/benefit analysis, the procedure is to determine the benefits and savings that are expected from a candidate system and compare them with costs. If benefits outweigh costs, then the decision is made to design and implement the system. Otherwise, further justification or alterations in the proposed system will have to be made if it is to have a chance of being approved. This is an ongoing effort that improves in accuracy at each phase of the system life cycle. More on cost/benefit analysis is covered in Chapter 8.

Technical Feasibility

Technical feasibility centers around the existing computer system (hardware, software, etc.) and to what extent it can support the proposed addition. For example, if the current computer is operating at 80 percent capacity—an arbitrary ceiling—then running another application could overload the system or require additional hardware. This involves financial considerations to accommodate technical enhancements. If the budget is a serious constraint, then the project is judged not feasible.

Behavioral Feasibility

People are inherently resistant to change, and computers have been known to facilitate change. An estimate should be made of how strong a reaction the user staff is likely to have toward the development of a computerized system. It is common knowledge that computer installations have

---

[1] Lois Zells, "A Practical Approach to a Project Expectation Document," *Computerworld (In-Depth)*, August 29, 1983, p. 1.

something to do with turnover, transfers, retraining, and changes in employee job status. Therefore, it is understandable that the introduction of a candidate system requires special effort to educate, sell, and train the staff on new ways of conducting business.

In our safe deposit example, three employees are more than 50 years old and have been with the bank over 14 years, four years of which have been in safe deposit. The remaining two employees are in their early thirties. They joined safe deposit about two years before the study. Based on data gathered from extensive interviews, the younger employees want the programmable aspects of safe deposit (essentially billing) put on a computer. Two of the three older employees have voiced resistance to the idea. Their view is that billing is no problem. The main emphasis is customer service—personal contacts with customers. The decision in this case was to go ahead and pursue the project.

Steps in Feasibility Analysis

Feasibility analysis involves eight steps:

1. Form a project team and appoint a project leader.
2. Prepare system flowcharts.
3. Enumerate potential candidate systems.
4. Describe and identify characteristics of candidate systems.
5. Determine and evaluate performance and cost effectiveness of each candidate system.
6. Weight system performance and cost data.
7. Select the best candidate system.
8. Prepare and report final project directive to management.

Form a Project Team and Appoint a Project Leader

The concept behind a project team is that future system users should be involved in its design and implementation. Their knowledge and experience in the operations area are essential to the success of the system. For small projects, the analyst and an assistant usually suffice; however, more complex studies require a project team. The team consists of analysts and user staff—enough collective expertise to devise a solution to the problem. In many cases, an outside consultant and an information specialist join the team until the job is completed.

Projects are planned to occupy a specific time period, ranging from several weeks to months. The senior systems analyst is generally appointed as project leader. He/she is usually the most experienced analyst in the team. The appointment is temporary, lasting as long as the project. Regular meetings take place to keep up the momentum and accomplish the mission—selection of the best candidate system. A record is kept of the progress made in each meeting.

Regarding the safe deposit case, since the whole user area consists of five employees, the analyst handled most of the work.

Prepare System Flowcharts

The next step in the feasibility study is to prepare generalized system flowcharts for the system. Information-oriented charts and data flow diagrams prepared in the initial investigation are also reviewed at this time. The charts bring up the importance of inputs, outputs, and data flow among key points in the existing system. All other flowcharts needed for detailed evaluation are completed at this point.

Enumerate Potential Candidate Systems

This step identifies the candidate systems that are capable of producing the outputs included in the generalized flowcharts. This requires a transformation from logical to physical system models. Another aspect of this step is consideration of the hardware that can handle the total system requirements. In the safe deposit case, it was found that virtually any microcomputer system with more than 128K-byte memory an dual disk drive will do the job. It was also learned that a microcomputer can be designed to interface with the bank's mainframe. In this design, actual processing is handled by the microcomputer, whereas information such as payments and credits are transmitted to the main computer files for proper adjustment through the customer's checking account. The question here is: Which microcomputer (IBM, Apple, Digital, etc.) should be selected? This is taken up in step 6 of the study.

An important aspect of hardware is processing and main memory. There are a large number of computers with differing processing sizes, main memory capabilities, and software support. The project team may contact vendors for information on the processing capabilities of the system available.

Describe and Identify Characteristics of Candidate Systems

From the candidate systems considered, the team begins a preliminary evaluation in an attempt to reduce them to a manageable number. Technical knowledge and expertise in the hardware/software area are critical for determining what each candidate system can and cannot do. In our safe deposit example, a search for the available microcomputers and safe deposit billing packages revealed the information summarized in Table 7-1.

These packages were the result of a preliminary evaluation of more than 15 other packages—all purporting to meet the requirements of the safe deposit billing system. When the number is reduced to three key packages, the next step is to describe in some detail the characteristics of each package. For example, the first candidate system runs on an IBM PC with a minimum of 128K-bytes of memory. The software is written in Pascal, a relatively new language. In case of enhancements, change has to be made through the software house, since the source code is not available to the

TABLE 7-1 Safe Deposit Billing Packages and Selected Characteristics

Characteristics	IBM PC	HP 100	Apple III
Memory required (K bytes)	128	64	264
Source language	Pascal	Basic	Basic compiler
Source code available	No	Yes	No
Purchase terms	Purchase (license)	Purchase (license)	Purchase (license)
Purchase price	$995	$800	$1,095
Number installed to date	200	30	50
Date of first installation	1/82	3/81	1980

user. The first package was installed in January 1982. More than 200 packages have been installed to date.

The next two candidate systems are similarly described. The information along with additional data available through the vendor highlight the positive and negative features of each system. The constraints unique to each system are also specified. For example, in the IBM PC package, the lack of an available source code means that the user has to secure a maintenance contract that costs 18 percent of the price of the package per year. In contrast, the HP 100 package is less expensive and offers a source code to the user. A maintenance contract (optional) is available at 8 percent of the price of the package.

Determine and Evaluate Performance and Cost Effectiveness of Each Candidate System

Each candidate system's performance is evaluated against the system performance requirements set prior to the feasibility study. Whatever the criteria, there has to be as close a match as practicable, although trade-offs are often necessary to select the best system. In the safe deposit case, the criteria chosen in advance were accuracy, growth potential, response time less than five seconds, expandable main and auxiliary storage, and user-friendly software. Often these characteristics do not lend themselves to quantitative measures. They are usually evaluated in qualitative terms (excellent, good, etc.) based on the subjective judgment of the project team (see Table 7-2).

The cost encompasses both designing and installing the system. It includes user training, updating the physical facilities, and documenting. System performance criteria are evaluated against the cost of each system to determine which system is likely to be the most cost effective and also meets the performance requirements. The safe deposit problem is easy. The analyst can plot performance criteria and costs for each system to determine how each fares. Table 7-3 summarize the outcome of the comparison.

TABLE 7–2 Candidate Qualitative Evaluation Matrix

Evaluation Criteria	IBM PC	HP 100	Apple III
Performance			
System accuracy*	Excellent	Excellent	Excellent
Growth potential‡	Very good	Good	Good
Response time†	Very good	Very good	Very good
User-friendly	Excellent	Very good	Very good
Costs			
System development	Good	Very good	Good
User training	Excellent	Good	Good
System operation	Very good	Fair	Very good
Payback‡	Very good	Good	Excellent

* Protection from mistakes.

† Elapsed time between preparation of data for input and their return as computer output.

‡ Time taken to recover investment in candidate system.

Costs are more easily determined when the benefits of the system are tangible and measurable. An additional factor to consider is the cost of the study design and development. As shown in Table 7–4, the cost estimate of each phase of the safe deposit project was determined for the candidate system (IBM PC). In many respects, the cost of the study phase is a "sunk cost" (fixed cost). Including it in the project cost estimate is optional.

Weight System Performance and Cost Data

In some cases, the performance and cost data for each candidate system show which system is the best choice. This outcome terminates the feasibility study. Many times, however, the situation is not so clear-cut. The performance/cost evaluation matrix in Table 7–3 does not clearly identify the best system, so the next step is to weight the importance of each

TABLE 7–3 Candidate System Performance/Cost Evaluation Matrix

Criteria	IBM PC	HP 100	Apple III
Performance			
System accuracy	99% (rounded)	97% (rounded)	97% (rounded)
Growth potential	To 500K bytes	To 250K bytes	To 250K bytes
Response time less than five seconds	Yes	Yes	Yes
User-friendly	Menu driven, interactive	Command driven	Menu driven
Costs			
System development	$2,500–4,000	$1,350–4,050	$2,100–4,900
User training	1–2 days	2–4 days	2–3 days
System operations	$0.27/box	$0.64/box	$0.24/box
Payback	4 months	4.8 months	3.6 months

TABLE 7–4 Project Cost Estimate of Final Candidate System

	Charges (per day)		Total
Study phase (4 days)			
Analyst	$400		$1,600
Design phase (6 days)			
Analyst	400	$2,400	
Programmer	250	1,500	3,900
Development phase (9 days)			
Data entry operator	80	720	
Programmer	250	2,250	2,970
			$8,470

criterion by applying a rating figure. Then the candidate system with the highest total score is selected.

The procedure for weighting candidate systems is simple:

1. Assign a weighting factor to each evaluation criterion based on the criterion's effect on the success of the system. For example, if the usability criterion is twice as important as the accuracy factor, usability is assigned weight 4 and accuracy is assigned weight 2.
2. Assign a quantitative rating to each criterion's qualitative rating. For example, ratings (poor, fair, good, very good, excellent) may be assigned respective values (1, 2, 3, 4, 5).
3. Multiply the weight assigned to each category by the relative rating to determine the score.
4. Sum the score column for each candidate system.

Table 7–5 is a weighted candidate evaluation matrix from the four steps and the data.

Select the Best Candidate System

The system with the highest total score is judged the best system. This assumes the weighting factors are fair and the rating of each evaluation criterion is accurate. According to our safe deposit example in Table 7–4, the IBM PC is the best system for the job. Growth potential was the criterion that had the greatest effect on the total score. The HP 100 and the Apple III were given lower ratings than the IBM PC because they were not judged to grow as easily and quickly. Additionally, user training was judged superior for the PC than for other candidate systems.

Most feasibility studies select from more candidate systems than we used in our safe deposit example. The criteria chosen and the constraints are also more complex. In any case, management should not make the

TABLE 7-5 Weighted Candidate Evaluation Matrix

Evaluation Criteria	Weighting Factor	IBM PC		HP100		Apple III	
		Rating	Score	Rating	Score	Rating	Score
Performance							
System accuracy	3	5	15	5	15	5	15
Growth potential	4	4	16	3	12	3	12
Response time	2	4	8	4	8	4	8
User-friendly	2	5	10	4	8	4	8
Costs							
System development	5	3	15	4	20	4	15
User training	3	5	15	3	9	3	9
System operation	2	4	8	2	4	4	8
Payback	3	4	12	3	9	5	15
			99		85		90

selection without having the experience to do so. Management cooperation and comments, however, are encouraged.

Feasibility Report

The culmination of the feasibility study is a feasibility report directed to management; it evaluates the impact of the proposed changes on the area(s) in question. The report is a formal document for management use, brief enough and sufficiently nontechnical to be understandable, yet detailed enough to provide the basis for system design.

There is no standard format for preparing feasibility reports. Analysts usually decide on a format that suits the particular user and system. Most reports, however, begin with a summary of findings and recommendations, followed by documented details. Starting with summary information highlights the essence of the report, giving management the option of reviewing the details later. The report contains the following sections:

1. *Cover letter* formally presents the report and briefly indicates to management the nature, general findings, and recommendations to be considered.
2. *Table of contents* specifies the location of the various parts of the report. Management quickly refers to the sections that concern them.
3. *Overview* is a narrative explanation of the purpose and scope of the project, the reason for undertaking the feasibility study, and the department(s) involved or affected by the candidate system. Also included are the names of the persons who conducted the study, when it began, and other information that explains the circumstances surrounding the study.
4. *Detailed findings* outline the methods used in the present system. The

system's effectiveness and efficiency as well as operating costs are emphasized. The section also provides a description of the objectives and general procedures of the candidate system. A discussion of output reports, costs, and benefits gives management a feel for the pros and cons of the candidate system.

5. *Economic justification* details point-by-point cost comparisons and preliminary cost estimates for the development and operation of the candidate system. A return on investment (ROI) analysis of the project is also included.
6. *Recommendations and conclusions* suggest to management the most beneficial and cost-effective system. They are written only as a recommendation, not a command. Following the recommendations, any conclusions from the study may be included.
7. *Appendixes* document all memos and data compiled during the investigaiton. They are placed at the end of the report for reference.

Disapproval of the feasibility report is rare if it has been conducted properly. When a feasibility team has maintained good rapport with the user and his/her staff it makes the recommendations easier to approve. Technically, the report is only a recommendation, but it is an authoritative one. Management has the final say. Its approval is required before system design is initiated. Chapter 9 covers in detail the design phase of the system life cycle.

Oral Presentation

The feasibility report is a good written presentation documenting the activities involving the candidate system. The pivotal step, however, is selling the proposed change. Invariably the project leader or analyst is expected to give an oral presentation to the end user. Although it is not as polished as the written report, the oral presentation has several important objectives. The most critical requirements for the analyst who gives the oral presentation are (1) communication skills and knowledge about the candidate system that can be translated into language understandable to the user, and (2) the ability to answer questions, clarify issues, maintain credibility, and pick up on any new ideas or suggestions.

The substance and form of the presentation depend largely on the purposes sought. Figure 7–1 suggests a general outline. The presentation may aim at informing, confirming, or persuading.

1. *Informing*. This simply means communicating the decisions already reached on system recommendations and the resulting action plans to those who will participate in the implementation. No detailed findings or conclusions are included.
2. *Confirming*. A presentation with this purpose verifies facts and recommendations already discussed and agreed upon. Unlike the persuading

FIGURE 7-1 Oral Presentation—Suggested Outline

I. Introduction.
    A. Introduce self.
    B. Introduce topic.
    C. Briefly describe current system.
        1. Explain why it is not solving the problem.
        2. Highlight user dissatisfaction with it.
        3. Briefly describe scope, objectives, and recommendations of the proposed system.
II. Body of presentation.
    A. Highlight weaknesses of current system.
    B. Describe proposed system. How is it going to solve the problem?
    C. Sell proposed system.
        1. Specify savings and benefits, costs and expenses.
        2. Use visual aids to justify project and explain system.
    D. Summarize implementation plan and schedule.
    E. Review human resources requirements to install system.
III. Conclusion.
    A. Summarize proposal.
    B. Restate recommendations and objectives of proposal.
    C. Summarize benefits and savings.
    D. Ask for top-level management support. Solicit go-ahead for project.
IV. Discussion period—Answer questions convincingly.

approach, no supportive evidence is presented to sell the proposed change, nor is there elaborate reasoning behind recommendations and conclusions. Although the presentation is not detailed, it should be complete. Confirming is itself part of the process of securing approval. It should reaffirm the benefits of the candidate system and provide a clear statement of results to be achieved.

3. *Persuading*. This is a presentation pitched toward selling ideas—attempts to convince executives to take action on recommendations for implementing a candidate system.

Regardless of the purpose sought, the effectiveness of the oral presentation depends on how successful the project team has been in gaining the confidence of frontline personnel during the initial investigation. How the recommendations are presented also has an impact. Here are some pointers on how to give an oral presentation:

1. Rehearse and test your ideas before the presentation. Show that you are in command. Appear relaxed.
2. Final recommendations are more easily accepted if they are presented as *ideas for discussion*, even though they seem to be settled and final.
3. The presentation should be brief, factual, and interesting. Clarity and

persuasiveness are critical. Skill is needed to generate enthusiasm and interest throughout the presentation.

4. Use good organization. Distribute relevant material to the user and other parties in advance.
5. Visual aids (graphs, charts) are effective if they are simple, meaningful, and imaginative. An effective graph should teach or tell what is to be communicated.
6. Most important, present the report in an appropriate physical environment where the acoustics, seating pattern, visual aid technology, and refreshments are all available.

The most important element to consider is the length of the presentation. The duration often depends on the complexity of the project, the interest of the user group, and the competence of the project team. A study that has companywide applications and took months to complete would require hours or longer to present. The user group that was involved at the outset would likely permit a lengthy presentation, although familiarity with the project often dictates a brief presentation. Unfortunately, many oral presentations tend to be a rehash of the written document, with little flare or excitement. Also, when the analyst or the project leader has a good reputation and success record from previous projects, the end user may request only a brief presentation.

Summary

1. A feasibility study is conducted to select the best system that meets performance requirements. This entails an identification description, an evaluation of candidate systems, and the selection of the best system for the job.
2. A system's required performance is defined by a statement of constraints, the identification of specific system objectives, and a description of outputs. The analyst is then ready to evaluate the feasibility of candidate systems to produce these outputs.
3. Three key considerations are involved in feasibility analysis: economic, technical, and behavioral. Economic analysis (known as cost/benefit analysis) determines whether the adoption of a system can be cost-justified. Technical considerations evaluate existing hardware and software. Behavioral feasibility determines how much effort will go into educating, selling, and training the user staff on a candidate system.
4. There are eight steps in a feasibility study:
   a. Form a project team and appoint a project leader.
   b. Prepare system flowcharts.
   c. Enumerate potential candidate systems.
   d. Describe and identify characteristics of candidate systems.

e. Determine and evaluate performance and cost effectiveness of each candidate system.
f. Weight system performance and cost data.
g. Select the best candidate system.
h. Prepare and report project directive to management.

Key Words

Candidate System
Cost/Benefit Analysis
Feasibility Study
Response Time
Source Code
Source Language

Review Questions

1. What makes up a system performance definition? Select a situation with which you are familiar and explain the steps to prepare the definition.
2. "Many feasibility studies produce disillusions to users and analysts." Do you agree? Why? Explain.
3. What considerations are involved in feasibility analysis? Which consideration do you think is the most crucial? Why?
4. Elaborate on the steps in feasibility analysis. If you were to shorten them to four steps, which ones would you pick? Why?
5. How important is a project team in feasibility analysis? Is it mandatory in every study? Where are the exceptions?
6. Use Table 7–3 as the basis for determining alternative performance/cost factors between the IBM PC jr. and Apple's MacIntosh systems. Which one would you recommend for the safe deposit department of the bank?
7. What makes up a feasibility report? How would you change it? Explain.

Application Problems

SAFE DEPOSIT TRACKING SYSTEM

Introduction

The First National Bank of South Miami is a full-service bank and a member of the Federal Reserve and the Federal Deposit Insurance

Corporation. At the close of 1983, the bank had assets totaling $159 million, capital of more than $19 million, deposits of $113 million, and a loan base of $37 million. Rated the 37th safest bank in the United States, First National has a highly successful management team and 119 employees in a single-story building in the heart of downtown South Miami.

The bank operates two remote automatic teller machine (ATM) locations and a drive-in facility. The original orientation was toward the community, serving the people in the immediate area. Although still primarily a community bank, First National is becoming increasingly commerical as more industrial firms move into South Miami, especially around a major university where the bank is located. Available are complete banking and trust services ranging from NOW accounts and money market certificates to commercial, installment, and auto loans to safe deposit box rentals.

The Organization Structure—Safe Deposit

In 1978, the bank's safe deposit department anticipated a sharp increase in demand. The safe deposit department is in the lobby area and operates under the operations department. It is staffed by three customer service representatives and one security guard (Exhibit 7–1). Although no single individual in the department has been given the official title of supervisor, in practice, one of the employees is in charge. Denise Aguiar, the "supervisor," began working in August 1981. Since then she has made several improvements in procedures as we shall discuss later. Dorothy Garrison and Linda Hoppe are the other two safe deposit clerks. They began employment in June 1981 and March 1983, respectively. None of the three women has prior computer experience. Humberto Tamayo joined safe deposit in January 1982 as the vault security guard. The bank currently has 6,195 safe deposit boxes and the capacity to increase the number to more than 10,000.

The safe deposit boxes range in size from 3 by 5 inches to 38 by 21 inches and are rented on an annual contractual basis. Exhibit 7–2 lists the most recent schedule of service charges by box size. Although rental charges are fixed, in practice the department must keep track of three fee schedules. First, the safe deposit box lessees whose contracts have not yet expired are charged by the older rental fee schedule. A second schedule holds for some current customers who are unhappy with the hefty price increases and have large account balances, borrowings, or utilization of other bank services. These customers are enticed into keeping their boxes by various discount allowances that are approved by a vice president. Third, the employees are offered safe deposit boxes at the old fee rates less $10, making the 3 by 5 box free.

EXHIBIT 7-1 Organization Chart

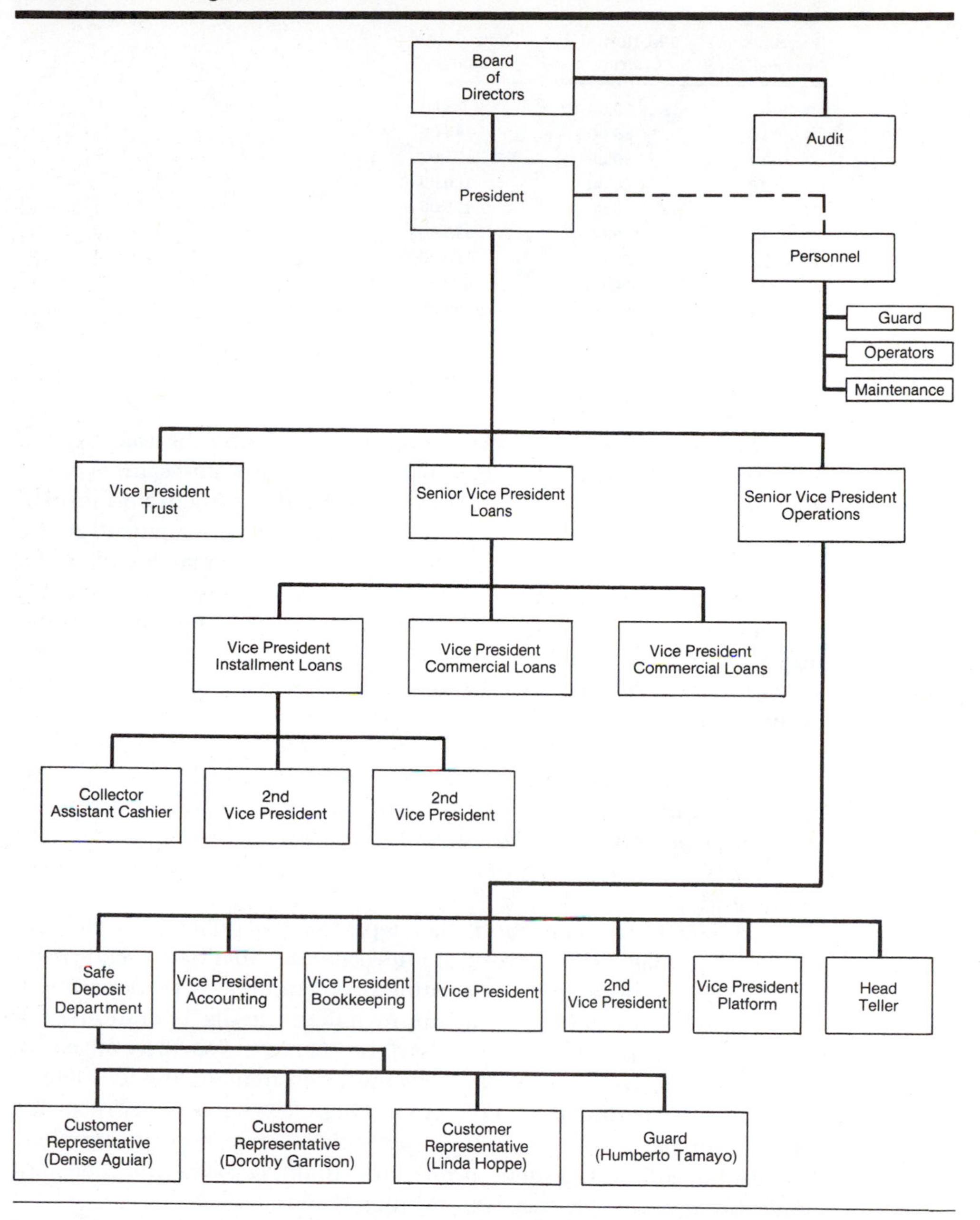

EXHIBIT 7–2 List of Rental Charges

Box Size (inches)	Old Rental Charge	New Rental Charge
3 × 5	$ 10.00	$ 25.00
5 × 5	25.00	45.00
3 × 10	30.00	75.00
5 × 10	40.00	100.00
10 × 10	70.00	175.00
15 × 10	125.00	225.00
13 × 21	350.00	350.00
26 × 21	600.00	600.00
38 × 21	1,000.00	1,000.00

Data Collection

The data collection process began by examining the safe deposit department's Manual of Instructions and Procedures and sample documents. To gather more information about the department and investigate some questions raised, staff members were interviewed at the bank. Denise provided most of the data. A tour of the facility offered a firsthand look at the layout of the vault and lobby area where business is conducted. A systems flowchart of procedures was drawn up based on the Manual of Instructions and Procedures and the rules and regulations booklet (Exhibit 7–3). This helped to clarify the information and indicate where changes might be needed. Any further questions were readily answered by the supervisor or a member of the board of directors who is quite familiar with the bank's operations. There was a good deal of communication between the bank and the project team—by phone, in person, and in writing. Management's encouragement and enthusiasm for the project were extremely helpful throughout the feasibility study.

Following data collection, the department's policies and procedures were analyzed by looking at improvements that had already been implemented and identifying existing problem areas. The department has recently instituted several improvements in its operations. For example, a previous problem involved the absence of security measures when the keys were sent to the locksmith for changing the lock after a customer discontinued the box rental. These keys were never recorded and could be missing, causing a potentially large security problem. Denise established a new procedure that required the constant monitoring of the whereabouts of all keys.

Another area in which changes have occurred is record-keeping and documentation. In the past, vital documents such as birth and death certificates, court orders, and records on customers and payments were not always readily available. Denise has organized this information and filed it so that it can be obtained more easily.

EXHIBIT 7–3A Initial Visit

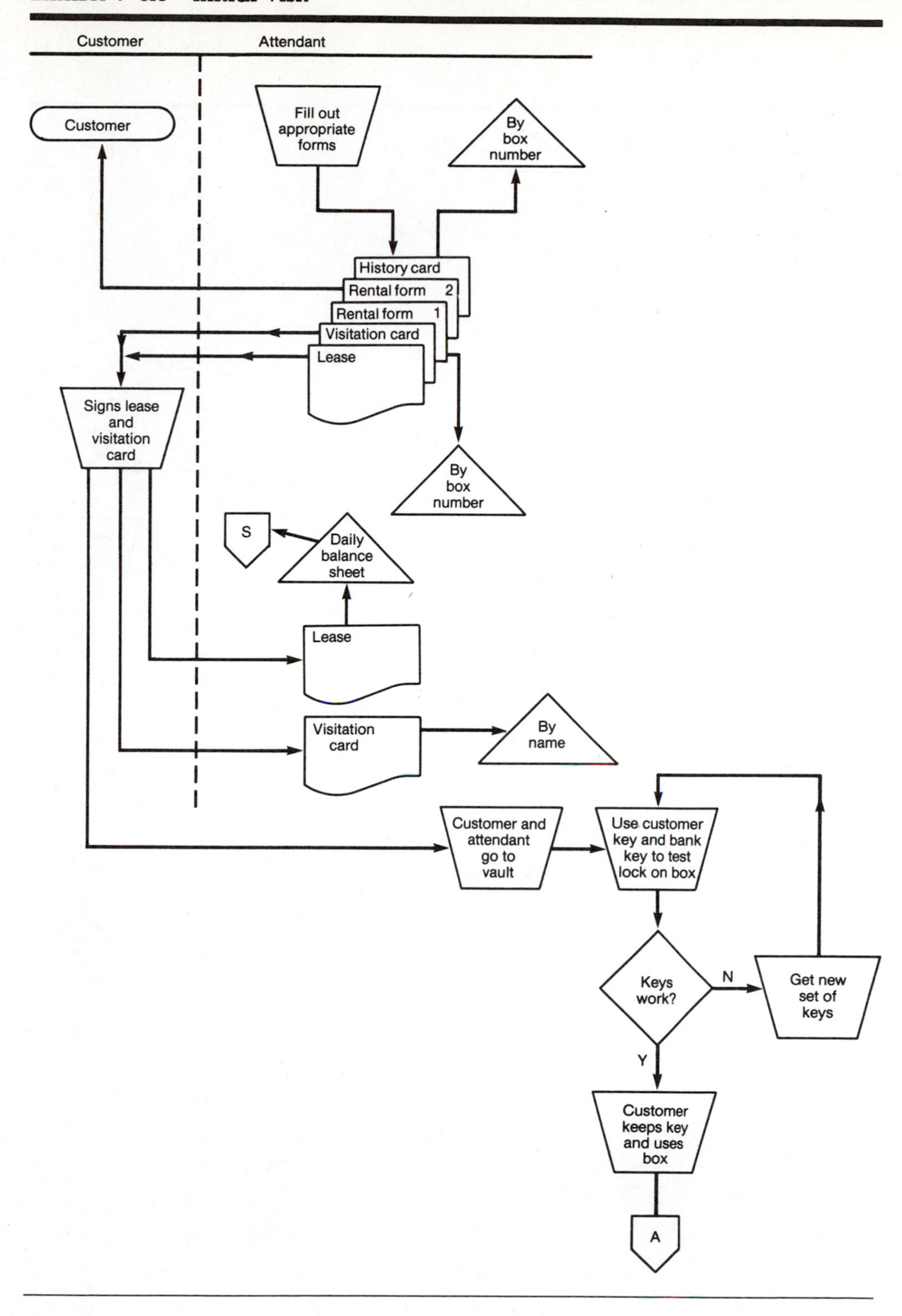

Customer
Attendant
Customer
Fill out appropriate forms
By box number
History card
Rental form 2
Rental form 1
Visitation card
Lease
Signs lease and visitation card
By box number
S
Daily balance sheet
Lease
Visitation card
By name
Customer and attendant go to vault
Use customer key and bank key to test lock on box
Keys work?
N
Get new set of keys
Y
Customer keeps key and uses box
A

EXHIBIT 7–3B Daily Activity

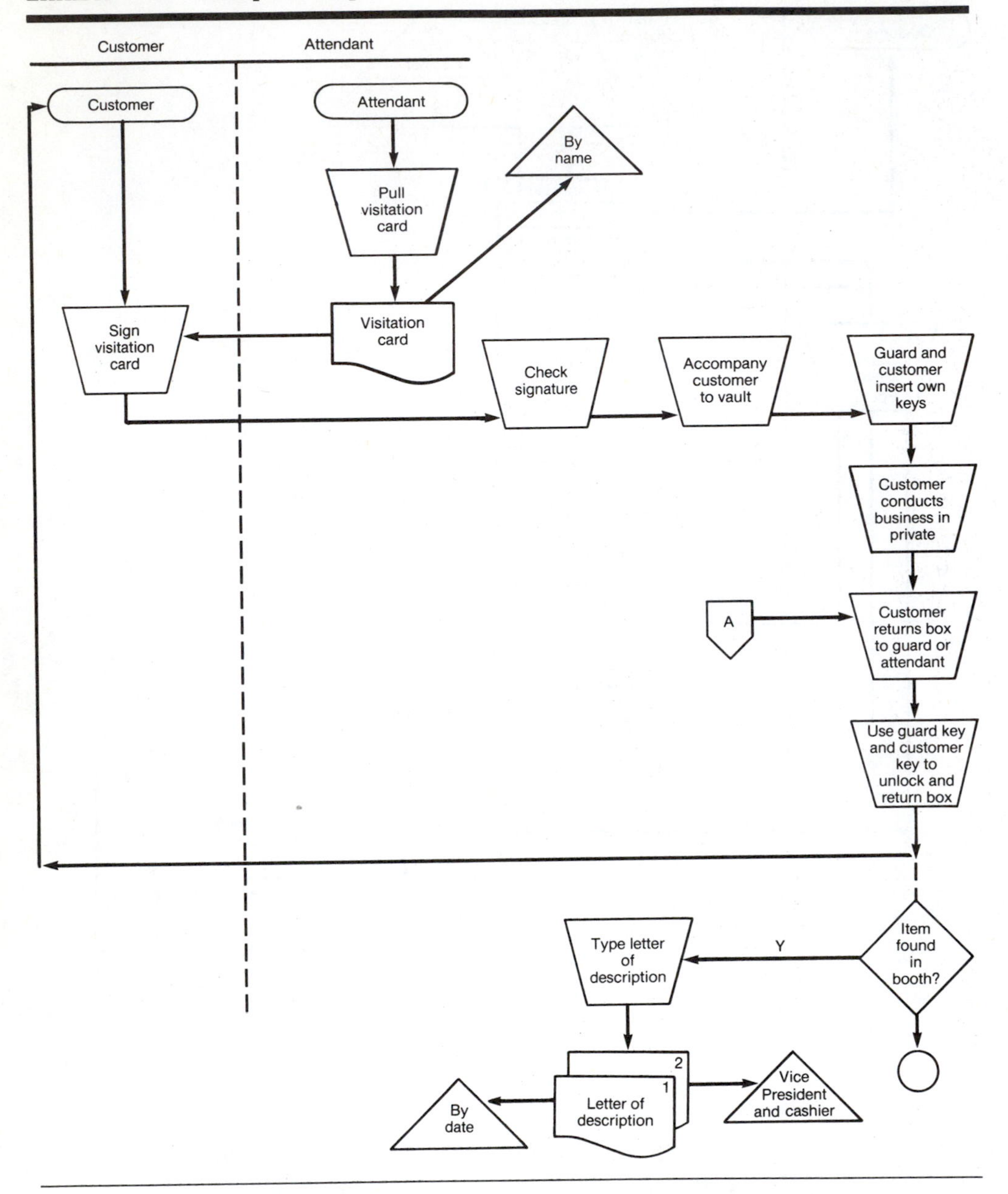

EXHIBIT 7-3C Billing

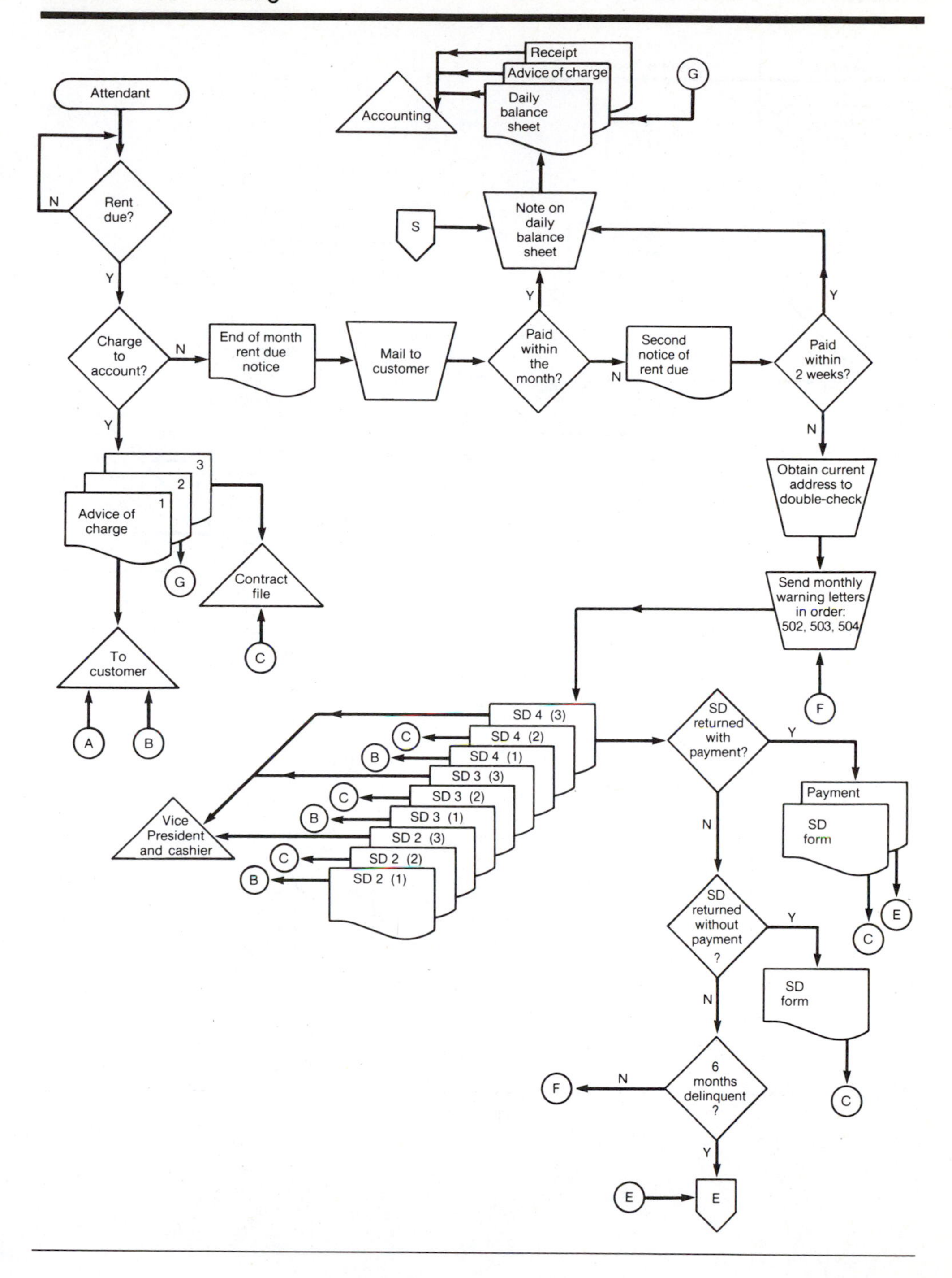

EXHIBIT 7–3D Six-Month Delinquent Rental

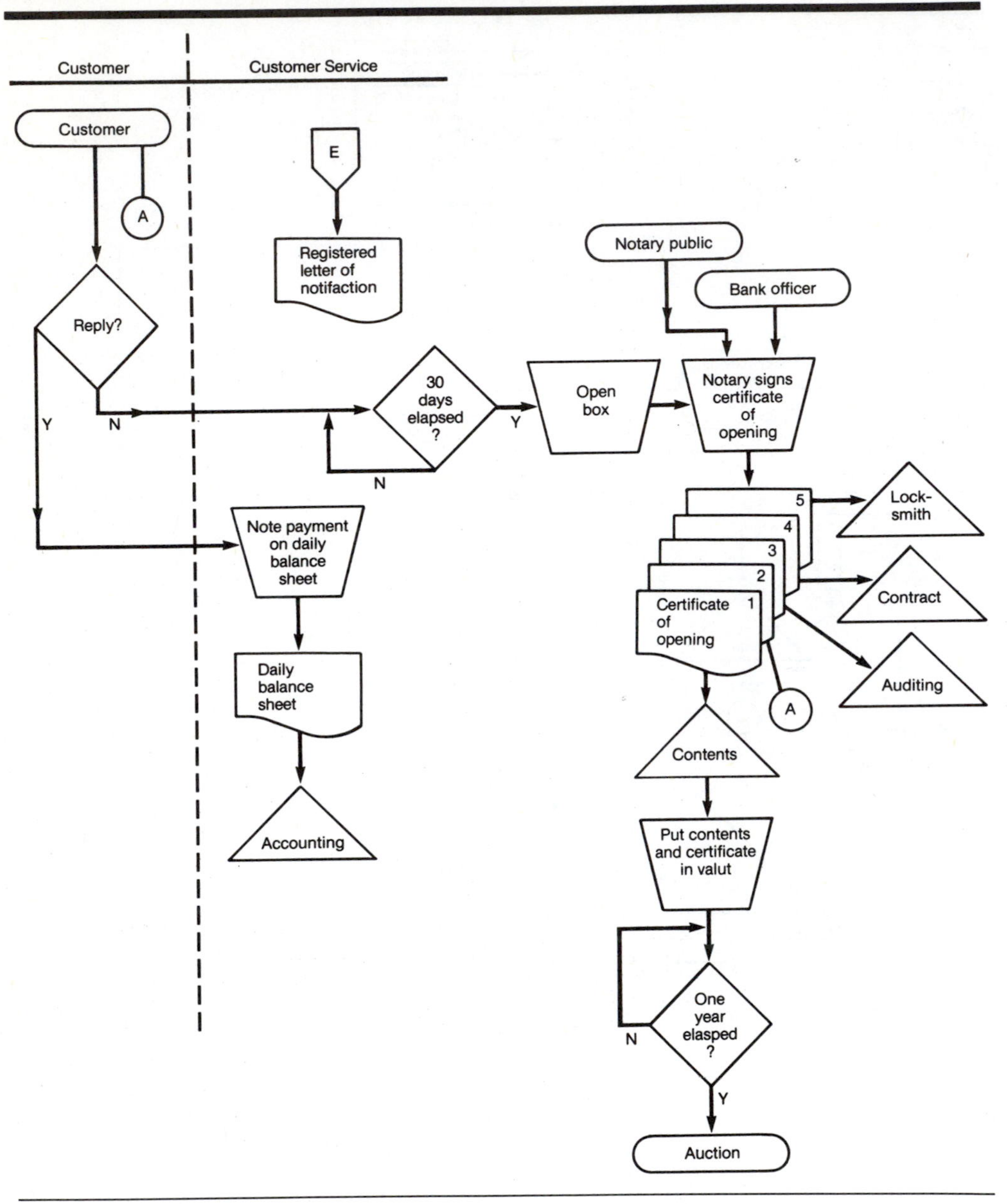

A new problem arose when the department was questioned regarding a purported loss of cash from a customer's safe deposit box; which spurred an investigation into the procedures of its operations. There was one instance where a change in a contract's terms regarding the rental of a box was improperly recorded. Consequently, a new procedure was established whereby contract changes are now immediately recorded as a new contract. The old contractual agreement on which the change is noted is filed as a duplicate contract. A situation that requires such action, for instance, would be a change in rental by two individuals to a single lessee. Furthermore, a manual of procedures was compiled. As shown in Exhibit 7–4, it specifies and documents the procedural requirements and information necessary for opening new accounts, conducting daily operations, and collecting delinquent rentals. The guidelines improve efficiency and minimize legal liabilities.

The Problem

Despite these recent procedural changes, problems still exist for safe deposit that indicate the need for further, more dramatic changes in the current system. The main problems center around the present labor-intensive manual record-keeping procedures. All filing and record-keeping are done manually by several people. Consequently, documents are easily lost or misfiled. The enormous volume of paperwork generates inefficiency and disorganization. Because the department has grown so large and the nature of the manual work is so tedious, the potential for errors is great. As stated by one employee, "the idea of future growth under this system is frightening."

Problems with financial record-keeping also attribute to the manual system, particularly with accuracy in the present billing method. One person is solely responsible for pulling contracts due for billing and for typing up the bills. This, along with manually noting payments on contracts, leaves room for inaccurate reporting and misfiling. One result, for example, is that delinquent accounts sometimes go unbilled.

Another area of concern with the present system is incomplete and inaccurate documentation. This is evident in the difficulty of tracing misplaced documents. Similar documents are not necessarily filed together and standard forms are not prenumbered. These all contribute to filing problems.

Documentation problems were found to stem from the *Manual of Instructions and Procedures.* First, there is no description of which files exist, how they are organized, what is kept where, and how often each file is updated. Second, box prices are somewhat arbitrary. Only one of the four pricing schedules is documented. Third, there are no formal job descriptions or recorded assignments of duty.

Related to record-keeping deficiencies are inventory procedures. Neither auditing nor safe deposit knows exactly how many boxes there

EXHIBIT 7–4

SUMMARY OF CURRENT PROCEDURES

Following is an outline of the procedures presently followed by the department, as documented in the Manual of Instructions and Procedures.

PROCEDURE FOR OPENING NEW ACCOUNTS

Forms to be completed upon the opening of a new account before the customer enters the vault:

I. Lease.
   A. Necessary identification information to be obtained.
      1. Name.
      2. Firm.
      3. Address.
      4. Phone number.
      5. Hair color.
      6. Eye color.
      7. Height.
      8. Weight.
   B. Lease agreement (contract) is read and signed by the person(s) opening the account. A key deposit is made and the deposit amount is recorded on the lease.
   C. If additional persons are to have access to the safe desposit box, a deputy must be appointed. This appointment is noted on the contract and the signature of the deputy is needed.
   D. Contract filed by box number.

II. Visitation card.
   A. Signed upon rental; this original signature is used for comparison purposes during each visit.
   B. Signed upon each visit by authorized customer.
      1. Signature compared to original signature.
      2. Initialed by attendant.
      3. Date and time of entrance noted.
   C. After needed information is obtained, customer is granted entrance.
   D. Filed by lessee name.

III. Rental form.
   A. Receipt for the initial rental fee (payable in advance).
   B. Receipt for key deposit.
   C. Lessee is requested to read safe deposit rules and regulations printed on reverse side.
   D. Filed by box number.
   E. Lessee receives copy.
   F. Transaction entered on daily balance sheet; all receipts filed in daily envelope.

EXHIBIT 7-4 *(continued)*

IV. History card.
    A. Information to be recorded:
        1. Box number.
        2. Person(s) renting box.
        3. Date rented.
        4. Attendant's initials.
    B. Filed by box number.

Customer is granted entrance to the vault:
I. Customer is escorted by safe deposit attendant to safe deposit box.
II. Customer's keys are tested.
III. Box is checked to ensure it is empty.
IV. Customer and box are taken to a private booth if personal belongings are to be placed in box at this time.

PROCEDURES FOR DAILY WORK

Customer wants access to safe deposit box.
I. Attendant pulls customer's visitation card.
    A. Customer signs.
    B. Attendant checks signature, initials card, and notes time and date.
II. Customer is accompanied into vault and to safe deposit box.
III. Attendant and customer each insert own keys; door is opened and box is removed.
IV. Customer conducts business with box in private booth.
V. Customer returns box to attendant.
VI. Attendant key and customer key are used to unlock door; box is returned and door locked.

Inventory of contents of safe deposit box.
I. Inventory must be authorized by
    A. Court order or
    B. Power of attorney.
II. Approval from President, VP and Cashier, or VP and Senior Loan Officer needed before entrance.
III. Customer is accompanied by officer or safe deposit custodian of the bank.
IV. Visitation card is signed and initialed by bank officer and safe deposit custodian and circumstance of entry noted.
V. Inventory of safe deposit box contents is made.
VI. Copy of inventory contents is filed by date.

Customer surrenders safe deposit box.
I. Release portion of lease agreement signed.
II. History card notations.
    1. Closing date.
    2. Number of keys returned.
III. All account documents are marked with closing date and filed with inactive documents.
IV. Key deposit returned.

EXHIBIT 7-4 *(concluded)*

BILLING PROCEDURE

- I. Safe deposit attendant manually determines for which accounts rental fees are due.
- II. Notices mailed.
  - A. If fee is charged to customer's checking account, an advice of charge is created in triplicate.
    - 1. Copies to
      - *a.* Customer.
      - *b.* Accounting department.
      - *c.* Contract file.
    - 2. Credit noted on daily balance sheet.
  - B. If fee is not charged to customer's account, a rental due notice is mailed to customer at month's end.
- III. If rent not received by the end of the month in which it was due, second notice will be mailed.
- IV. If rent not received within two weeks after second notice, current address must be verified.
- V. If rent still not paid, three monthly warning letters are mailed (SD2, SD3, SD4).
  - A. Two copies typed.
    - 1. Original signed by VP and Cashier and mailed to customer.
    - 2. Second copy filed in contract file.
- VI. If rent is not renewed within three months after expiration of lease term, the safe deposit corporation may deny admission to safe deposit box.
- VII. If rent is six months delinquent, delinquent procedures are enacted.
- VIII. When rent received, payment noted on daily balance sheet.
- IX. All cash and daily balance sheet sent to Accounting at end of day; receipts filed in daily envelope.

SIX-MONTH DELINQUENT RENTAL PROCEDURE

- I. Registered letter mailed to customer's last known address, advising that the box will be opened and contents stored at renter's expense unless rental payment is made within 30 days.
- II. If payment is not received, box is opened in presence of Bank Officer and Notary Public.
  - A. Contents sealed in package.
  - B. Notary Public executes certificate reciting.
    - 1. Name of renter.
    - 2. Date of box opening.
    - 3. List of contents.
  - C. Five copies of certificate of opening.
    - 1. Included in package holding contents.
    - 2. Last known address of renter.
    - 3. Auditing department.
    - 4. Safe deposit department.
    - 5. Locksmith.
- III. Package placed in bank vault; same rent charged as for safe deposit box.
- IV. If one year elapses and package is not claimed, a public auction is held to sell contents. Proceeds used to pay for rent and expenses.

are. Discrepancies exist between the two departments (auditing and safe deposit) records. Boxes are manually counted. Their numbers do not follow a logical sequence corresponding with locations in the vault. In addition, some of the boxes cannot be matched with a contract, and some contracts are missing altogether.

Finally, two important security deficiencies have been identified. First, contracts filed in the vault are not locked up, which means they are accessible to customers in the vault at any time. Second, past-due contracts are filed in the lobby in an unlocked desk that is not fireproof. These conditions jeopardize the interests of both the customer and the bank.

In summary, specific deficiencies with the present system have been identified in areas such as manual record-keeping and filing procedures, incomplete and inaccurate documentation, and inadequate security. Despite recent procedural and policy changes, a computerized safe deposit tracking and billing system could result in significant improvements. Such a system would provide needed up-to-date information, improve accuracy and file organization, and increase security by making errors more readily detectable. With a user's manual and a more specific policy manual, documentation would be complete.

The Alternatives

In determining feasibility, the first step was to evaluate to what extent current needs can be met and problems solved. A quick search of existing software packages revealed specific improvements, including:

a. Improved billing accuracy and efficiency.

b. Provision for updated box tracking.

c. Reduction of record misfiling.

d. Elimination of monthly manual check for box occupancy.

After various microcomputers were reviewed, it became obvious that an IBM PC would be appropriate. For a relatively low cost, such a system would meet the department's immediate needs, has enough memory capacity for expansion, and is flexible enough to be used for other functions such as word processing, general ledger, payroll, and financial analysis. The system has an excellent reputation for performance and maintenance and could later be networked into a large system. Even though department personnel are completely unfamiliar with computers, the system would not be intimidating and can be installed with limited training. Politically, the IBM PC would be a good choice. The vice president in charge of accounting had a three-day seminar on the system with a previous employer. His biases could not be overlooked.

The first major task undertaken was a search for a software package that is capable of meeting the user's needs, keeping in mind the following:

a. Automated billing.
b. Box tracking.
c. Online safe deposit maintenance.
d. User-friendly documentation and easy training.
e. Loading the software on the IBM PC.

Although one option was to do the programming, it was soon found that time and cost would be prohibitive. After various directories of financial software packages were searched, a package was found to fit the overall requirements.

The Package

There were three limitations concerning the proposed safe deposit package:

a. There was no provision for a history card file to keep track of the present and previous customers of each box for reference.
b. The package handles a maximum of 8,000 boxes on the PC and 32,000 boxes on the XT (hard disk) model.
c. Being less than two years old, the package has flaws such as limited editing capability and poor documentation.

To compensate for these drawbacks, the history card file problem was solved by setting up a dBASE II file on the PC to work in conjunction with the software package. The capacity constraint was settled when it was learned that management does not expect SDC to exceed that limit in box rentals before 1990. In the long run, this would be more cost effective with the PC than underutilizing an XT system. Finally, the vendor assured the bank that all program flaws will be eliminated in a forthcoming updated edition.

Costs and Benefits

Computerizing safe deposit operations will involve new costs that can be justified by saving one full-time clerk and substantially improving customer service and overall performance. Listed here are the costs and savings associated with the candidate system:

Hardware/software

Hardware

IBM PC with 64K memory	$2,630
Three 64K-byte memory chips	490
Monochrome display and printer adapter board	335
Monochrome display (screen)	345
Epson printer—letter quality	1,100
	$4,900

Software	
DOS 2.0 operating system	65
dBASE II	385
Safe deposit tracking/billing system	950
Maintenance (one-time fee)	150
Supplies (disks, paper, etc.)	250
	$1,800
Total costs	$6,700

Direct savings

One full-time employee (salary + 25% benefits) $16,750

Net direct benefits $16,750 − 6,700 = $ 10,050

Break-even point 6,700/16,750 = 40 percent or 4.8 months

Assignment

a. Evaluate the feasibility study undertaken by the project team—its strengths, weaknesses, procedures used, and expected outcome.

b. If you were doing the study, would you have considered hardware before software? Why? Elaborate.

2 JEFFERSON CREDIT CENTER

This case is a continuation of the company's background presented in Chapter 4. In this section, two alternative systems are described and a preliminary choice is made for implementation.

Problem Definition

As described in the first part of the case (see Chapter 4), the current problem in the operations of the Jefferson credit center is inefficient storage and retrieval. In the present file system, both customer inquiries and payment slips are stored in physical paper files under date indexes. Consequently, misfiles and the tertiary relevance of the account number and document type make the search process highly inefficient. This reduces the credit center's ability to properly respond to customer inquiries.

Goals and Considerations

With the preceding problem in mind, the study was oriented toward (1) locating and evaluating microfilm processing and data storage and retrieval systems capable of meeting the Jefferson credit department's needs and (2) recommending the system best suited to the unique needs and limitations of credit center operations.

In the system feasibility study, the following goals were expected to be achieved by the candidate system:

a. File control—All out-of-file or misfiled conditions should be eliminated.

b. Multiple access—Multiple users should have access to the same information simultaneously.

c. Labor savings—Fewer personnel should be able to retrieve more information in less time, providing greater productivity per employee.

d. Storage capacity—The system should be large enough to allow unlimited expansion.

e. Document organization—Retrieval of documents can be specified in any order regardless of date or order filmed.

f. Information revision—The system must be capable of accepting additional information via the CRT terminal.

g. Future flexibility—The system should be capable of adding more files (readers if needed) in the event of future growth.

In addition to these goals, each system was evaluated in light of the following considerations:

a. Compatibility with the present credit management system on the IBM 3031 and 4341.

b. Feasibility of in-house (versus contracting) conversion and updating of existing paper file records.

c. Ease of transition, which involves implementation procedures and employee training.

d. Affordability.

Of the systems evaluated, two top candidates were chosen. Each system is briefly described here.

Kodak Microimagination System

The Kodak Microimagination System is delineated into two separate areas: filming is performed by the Reliant 800 Microfilmer, and storage and retrieval functions are performed by the IMT-150 Intelligent Terminal. The Reliant 800 microfilms checks at the rate of 700 per minute. The machine has reduction capabilities and a film capacity of either two 100-feet rolls of 5.4-mm film or two 215-feet rolls of 2.5-mm film. The unique aspects of the system are its flexibility and speed. Accessories can be added with ease. The indexing system developed while microfilming the document allows for a retrieval time of less than five seconds.

The Reliant 800 is equipped with an intelligent controller that receives program signals and translates them into operations that suit the indexing needs. The client can select any four indexing options from 13 available programs. These indexing choices can be changed as the user's needs change.

The second component of the system is basically a microcomputer that has a built-in memory and short-term (temporary) storage that can be erased when not in use. The unit is capable of searching through several varieties of film formats. When the document is brought to screen, the IMT-150 has an automatic image position feature that "locks in" a clear, complete picture on the viewing screen. Once on the viewing screen, a print can be obtained in 12 seconds.

An aspect of the Kodak system that is of vital importance to Jefferson stores is its easy adaptability and compatibility with Jefferson's existing mainframe computer. The Computer-Assisted Retrieval (CAR) allows the user to keep the images on inexpensive, easy-to-use microfilm magazines, while the computer data base maintains an index of the location of each microfilm image. With the CAR, the computer does all the sorting and indexing of randomly filmed documents while the IMT-150 retrieves the document.

CARMS/11 Microimage System

The second candidate system considered was the CARMS/11 system. The California-based vendor is one of the leading suppliers of rapid-access information-retrieval systems using micrographic and computer technologies. The proposed system is a fully automated, computer-controlled record management system. It is designed to:

a. Eliminate misfiles and out-of-file documents.

b. Provide instant retrieval of payment and customer inquiries.

c. Increase productivity by creating instant access to data.

d. Allow for incremental expansion into other areas within the Jefferson credit center.

When the operator initiates a search by depressing a function key, he/she receives prompts querying as to what functions are to be performed. In addition to system status, the system can be used to update, amend, or append information to a file, thus providing the operator with complete information whenever it is needed.

Comparative Analysis

The next step in this project was to evaluate the pros and cons of the two candidate systems. The present system used by Jefferson's credit center is already obsolete. Both systems considered, the Kodak Reliant and the CARMS/11 are completely compatible with the IBM 3031. In the case of the Kodak system, a small software package is all that is required to integrate the two systems. On the other hand, the

CARMS/11 system would require extensive data base and file control software to control indexing, storage, and retrieval of large amounts of information.

A second consideration is storage requirements. Jefferson's credit center receives and microfilms between 12 and 25 batches of 250 checks per day. It also receives 725 customer inquiries per week. The proposed system must be capable of storing two years of such data. This amounts to 2 million to 3 million documents. The CARMS/11 system, with a storage capacity of 100,000 documents per ultrastrip cassette, would require 20 to 25 cassettes. On the other hand, the Kodak system would require 150 cassettes to store two years of documentation.

Related to storage requirements is quick retrieval time. The maximum time for accessing a document should be 15 seconds; 25 seconds if a hard copy is required. This criterion favors the CARMS/11 system. With 20 to 25 total storage cassettes, the CARMS/11 access time is between 2 and 12 seconds. The Kodak system, with more than five times as many film cassettes, would require more search time for the cassette. In addition, the chance for misplacing the cassette is greatly increased with the Kodak system.

A third criterion to consider is document filming and indexing. Both systems are comparable. Filming and indexing take five to eight hours per day.

A final criterion used in the evaluation is vendor service and support. Kodak's vendor is in Lynchburg (home of the credit center), whereas the CARMS/11 dealer operates out of the Washington, D.C., area—140 miles away. A system engineer must be within easy reach to remedy malfunctions and provide an ongoing training program.

Cost Comparison

Most of Kodak's cost is for hardware. Jefferson would have to purchase a new microfilmer, two intelligent terminals, some peripheral accessories, and a fairly inexpensive software package (see Exhibit 7–5). On the other hand, the CARMS/11 system's cost is mainly the software package and its two work stations. Compared to Kodak's software package, which merely provides the interface between the IMT-150 and the IBM 3031, the CARMS/11 software package provides data base management services, but also redundant operating system services (see Exhibit 7–6).

Development costs present some interesting features. The CARMS/11 system requires over twice the processing and development costs incurred by the Kodak system. This is best shown by Exhibits 7–7 and 7–8, which represent the processing and development costs for the next two years' business volume. The increased cost of CARMS/11 can be attributed solely to the ultrastrip conversion. The ultrastrip's benefits seem to outweigh its rather expensive price tag, however.

EXHIBIT 7–5 Kodak Microimagination System Costs

Item	Cost
Reliant 800 Microfilmer	$ 8,600
Intelligent controller	3,025
Imprinter	6,400
Imprinter keyboard	2,550
Patch sensor assembly	1,760
Image mark counter	535
Bench work base	585
Shelf	155
Receiving hopper	165
Film unit for inquiries	2,190
Film unit for checks	2,190
Auto-feeder	1,325
Stacker	985
Image marker	905
Two IMT-150 terminals	29,570
Interface with computer	2,800
Total	$63,740
Development and processing costs of past documents	11,305
Total to be current	$75,045

EXHIBIT 7–6 Purchase Cost for CARMS/11 Microimage System

Two CARMS/11 work stations	$ 38,000
Auto-feeder	1,325
50 Strip removable cartridges	1,000
Software	35,000
Total system price	$ 75,325
Conversion of present microfilm to ultrastrips—two years	35,000
Total system and conversion costs	$110,325

EXHIBIT 7–7 CARMS/11 Developing Costs

Description	Costs	
Cost of developing one image microfilm and ultrastrip	0.05/image	
1,500,000 checks/year, 3 checks/image		$25,000/year
600 customer inquiries/week, 1 inquiry/page		1,440/year
Total microfilm and ultrastrip development costs/year		$26,440

EXHIBIT 7-8 Kodak Developing Costs

1,500,000 checks/year,	
20,000 checks/cartridge,	
75 cartridges required/year,	
two years' worth = 150 cartridges,	
$7.00/cartridge	$1,050
600 customer inquiries,	
24,000/year,	
two cartridges required/year,	
two years' worth	148
Total	$1,198

In comparing the two systems, it is necessary to weigh the cost, vendor reputation, nearby service center, and support of the Kodak system against the greater flexibility, speed, and storage capabilities of the CARMS/11 system. Because such a large portion of the credit center's activity centers around customer service, speed and efficiency are of prime importance. Consequently, the CARMS/11's ability to increase the speed of access/retrieval was the deciding advantage. Although Kodak operates a service center in Lynchburg, CARMS's dealer can provide same-day service as well.

Finally, although the CARMS/11 system costs $30,000 more than the Kodak system, no budget constraints were provided. Furthermore, the system's unique attributes justify the additional cost.

Assignment

Evaluate the feasibility study. What were the strong points of the study? The weak points? What additional information is needed to do a complete study? Elaborate.

Selected References

Andrews, Wm. C. "The Business Systems Proposal." *Journal of Systems Management*, February 1978, pp. 39–41.

Gore, Marvin, and John Stubbe. *Elements of Systems Analysis*. 3rd ed. Dubuque, Iowa: Wm. C. Brown, 1983, pp. 240–67.

Neuschel, Richard F. "Presenting and Selling Systems Recommendations." *Journal of Systems Management*, March 1982, pp. 5–13.

Powers, Michael; Davis Adams; and Harlan D. Mills. *Computer Information System Development: Analysis & Design*. Cincinnati: South-Western Publishing, 1984, pp. 120–47.

Zells, Lois A. "A Practical Approach to a Project Expectation Document." *Computerworld (In-Depth)*, August 29, 1983, p. 1.

Chapter 8

Cost/Benefit Analysis

At a Glance

Data gathering, traditional or structured, is only one part of systems analysis. The next steps are examining the data gathered, assessing the situation, looking at the alternatives, and recommending a solution. The costs and benefits of each alternative guide the selection of the best system for the job.

Cost and benefits may be tangible or intangible, direct or indirect, fixed or variable. Cost estimates also take into consideration hardware, personnel, facility, and supply costs for final evaluation. Cost/benefit analysis, then, identifies the costs and benefits of a given system and categorizes them for analysis. Then a method of evaluation is selected and the results are interpreted for action. The evaluation methods range from the simple net benefit analysis to more complex methods such as present value and payback analyses.

By the end of this chapter, you should know:

1. What is involved in data analysis.
2. Cost and benefit categories.
3. How to identify and classify costs and benefits.
4. The various evaluation methods for cost/benefit analysis.

INTRODUCTION

In Chapters 5 and 6, we discussed various tools analysts use for gathering details about the system under study. Data gathering is only one part of systems analysis, however. The next steps are to examine the data, assess the situation, look at the alternatives, and recommend a candidate system. The costs and benefits of each alternative guide the selection of one alternative over the others. Since this aspect of analysis is so important, it will take up most of the chapter.

This chapter discusses approaches to developing design recommendations for the end user. Each approach has costs and benefits that are compared with those of other approaches before a final recommendation is made. The outcome is a system proposal (also called a project proposal) that summarizes the findings of the analysis and states the recommendations for design. By the end of this chatper, you should be able to evaluate how current operations are performed, the categories of costs and benefits, key methods for cost/benefit analysis, and how a system proposal is developed.

DATA ANALYSIS

Data analysis is a prerequisite to cost/benefit analysis. System investigation and data gathering lead to an assessment of current findings. Our interest is in determining how efficiently certain steps are performed, how they contribute to achieving the intended goals, and the cost of making improvements. Let us return to our safe deposit scenario (from Chapter 4) to illustrate the point.

The safe deposit department was authorized to double its capacity from 4,000 to 8,000 boxes in an effort to meet increased demand. Consequently, the number of employees changed from three to five, with one employee assigned full-time to billing. Analysis of the data collected made it obvious that customers were frequently billed too late, too often, or not at all. Access to customer information or status of vacant boxes was a nightmare. Customer lines were long, and service was jeopardized.

The representative facts for the safe deposit department are shown in Figure 8–1. The system profile summarizes the operating characteristics of the safe deposit system, such as the volume of work, nature of processes, physical facilities, and personnel. From the analysis, the system design requirements are identified. These features must be incorporated into a candidate system to produce the necessary improvements. The system requirements are:

1. Better customer service.
2. Faster information retrieval.
3. Quicker notice preparation.
4. Better billing accuracy.

FIGURE 8–1 Representative Facts and Candidate System Design Objectives

Analysis	Current System Facts	Objectives (System Design Requirements)
What is done? (processes)	Open customer account Assign safe deposit box Issue key Receive annual rent	Better customer service Faster information retrieval Quicker notice preparation
How is it done? (processing detail)	Some 40 boxes opened or closed daily Master card file located too far from customer inquiry station Heaviest activity on Fridays and before holidays Too many steps taken with new customers Delay in billing—all manual Some 80 renewal payment notices prepared daily Cash received given to head teller at end of each day Poorly designed application forms Accounting gets daily summary Procedure for customer access to boxes is neither documented nor consistent	Better billing accuracy Lower processing and operating cost Improved staff efficiency More consistent billing procedures to eliminate errors
Who does it? (personnel)	One person handles billing (full-time) One person handles security Three persons process customers into and out of safe deposit area Except for two persons, rest of staff is poorly trained Communication among staff is adequate	Better trained personnel Experience in computer use for other applications
Where is it done? (physical location/facility)	Location allows privacy and security Billing carried out close to customer counter	Allocate quiet space for computer Provide security measure for information access
Assessment of processing	Time to prepare a renewal notice is 10 minutes Time to process a customer is 3.5 minutes 15 percent of billing is erroneous in amount, box number, or amount of rent 28 percent of vacant boxes cannot be located on existing books Frequent notices regarding "to be renewed" boxes cost $8,000 for mailing Employee payroll is as high as junior officers in operations area	

5. Lower processing and operating costs.
6. Improved staff efficiency.
7. Consistent billing procedure to eliminate errors.

To achieve these design objectives, several alternatives must be evaluated; there is seldom just one alternative. The analyst then selects those that are feasible economically, technically, and operationally. The approach may emphasize the introduction of a computerized billing system, replacement of staff, improved billing practices, changes in operating procedures, or a combination of several options.

As you can imagine, each approach has its benefits and drawbacks. For example, one alternative is to introduce a computer-based safe desposit tracking and billing system designed to improve billing accuracy and notice preparation and lower processing and operating costs. It would also promote staff efficiency by allowing the existing staff to concentrate on customer service and provide online information on box availability and the like. The drawbacks include laying off the billing clerk who recently got married and strong resistance by the majority of the staff to a computerized environment.

Another alternative might be simply to devise a semiautomatic (ferris-wheel type) system that organizes master cards and customer records and improves their access. A word processing system might be introduced to speed the preparation of billing notices. The edit feature of word processors would improve the accuracy in billing preparation. If these were the only two alternatives available, which alternative must be selected? An analysis of the costs and benefits of each alternative guides the selection process. Therefore, the analyst needs to be familiar with the cost and benefit categories and the evaluation methods before a final selection can be made. Details on these topics are given in the next section.

COST/BENEFIT ANALYSIS

Cost and Benefit Categories

In developing cost estimates for a system, we need to consider several cost elements. Among them are hardware, personnel, facility, operating, and supply costs.

1. *Hardware costs* relate to the actual purchase or lease of the computer and peripherals (for example, printer, disk drive, tape unit). Determining the actual cost of hardware is generally more difficult when the system is shared by various users than for a dedicated stand-alone system. In some cases, the best way to control for this cost is to treat it as an operating cost.

2. *Personnel costs* include EDP staff salaries and benefits (health insurance, vacation time, sick pay, etc.) as well as pay for those involved in developing the system. Costs incurred during the development of a system

are one-time costs and are labeled developmental costs. Once the system is installed, the costs of operating and maintaining the system become recurring costs.

3. *Facility costs* are expenses incurred in the preparation of the physical site where the application or the computer will be in operation. This includes wiring, flooring, acoustics, lighting, and air conditioning. These costs are treated as one-time costs and are incorporated into the overall cost estimate of the candidate system.

4. *Operating costs* include all costs associated with the day-to-day operation of the system; the amount depends on the number of shifts, the nature of the applications, and the caliber of the operating staff. There are various ways of covering operating costs. One approach is to treat operating costs as overhead. Another approach is to charge each authorized user for the amount of processing they request from the system. The amount charged is based on computer time, staff time, and volume of the output produced. In any case, some accounting is necessary to determine how operating costs should be handled.

5. *Supply costs* are variable costs that increase with increased use of paper, ribbons, disks, and the like. They should be estimated and included in the overall cost of the system.

A system is also expected to provide benefits. The first task is to identify each benefit and then assign a monetary value to it for cost/benefit analysis. Benefits may be tangible and intangible, direct or indirect, as we shall see later.

The two major benefits are improving performance and minimizing the cost of processing. The performance category emphasizes improvement in the accuracy of or access to information and easier access to the system by authorized users. Minimizing costs through an efficient system—error control or reduction of staff—is a benefit that should be measured and included in cost/benefit analysis.

Procedure for Cost/Benefit Determination

There is a difference between expenditure and investment. We spend to get what we need, but we invest to realize a return on the investment. Building a computer-based system is an investment. Costs are incurred throughout its life cycle. Benefits are realized in the form of reduced operating costs, improved corporate image, staff efficiency, or revenues. To what extent benefits outweigh costs is the function of cost/benefit analysis.

Cost/benefit analysis is a procedure that gives a picture of the various costs, benefits, and rules associated with a system. The determination of costs and benefits entails the following steps:

1. Identify the costs and benefits pertaining to a given project.
2. Categorize the various costs and benefits for analysis.

3. Select a method of evaluation.
4. Interpret the results of the analysis.
5. Take action.

Costs and Benefits Identification

Certain costs and benefits are more easily identifiable than others. For example, direct costs, such as the price of a hard disk, are easily identified from company invoice payments or canceled checks. Direct benefits often relate one-to-one to direct costs, especially savings from reducing costs in the activity in question. Other direct costs and benefits, however, may not be well defined, since they represent estimated costs or benefits that have some uncertainty. An example of such costs is reserve for bad debt. It is a discerned real cost, although its exact amount is not so immediate.

A category of costs or benefits that is not easily discernible is *opportunity costs* and *opportunity benefits*. These are the costs or benefits forgone by selecting one alternative over another. They do not show in the organization's accounts and therefore are not easy to identify.

Classifications of Costs and Benefits

The next step in cost and benefit determination is to categorize costs and benefits. They may be tangible or intangible, direct or indirect, fixed or variable. Let us review each category.

Tangible or Intangible Costs and Benefits. Tangibility refers to the ease with which costs or benefits can be measured. An outlay of cash for a specific item or activity is referred to as a *tangible* cost. They are usually shown as disbursements on the books. The purchase of hardware or software, personnel training, and employee salaries are examples of tangible costs. They are readily identified and measured.

Costs that are known to exist but whose financial value cannot be accurately measured are referred to as *intangible* costs. For example, employee morale problems caused by a new system or lowered company image is an intangible cost. In some cases, intangible costs may be easy to identify but difficult to measure. For example, the cost of the breakdown of an online system during banking hours will cause the bank to lose deposits and waste human resources. The problem is by how much? In other cases, intangible costs may be difficult even to identify, such as an improvement in customer satisfaction stemming from a real-time order entry system.

Benefits are also classified as tangible or intangible. Like costs, they are often difficult to specify accurately. Tangible benefits, such as completing jobs in fewer hours or producing reports with no errors, are quantifiable. Intangible benefits, such as more satisfied customers or an improved corporate image, are not easily quantified. Both tangible and intangible costs and benefits, however, should be considered in the evaluation process.

Management often tends to deal irrationally with intangibles by ignor-

FIGURE 8-2 Tangible and Intangible Costs and Benefits for a Given Project

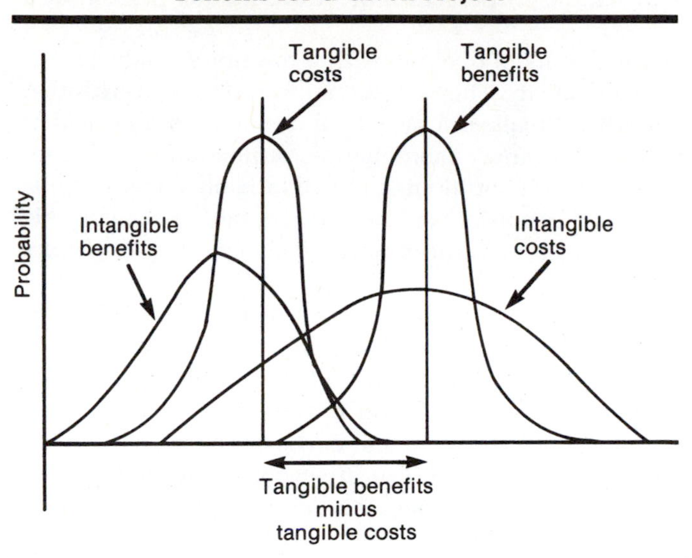

ing them. According to Oxenfeldt, placing a zero value on intangible benefits is wrong.[1] Axelrod reinforces this point by suggesting that if intangible costs and benefits are ignored, the outcome of the evaluation may be quite different from when they are included.[2] Figure 8-2 is a hypothetical representation of the probability distribution of tangible and intangible costs and benefits. It indicates the degree of uncertainty surrounding the estimation of costs and benefits. If the project is evaluated on a purely tangible basis, benefits exceed costs by a substantial margin; therefore, such a project is considered cost effective. On the other hand, if intangible costs and benefits are included, the total tangible and intangible costs exceed the benefits, which makes the project an undesirable investment. Furthermore, including all costs increases the spread of the distribution (compared with the tangible-only distribution) with respect to the eventual outcome of the project.

Direct or Indirect Costs and Benefits. From a cost accounting point of view, costs are handled differently depending on whether they are direct or indirect. *Direct costs* are those with which a dollar figure can be directly associated in a project. They are applied directly to the operation. For example, the purchase of a box of diskettes for $35 is a direct cost because we can associate the diskettes with the dollars expended. *Direct*

[1] A. R. Oxenfeldt, *Cost-Benefit Analysis for Executive Decision Making* AMACOM, American Management Association, 1979), p. 51.

[2] C. Warren Axelrod, *Computer Productivity* (New York: John Wiley & Sons, 1982), pp. 61–89.

benefits also can be specifically attributable to a given project. For example, a new system that can handle 25 percent more transactions per day is a direct benefit.

Indirect costs are the results of operations that are not directly associated with a given system or activity. They are often referred to as overhead. A system that reduces overhead realizes a savings. If it increases overhead, it incurs an additional cost. Insurance, maintenance, protection of the computer center, heat, light, and air conditioning are all tangible costs, but it is difficult to determine the proportion of each attributable to a specific activity such as a report. They are overhead and are allocated among users according to a formula.

Indirect benefits are realized as a by-product of another activity or system. For example, our proposed safe deposit billing system that provides profiles showing vacant boxes by size, location, and price, will help management decide on how much advertising to do for box rental. Information about vacant boxes becomes an indirect benefit of the billing even though it is difficult to specify its value. Direct and indirect costs and benefits are readily identified for tangible costs and benefits, respectively.

Fixed or Variable Costs and Benefits. Some costs and benefits are constant, regardless of how well a system is used. *Fixed costs* (after the fact) are *sunk* costs. They are constant and do not change. Once encountered, they will not recur. Examples are straight-line depreciation of hardware, exempt employee salaries, and insurance. In contrast, *variable costs* are incurred on a regular (weekly, monthly) basis. They are usually proportional to work volume and continue as long as the system is in operation. For example, the costs of computer forms vary in proportion to the amount of processing or the length of the reports required.

Fixed benefits are also constant and do not change. An example is a decrease in the number of personnel by 20 percent resulting from the use of a new computer. The benefit of personnel savings may recur every month. *Variable benefits*, on the other hand, are realized on a regular basis. For example, consider a safe deposit tracking system that saves 20 minutes preparing customer notices compared with the manual system. The amount of time saved varies with the number of notices produced.

Savings versus Cost Advantages. Savings are realized when there is some kind of cost advantage. A cost advantage reduces or eliminates expenditures. So we can say that a true savings reduces or eliminates various costs being incurred. Figure 8–3 is a summary of savings from the use of a new online teller system. In this installation, $131,870 was saved through a reduction in personnel, handling charges, and proof machine rental. After deducting processing charges of $90,990, the net savings from the online system was $40,880. This is a savings that provides relief from current costs. It is realized specifically as a result of the additional processing costs incurred in the new system.

FIGURE 8–3 An Example of Savings That Reduce Current Costs

Summary of Savings
from an Online Teller System

A. Reduction in personnel and payroll:

Position	N	Average Annual Pay (includes 25 percent benefits)	Total
Collections teller	1	$12,400	$ 12,400
Savings teller	5	11,610	58,050
Bookkeeper	3	9,940	29,820
Proof operator	1	10,900	10,900
Subtotal	10		$111,170
B. Reduction in handling charges			6,840
C. Reduction in proof machine rental:			
Previous units (4 @ $4,500)		$18,000	
Present units (3 @ $1,380)		4,140	
Net savings from rentals			13,860
Total gross savings			$131,870
Less processing charges:			
Online demand deposit processing		$33,660	
Proof of deposit reporting		27,000	
Online savings processing		5,100	
Teller machine rental		25,230	
Total processing charges			90,990
Net savings/year			$ 40,880

There are savings, however, that do not directly reduce existing costs. To illustrate, examine the following case:

> A systems analyst designed an online teller system that requires 14 new terminals. No reduction in personnel is immediately planned. Renovation of the bank lobby and the teller cages will be required. The primary benefits are:
>
> 1. Savings in tellers' time to update accounts and post transactions.
> 2. Faster access and retrieval of customer account balances.
> 3. Availability of additional data for tellers when needed.
> 4. Reduction of transaction processing errors.
> 5. Higher employee morale.
> 6. Capability to absorb 34 percent of additional transactions.

This is a case where no dollars can be realized as a result of the costs incurred for the new installation. There might be potential savings if additional transactions help another department reduce its personnel. Similarly, management might set a value (in terms of savings) on the improved accuracy of teller activity, on quicker customer service, or on the psychological benefits from installing an online teller system. Given the profit motive, savings (or benefits) would ultimately be tied to cost reductions. Management has the final say on how well the benefits can be cost-justified.

Select Evaluation Method

When all financial data have been identified and broken down into cost categories, the analyst must select a method of evaluation. Several evaluation methods are available, each with pros and cons. The common methods are:

1. Net benefit analysis.
2. Present value analysis.
3. Net present value.
4. Payback analysis.
5. Break-even analysis.
6. Cash-flow analysis.

Net Benefit Analysis. Net benefit analysis simply involves subtracting total costs from total benefits. It is easy to calculate, easy to interpret, and easy to present. The main drawback is that it does not account for the time value of money and does not discount future cash flow. Figure 8–4 illustrates the use of net benefit analysis. Cash flow amounts are shown for three time periods: Period 0 is the present period, followed by two succeeding periods. The negative numbers represent cash outlays. A cursory look at the numbers shows that the net benefit is $550.

The time value of money is extremely important in evaluation processes. Let us explain what it means. If you were faced with an opportunity that generates $3,000 a year, how much would you be willing to invest? Obviously, you'd like to invest less than the $3,000. To earn the same money five years from now, the amount of investment would be even less. What is suggested here is that money has a time value. Today's dollar and tomorrow's dollar are not the same. The time lag accounts for the time value of money.

The time value of money is usually expressed in the form of interest on the funds invested to realize the future value. Assuming compounded interest, the formula is:

$$F = P(1 + i)^n$$

FIGURE 8–4 Net Benefit Analysis—An Example

Cost/Benefit	Year 0	Year 1	Year 2	Total
Costs	$−1,000	$−2,000	$−2,000	$−5,000
Benefits	0	650	4,900	5,550
Net benefits	$−1,000	$−1,350	$+2,900	$ 550

where

F = Future value of an investment.
P = Present value of the investment.
i = Interest rate per compounding period.
n = Number of years. periods

For example, $3,000 invested in Treasury notes for three years at 10 percent interest would have a value at maturity of:

$$F = \$3{,}000(1 + 0.10)^3$$
$$= 3{,}000(1.33)$$
$$= \$3{,}993$$

Present Value Analysis. In developing long-term projects, it is often difficult to compare today's costs with the full value of tomorrow's benefits. As we have seen, the time value of money allows for interest rates, inflation, and other factors that alter the value of the investment. Furthermore, certain investments offer benefit periods that vary with different projects. Present value analysis controls for these problems by calculating the costs and benefits of the system in terms of today's value of the investment and then comparing across alternatives.

A critical factor to consider in computing present value is a discount rate equivalent to the forgone amount that the money could earn if it were invested in a different project. It is similar to the opportunity cost of the funds being considered for the project.

Suppose that $3,000 is to be invested in a microcomputer for our safe deposit tracking system, and the average annual benefit is $1,500 for the four-year life of the system. The investment has to be made today, whereas the benefits are in the future. We compare present values to future values by considering the time value of money to be invested. The amount that we are willing to invest today is determined by the value of the benefits at the end of a given period (year). The amount is called the *present value* of the benefit.

To compute the present value, we take the formula for future value ($F = P(1 + i)^n$) and solve for the present value (P) as follows:

$$P = \frac{F}{(1 + i)^n}$$

So the present value of $1,500 invested at 10 percent interest at the end of the fourth year is:

$$P = \frac{1{,}500}{(1 + 0.10)^4}$$

$$= \frac{1{,}500}{1.61} = \$1{,}027.39$$

That is, if we invest $1,027.39 today at 10 percent interest, we can expect to

FIGURE 8-5 Present Value Analysis Using 10 Percent Interest Rate (Discounted)

Year	Estimated Future Value		Discount Rate*		Present Value†	Cumulative Present Value of Benefits
1	$1,500	×	0.908	=	$1,363.63	$1,363.63
2	1,500	×	0.826	=	1,239.67	2,603.30
3	1,500	×	0.751	=	1,127.82	3,731.12
4	1,500	×	0.683	=	1,027.39	4,758.51

* $1/[(1 + i)^n]$
† $P = F/[(1 + i)^n]$

have $1,500 in four years. This calculation can be represented for each year where a benefit is expected. For a four-year summary, see Figure 8-5.

Net Present Value. The net present value is equal to discounted benefits minus discounted costs. Our $3,000 microcomputer investment yields a cumulative benefit of $4,758.51, or a net present gain of $1,758.51. This value is relatively easy to calculate and accounts for the time value of money. The net present value is expressed as a percentage of the investment—in our example:

$$\frac{1,758.51}{3,000} = 0.55 \text{ percent}$$

Payback Analysis. The payback method is a common measure of the relative time value of a project. It determines the time it takes for the accumulated benefits to equal the initial investment. Obviously, the shorter the payback period, the sooner a profit is realized and the more attractive is the investment. The payback method is easy to calculate and allows two or more activities to be ranked. Like the net profit, though, it does not allow for the time value of money.

The payback period may be computed by the following formula:

$$\frac{\text{Overall cost outlay}}{\text{Annual cash return}} = \frac{(A \times B) + (C \times D)}{5 \quad + \quad 2} = \frac{\text{Years + Installation time } (G)}{\text{Years to recover}}$$

Elements of the formula:

(A) Capital investment (includes escalation costs)
(B) Investment credit (i.e., 1.00 − 0.08 = 0.92; must use current rate)
(C) Cost investment (i.e., site preparation—includes escalation)
(D) Company's federal income tax bracket
(E) State and local taxes
(F) Life of capital (expected)
(G) Time to install system
(H) Benefits and savings

(1) Projects benefits (includes escalation)
(2) Depreciation (Capital investment − Salvage ÷ Expected Life)
(3) State and local taxes (percent)
(4) Benefits before FIT (federal income tax): (1) − (2) − (3) = (4)
(5) Benefits after FIT: (4) − [(4) × (D)]

Example of Calculation

Elements

(A)	Capital investment in a new computer	$200,000
(B)	Investment credit difference (100% − 8% investment credit)	92%
(C)	Cost investment (site preparation)	$ 25,000
(D)	Company's income tax bracket difference (100% − 46%)	54%
(E)	State and local taxes	2%
(F)	Life of capital (no salvage value)	5 years
(G)	Time to install system	1 year
(H)	Benefits (include escalation or inflation)	$250,000

Calculation

(1)	Benefits before depreciation and taxes *(H)*		$250,000
(2)	Less depreciation {$200,000*(A)*/5[Life*(F)*]}	$40,000	
(3)	Less state and local taxes [$200,000 × 0.02*(E)*]	4,000	44,000
(4)	Benefits before FIT		$206,000
	Less tax difference ($206,000 × 0.46)		94,760
(5)	Benefits after FIT		$111,240

Formula calculation

$$\frac{[\$200{,}000(A) \times 0.92(B)] + [\$25{,}000(C) \times 0.54(D)]}{\underset{(5)}{\$111{,}240} + \underset{(2)}{\$40{,}000}} =$$

$$(\$184{,}000 + 13{,}500) \text{ or } \frac{197{,}500}{151{,}240}$$

$$\frac{\$197{,}500}{151{,}240} = 1.3 \text{ years plus installation time } (G)$$

$$= 2.3 \text{ years to recover investment}$$

2 years and 4 months is the payback period

Break-even Analysis. *Break-even* is the point where the cost of the candidate system and that of the current one are equal. Unlike the payback method that compares costs and benefits of the candidate system, break-even compares the costs of the current and candidate systems. When a candidate system is developed, initial costs usually exceed those of the current system. This is an *investment period.* When both costs are equal, it is break-even. Beyond that point, the candidate system provides greater benefit (profit) than the old one—a *return period.*

FIGURE 8–6 Break-Even Chart—An Illustration

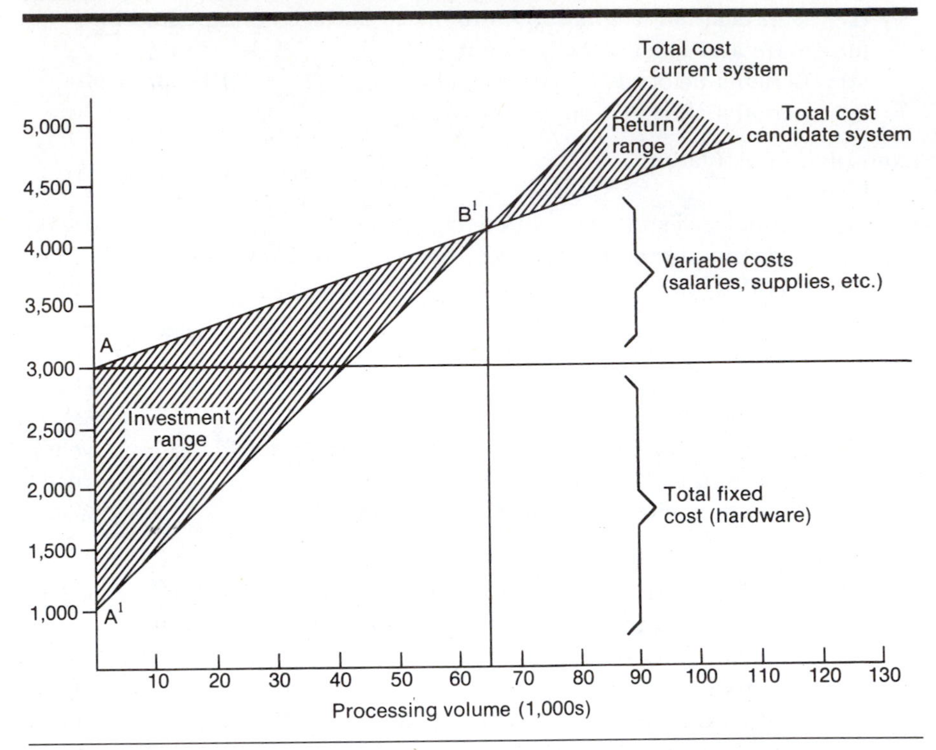

Figure 8–6 is a break-even chart comparing the costs of the current and candidate systems. The attributes are processing cost and processing volume. Straight lines are used to show the model's relationships in terms of the variable, fixed, and total costs of the two processing methods and their economic benefits. Intersection B′ indicates the point where the total cost of processing 65,000 transactions by the current system is equal to the total cost of using the candidate system. The shaded area beyond that point is the return period. The shaded area AB′A′ is the investment period. According to the chart, then, it would be more economical to process manually when volume is below 65,000 transactions during a given time period. Processing volume above B′ favors the candidate system.

Cash-Flow Analysis. Some projects, such as those carried out by computer and word processing services, produce revenues from an investment in computer systems. Cash-flow analysis keeps track of accumulated costs and revenues on a regular basis. The "spread sheet" format also provides break-even and payback information.

Figure 8–7 illustrates the performance of a new computer service over a one-year period. Revenues for the first month in operation (January) were $22,000. Operating expenses (including facility preparation) were $51,175,

FIGURE 8-7 Cash-Flow Analysis—An Example

	January	February	March	April	May	June	July	August	September	October	November	December
Revenues from computer service	$22,000	$22,000	$26,000	$27,100	$41,000	$48,000	$59,050	$59,010	$66,450	$ 64,040	$ 69,700	$ 71,040
Operating expenses:												
Facility preparation	$24,000	$ 9,000	$ 1,400									
Hardware lease	7,400	7,400	7,400	$ 7,400	$ 7,400	$ 7,400	$ 7,400	$ 7,400	$ 7,400	$ 7,400	$ 7,400	$ 7,400
Insurance	195	195	195	195	195	195	195	195	195	195	195	195
Salaries	11,100	11,100	11,100	10,600	10,600	10,600	12,010	12,010	12,010	12,010	12,010	12,010
Supplies	3,640	2,700	2,740	3,950	4,600	4,840	4,885	5,110	5,225	5,170	5,360	8,801
Telecommunication expenses	1,300	1,300	1,300	1,300	1,900	1,900	1,900	1,900	1,900	1,900	1,750	1,750
Travel/entertainment	2,000	1,600	900	800	900	600	400	400	500	500	200	700
User training	1,500	1,500	2,760	2,810	2,850	2,850	2,850	2,910	2,800	2,900	3,100	3,050
Total expenses	$51,175	$34,795	$27,805	$27,055	$28,445	$28,385	$29,640	$29,925	$28,030	$ 30,075	$ 30,015	$ 30,906
Cash flow (revenue − expense)	(29,175)	(12,795)	(1,805)	45	12,555	19,615	29,410	30,085	22,420	33,965	39,685	40,134
Accumulated cash flow	($29,175)	($41,970)	($43,775)	($43,730)	($31,180)	($11,565)	$17,845	$47,930	$70,350	$104,315	$144,000	$184,134

which resulted in a net expenditure of $29,175. Break-even occurs at the end of the fourth month (April). The cash flow then was $45, although the accumulated cash flow was $ −43,730. This was the result of excess expenses over revenues, facility preparation costs, and the like. Accumulated cash flow began to turn positive. This was the beginning of the payback period. The evaluation methods are summarized in Figure 8–8.

Interpret Results of the Analysis and Final Action

When the evaluation of the project is complete, the results have to be interpreted. This entails comparing actual results against a standard or the result of an alternative investment. The interpretation phase as well as the subsequent decision phase are subjective, requiring judgment and intuition. Depending on the level of uncertainty, the analyst may be confronted with a single known value or a range of values. In either case, simpler measures such as net benefit analysis are easier to calculate and present than other measures, although they do not discount future cash flows. If it can be modified to include the time value of money, the net benefit method would be comparable to the net present value method. More complex measures such as net present value account for the time value of money but are more difficult to evaluate and present.

The decision to adopt an alternative candidate system can be highly subjective, depending on the analyst's or end user's confidence in the estimated costs and benefits and the magnitude of the investment.

In summary, cost/benefit analysis is a tool for evaluating projects rather than a replacement of the decision maker. In real-life business situations, whenever a choice among alternatives is considered, cost/benefit analysis is an important tool. Like any tool, however, it has problems:

1. *Valuation problems.* Intangible costs and benefits are difficult to quantify, and tangible costs are generally more pronounced than tangible benefits. In most cases, then, a project must have substantial intangible benefits to be accepted.

2. *Distortion problems.* There are two ways of distorting the results of cost/benefit analysis. One is the intentional favoritism of an alternative for political reasons. The second is when data are incomplete or missing from the analysis.

3. *Completeness problems.* Occasionally an alternative is overlooked that compromises the quality of the final choice. Furthermore, the costs related to cost/benefit analysis may be on the high side or not enough costs may be considered to do a complete analysis. In either case, the reliability of the final choice is in doubt.

THE SYSTEM PROPOSAL

The final decision following cost/benefit analysis is to select the most cost-effective and beneficial system for the user. At this time, the analyst prepares a feasibility report on the major findings and recommendations. As

FIGURE 8–8 Evaluation Methods—A Summary

Evaluation Method	Procedure	Advantages	Limitations
Net benefit	Total benefits minus total costs	Easy to calculate Easy to interpret Easy to present	Does not account for the time value of money
Present value	$P = F/[(1 + i)^n]$ F = Future value of an investment $F = P(1 + i)^n$	Easy to calculate Equates different investment opportunities with various costs and benefits and discount rates Accounts for time value of money	It is only a relative (not absolute) measure of a project's return on investment
Net present value	Discounted benefits minus discounted costs	Relatively easy to calculate Accounts for time value of money	It is only a relative (not absolute) measure of a project's return on investment
Payback	Investment divided by yearly savings Compares costs and benefits of candidate system	Easy to calculate Has straightforward interpretation for choice between two or more alternatives for candidate systems	Conservative economic measure Applied to one opportunity at a time Does not compare profitability of multiple investment alternatives Dos not allow for time value of money.
Break-even	Compares costs of using present and candidate systems	Easy to understand	Does not allow for time factor and depreciation value of money
Cash-flow	Revenue minus expense on a period-by-period basis	Combines benefits of break-even and payback methods	Ignores time value of money For a limited time period, it does not take into account the profitability of the project Ignores behavioral implications of the numbers in the financial statement

FIGURE 8–9 Feasibility Report—An Outline

A written feasibility report should include the following:

I. TITLE PAGE Defines the name of the project and who it is for

II. TABLE OF CONTENTS List various parts, features, and exhibits, showing page numbers

III. SCOPE Present a brief explanation of the system boundaries

IV. STATEMENT OF PROBLEM Describe current system
Describe proposed system
Indicate how proposed system will solve the problem(s)

V. SUMMARY/ABSTRACT (optional) Give executive a summary of project, highlighting benefits

VI. COST/BENEFIT STATEMENT List benefits and savings in quantitative terms
Present dollar savings versus costs
Summarize cost of new equipment, one-time charges, etc.
Quantify net savings and expected returns

VII. IMPLEMENTATION SCHEDULE Submit implementation plan
Specify human resources requirements, systems and procedures, etc.
Include PERT-CPM or Gantt chart

VIII. HARDWARE CONFIGURATION (optional) Lay out computer configuration
Describe terminal network and equipment (CRTs, printers, etc.)
List communication equipment (data sets, lines, etc.)

IX. CREDITS Give credit to those who contributed to the project study

X. APPENDIX Include exhibits, correspondence on project, and other miscellaneous documentation

---

explained in Chapter 7, the report is a detailed summary of the investigation that has been carried out. It outlines the options and recommendations. It is presented to management for determining whether a candidate system should be designed. Effective reports follow carefully planned formats that management can understand and evaluate without having to read the entire document. The content and format of the feasibility report are summarized in Figure 8–9.

Summary

1. Data analysis is a prerequisite to cost/benefit analysis. From the analysis, system design requirements are identified and alternative systems evaluated. The analysis of the costs and benefits of each alternative guides the selection process. Therefore, a knowledge of cost and benefit categories and evaluation methods is important.

2. In developing cost estimates, we need to consider hardware, personnel, facility, operating, and supply costs. In addition, a system is expected to provide benefits. We need to identify each benefit and assign it a monetary value for cost/benefit analysis.
3. Cost/benefit analysis gives a picture of the various costs, benefits, and rules associated with each alternative system. The procedure entails:
   a. Identifying the costs and benefits pertaining to a project.
   b. Categorizing the various costs and benefits for analysis.
   c. Selecting a method of evaluation.
   d. Interpreting the results of the analysis.
   e. Taking action.
4. Costs and benefits are classified as tangible or intangible, direct or indirect, fixed or variable:
   a. *Tangible costs:* Outlays of cash for an item or activity.
   b. *Intangible costs:* Those that have financial values not easily measured.
   c. *Direct costs:* Those where a dollar figure can be directly associated with a project.
   d. *Indirect costs:* The results of operations not directly associated with a system or activity.
   e. *Fixed costs:* They are constant and do not change—nonrecurring.
   f. *Variable costs:* They are proportional to work volume.
5. When all financial data have been identified, the analyst must select a method of evaluation. There are several methods available:
   a. *Net benefit analysis:* Involves subtracting total costs from total benefits. It is easy to calculate, interpret, and present. The main drawback is not accounting for the time value of money and not discounting future cash flows.
   b. *Present value analysis:* Calculates the costs and benefits of the system in terms of today's value of the investment and then compares.
   c. *Net present value:* Discounted benefits minus discounted costs. It is relatively easy to calculate and accounts for the time value of money.
   d. *Payback analysis:* A common measure of the relative time value of a project. It determines the time it takes for the accumulated benefits to equal the initial investment. It is easy to calculate and allows the ranking of two or more activities.
   e. *Break-even analysis:* The point at which the cost of the candidate system and that of the current one are equal.
   f. *Cash-flow analysis:* Keeps track of accumulated costs and revenues on a regular basis.
6. Once the evaluation of the project is complete, actual results are compared against standards or alternative investments. The decision to adopt an alternative system can be highly subjective, depending on the

analyst's or user's confidence in the estimated cost and benefit values and the magnitude of the investment.

Key Words

Break-even Analysis
Cash-Flow Analysis
Cost/Benefit Analysis
Direct Cost
Fixed Cost
Future Value
Indirect Cost
Intangible Cost
Investment Period
Net Benefit Analysis
Net Pay Value
Opportunity Cost
Payback Analysis
Present Value
Return Period
Savings
Sunk Cost
Tangible Cost
Variable Cost

Review Questions

1. What cost elements are considered in cost/benefit analysis? Which element do you think is the most difficult to estimate? Why?
2. Define and explain the procedure for cost/benefit determination.
3. How easy is it to identify the costs and benefits of a system? Give examples of costs that are not easily identifiable.
4. Distinguish between the following:
   a. Opportunity and sunk costs.
   b. Direct and indirect costs.
   c. Fixed and tangible costs.
   d. Tangible and intangible benefits
5. "If tangible costs and benefits are ignored, the outcome of the evaluation may be quite different from one in which they are included." Do you agree? Illustrate your answer.
6. "Savings are realized when a cost advantage of some kind exists." Elaborate.
7. How do net present value and present value analyses differ? Illustrate.
8. What are the pros and cons of the following evaluation methods?
   a. Payback method.
   b. Cash-flow analysis.
   c. Break-even analysis.
9. If the evaluation methods used in cost/benefit analysis are seemingly quantitative, why are the interpretive phase and the subsequent decision phase subjective? Explain.
10. Briefly describe the essential elements of a project study report.

Application Problems

1 Suppose a firm went through a request for proposal, vendor proposals, benchmarking, and the final choice of a system that can be purchased or leased. Under the purchase option:

a. The price tag is $500,000.

b. The expected useful life is five years, and the salvage value is $40,000.

c. The vendor allows a trade-in on the user's old hardware of $100,000. The book value is $70,000, and there is a remaining life of one year.

d. Maintenance service is available at $8,000 per year.

Under the lease option:

a. Lease charges are $110,000 per year under a five-year contract.

b. Lessor's (vendor) maintenance fees are $8,000 per year.

c. There is no trade-in allowance, although a third party is offering $75,000 for the old equipment.

d. Cost of capital to the user is 10 percent, and the effective tax rate is 40 percent.

Assignment

Determine the net present value applied to the purchase/lease options. Keep in mind the following:

a. The benefits derived from either option are assumed to be equal.

b. When using the net present value method, you are looking for the net present value of either cash inflows or cash outflows.

c. Show all cash inflows and outflows net of their tax effect.

d. Proceeds from the sale of the old system are reduced by the tax on the gain from the sale.

e. Tax benefits that result from the deductibility of the service contract, the lease payments, and depreciation are taken into account in the analysis. The effect is a reduction in the cash outflows related to the expenditures.

f. Maintenance, lease payments, and depreciation are annuities, fixed amounts payable over a period of time.

2 A medium-sized bank has decided to automate its trust accounting service. The vice president of the trust department requested a cost/benefit analysis of a trust package for possible installation. The systems analyst first reviewed the cost of operating present trust activities. The vice president and three trust officers earned $64,000 in salaries. Inventory and supplies average $400 each year. Trust account statements, asset reviews, and other trust reports are produced on an outdated system at an annual cost of $400. Overhead (air conditioning, lighting, power, and maintenance) average $2,165 a year.

The analyst then evaluated three software packages designed for trust work. One package met the user's requirements. The purchase price was $13,980. When implemented, the computerized service will reduce payroll by $10,000—the salary of a junior trust officer. Inventory and supplies were priced at $1,900. Overhead was calculated at $2,660 for air conditioning, lighting, power, and maintenance.

When all the information was gathered, a report was drafted for the vice president. With these facts and figures, he could see the cost of a computerized trust package. The operation costs of both systems are summarized as follows:

	Proposed System	Present System
Salaries	$54,000	$64,000
Inventory and supplies	1,900	400
System charges	13,980	400
Overhead		
Air conditioning	900	780
Lighting	860	700
Power	610	400
Maintenance-janitorial	290	285
Total	$72,540	$66,965

In the event the proposed system is implemented, a brokerage house that deals with the trust department agreed to underwrite 25 percent of the system charge, or $3,495 per year. So the actual operations costs to the trust department are $69,045.

Assignment

a. Did the analyst project the correct salary costs for both systems? Explain.

b. Did the analyst provide the vice president with all the costs for the new system? Did he collect all the costs relating to the present system?

c. Given the operations costs, can the vice president cost-justify the proposed system? Explain.

3 The systems analyst of the MIS department was contacted one morning by the production manager of the candy bar division. The problem was that the 3-ounce candy bar is sticking to the wrapper. A large percentage of the bars begin to stick to their wrappers after four or five weeks on the grocer's shelf. Consequently, customer complaints have forced a supermarket chain to switch to a competitor's brand.

After a two-week observation of the mixing and wrapping processes and inspection of the wrapper, it was found that the three-year-old mixing unit overheated after 6.5 hours of operation. This caused a sugar ingredient to react slowly to room temperature three to four weeks after production. The mixing machine has five more years of projected life. The alternatives are to modify the present machine at a cost of $79,000 and to replace it with a new (but more reliable) machine that has been successfully used by a competitor. The new machine costs $150,000 plus $14,000 installation charge. It has no salvage value at the end of the eight years.

The next step was to compute the operation costs of both systems.

In a meeting with his supervisor, the analyst tried to show how the new system would increase sales of the 3-ounce candy bar and produce reliable production that would easily offset the initial investment in the new machine. The supervisor did not seem impressed. After a lengthy discussion, the analyst was asked to redo the computations and the analysis.

PROPOSED SYSTEM

Operating costs		
Salaries (three employees @ $5.40/hour or $648/week)		
Total salaries	$33,696	
Fringe benefits (@ 10%)	3,370	
Total payroll/year		$37,066
Service fees		12,000
Overhead (additional expenses due to new system)		
Maintenance (three hours/day @ $8.00/hour)	$ 8,736	
Insurance	955	
Utilities	$ 1,140	
Total		$10,831
Total operating costs/year		$59,897

PRESENT SYSTEM

Operating costs		
Salaries (four employees @ $4.50/hour or $720 week)		
Total salaries	$37,440	
Fringe benefits (@ 10%)	3,744	
Total payroll/year		$41,184
Overhead		
Maintenance	$ 5,824	
Insurance	675	
Utilities	1,008	
Total		7,507
Total operating costs/year		$48,691

Assignment

a. One problem that was pointed out was computing employee wages. What seems to be inaccurate about the salary section in both system? How would you correct the problem?

b. With respect to the types of costs incurred in operating either system, did the analyst include all relevant costs? Show where additional costs should be included. What other expenses should the report emphasize (if any)?

Selected References

Axelrod, C. Warren. *Computer Productivity*. New York: John Wiley & Sons, 1982, pp. 61–89.

Davis, Wm. *Systems Analysis and Design: A Structured Approach.* Reading, Mass.: Addison-Wesley Publishing, 1983, pp. 313–24.

Oxenfeldt, A. R. *Cost-Benefit Analysis for Executive Decision Making*. AMACOM, American Management Association, New York 1979.

Powers, Michael; Davis Adams; and Harlan D. Mills. *Computer Information System Development: Analysis & Design*. Cincinnati: South-Western Publishing, 1984, pp. 184–213.

Part Three

Systems Design

9 THE PROCESS AND STAGES OF SYSTEMS DESIGN

10 INPUT/OUTPUT AND FORMS DESIGN

11 FILE ORGANIZATION AND DATA BASE DESIGN

Chapter 9

The Process and Stages of Systems Design

At a Glance

System design is a solution, a "how to" approach to the creation of a new system. This important phase is composed of several steps. It provides the understanding and procedural details necessary for implementing the system recommended in the feasibility study. Emphasis is on translating the performance requirements into design specifications. Design goes through logical and physical stages of development. Logical design reviews the present physical system; prepares input and output specifications; makes edit, security, and control specifications; details the implementation plan; and prepares a logical design walkthrough. The physical design maps out the details of the physical system, plans the system implementation, devises a test and implementation plan, and specifies any new hardware and software.

Structured design methodologies are emphasized for design work. They include structure charts, HIPO and IPO charts, and structured walkthrough. Major development activities, audit considerations, and processing controls are also discussed.

By the end of this chapter, you should know:

1. The process of system design.
2. How logical design differs from physical design.
3. Top-down design and functional decomposition.
4. Forms-driven methodology.
5. The major development activities in structured design.
6. Audit considerations for system development.

Audit Considerations

PROCESSING CONTROLS AND DATA VALIDATION

AUDIT TRAIL AND DOCUMENTATION CONTROL

INTRODUCTION

The discussion so far brings us to a pivotal point in the system development life cycle. User requirements have been identified. Information has been gathered to verify the problem and evaluate the existing system. A feasibility study has been conducted to review alternative solutions and provide cost/benefit justification. The culmination of the study is a proposal summarizing the findings and recommending a candidate system for the user.

If the figures and the reasoning behind the candidate system make sense, management authorizes the proposed change. At this point in the systems life cycle, the design phase begins. The design is a solution, a "how to" approach, compared to analysis, a "what is" orientation. It translates the system requirements into ways of operationalizing them. In this chapter, we cover the process and stages of systems design, the tools used to design candidate systems, and the user's input to systems design. It is an overview chapter that outlines what follows. The next five chapters elaborate on various aspects of design—input/output and forms design (Chapter 10), file organization and data base design (Chapter 11), system testing and quality assurance (Chapter 12), and system implementation and software maintenance (Chapter 13).

THE PROCESS OF DESIGN

The design phase focuses on the detailed implementation of the system recommended in the feasibility study. Emphasis is on translating performance specifications into design specifications. The design phase is a transition from a user-oriented document (system proposal) to a document oriented to the programmers or data base personnel.

Logical and Physical Design

Systems design goes through two phases of development: *logical* and *physical* design. As we saw in Chapter 6, a data flow diagram shows the logical flow of a system and defines the boundaries of the system. For a candidate system, it describes the inputs (source), outputs (destination), data bases (data stores), and procedures (data flows)—all in a format that meets the user's requirements. When analysts prepare the logical system design, they specify the user needs at a level of detail that virtually determines the information flow into and out of the system and the required data resources. The design covers the following:

1. Reviews the current physical system—its data flows, file content, volumes, frequencies, etc.
2. Prepares *output* specifications—that is, determines the format, content, and frequency of reports, including terminal specifications and locations.

3. Prepares *input* specifications—format, content, and most of the input functions. This includes determining the flow of the document from the input data source to the actual input location.
4. Prepares edit, security, and control specifications. This includes specifying the rules for edit correction, backup procedures, and the controls that ensure processing and file integrity.
5. Specifies the implementation plan.
6. Prepares a logical design walkthrough of the information flow, output, input, controls, and implementation plan.
7. Reviews benefits, costs, target dates, and system constraints.

As an illustration, when a safe deposit tracking system is designed, system specifications include weekly reports, a definition of boxes rented and boxes vacant, and a summary of the activities of the week—boxes closed, boxes drilled, and so on. The logical design also specifies output, input, file, and screen layouts. In contrast, procedure specifications show how data are entered, how files are accessed, and how reports are produced (see Figure 9–1).

Following logical design is *physical* design. This produces the working system by defining the design specifications that tell programmers exactly what the candidate system must do. In turn, the programmer writes the necessary programs or modifies the software package that accepts input

FIGURE 9–1 Systems Design Goes through Logical and Physical Design

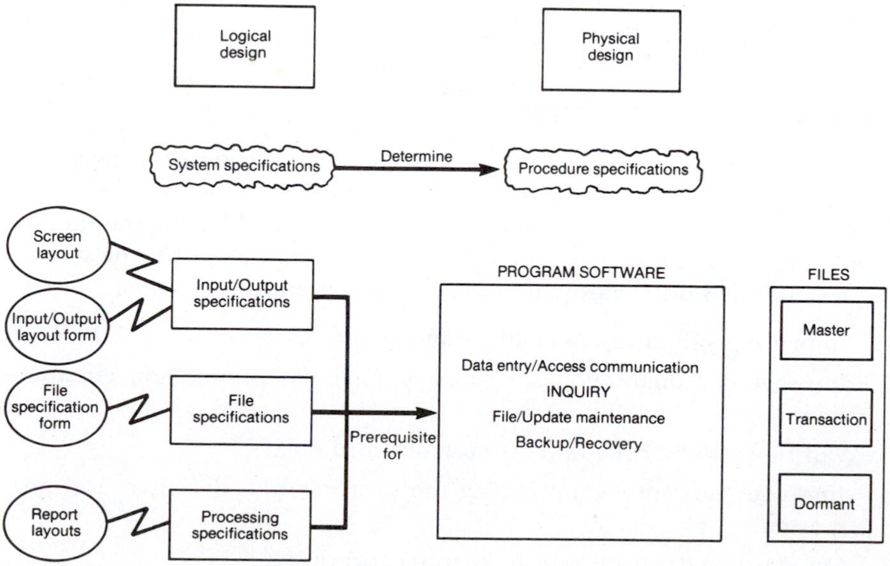

from the user, performs the necessary calculations through the existing file or data base, produces the report on a hard copy or displays it on a screen, and maintains an updated data base at all times. Specifically, physical system design consists of the following steps:

1. Design the physical system.
   a. Specify input/output media.
   b. Design the data base and specify backup procedures.
   c. Design physical information flow through the system and a physical design walkthrough.

2. Plan system implementation.
   a. Prepare a conversion schedule and a target date.
   b. Determine training procedure, courses, and timetable.

3. Devise a test and implementation plan and specify any new hardware/ software.
4. Update benefits, costs, conversion date, and system contraints (legal, financial, hardware, etc.).

The physical design for our safe deposit illustration is a software package written in Pascal (a programming language). It consists of program steps that accept new box rental information; change the number of boxes available with every new box rental; print a report by box type, box size, and box location; and store the information in the data base for reference. The analyst instructs the software programmer to have the package display a menu that specifies for the user *how* to enter a new box rental, produce a report, or display various information on the screen. These and other procedure specifications are tested and implemented as a working model of the candidate system.

DESIGN METHODOLOGIES

During the past decade, there has been a growing move to transform the "art" of systems analysis and design into an "engineering-type" discipline. The feeling that there has to be a more clearly defined logical method for developing a system that meets user requirements has led to new techniques and methodologies that fundamentally attempt to do the following:

1. Improve productivity of analysts and programmers.
2. Improve documentation and subsequent maintenance and enhancements.
3. Cut down drastically on cost overruns and delays.
4. Improve communication among the user, analyst, designer, and programmer.
5. Standardize the approach to analysis and design.
6. Simplify design by segmentation.

FIGURE 9–2 Data Flow Diagram—Safe Deposit Customer Master File Update Procedure

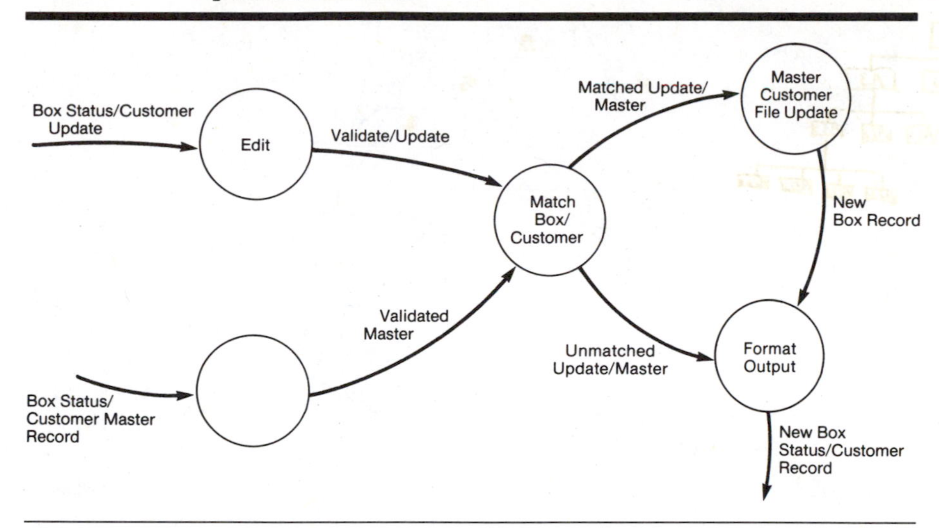

Structured Design

Structured design is a data-flow-based methodology. The approach begins with a system specification that identifies inputs and outputs and describes the functional aspects of the system. The system specifications, then are used as a basis for the graphic representation—data flow diagram(DFD)—of the data flows and processes (see Figures 9–2 and 9–3). From the DFD, the next step is the definition of the modules and their relationships to one another in a form called a *structure chart,* using a data dictionary and other structured tools.

Structured design partitions a program into small, independent modules. The are arranged in a hierarchy that approximates a model of the business area and is organized in a *top-down* manner with the details shown at the bottom. Thus, structured design is an attempt to minimize

FIGURE 9–3 The Structured Design Method

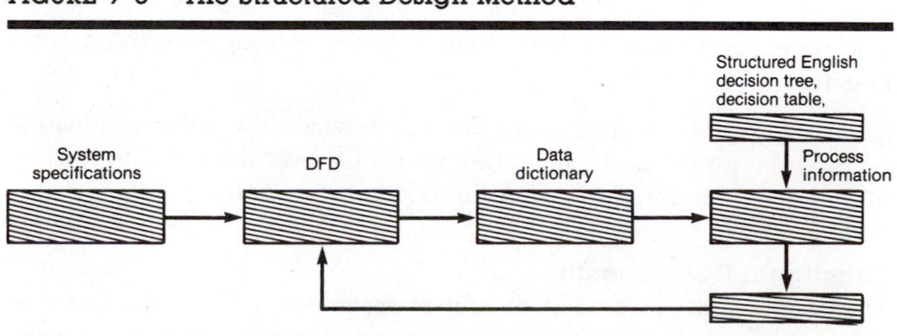

FIGURE 9–4 Decomposition—A Framework

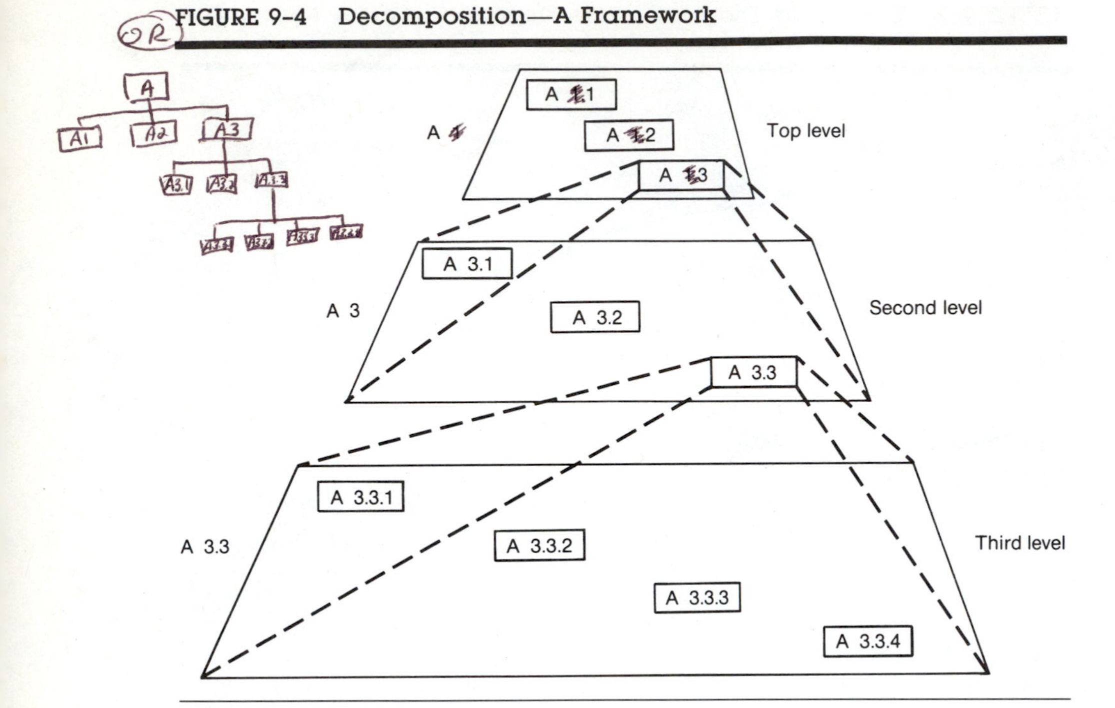

complexity and make a problem manageable by subdividing it into smaller segments, which is called *modularization* or *decomposition* (see Figure 9–4). In this way, structuring minimizes intuitive reasoning and promotes maintainable, provable systems.

A design is said to be *top-down* if it consists of a hierarchy of modules, with each module having a single entry and a single exit subroutine. The primary advantages of this design are as follows:

1. Critical interfaces are tested first.
2. Early versions of the design, though incomplete, are useful enough to resemble the real system.
3. Structuring the design, per se, provides control and improves morale.
4. The procedural characteristics define the order that determines processing.

So structured design arises from the hierarchical view of the application rather than the procedural view. The top level shows the most important division of work; the lowest level at the bottom shows the details.

Functional Decomposition

The documentation tool for structured design is the hierarchy or *structure chart.* It is a graphic tool for representing hierarchy, and it has three elements:

1. The *module* is represented by a rectangle with a name (see Figure 9–5). It is a contiguous set of statements.

FIGURE 9–5
A Module

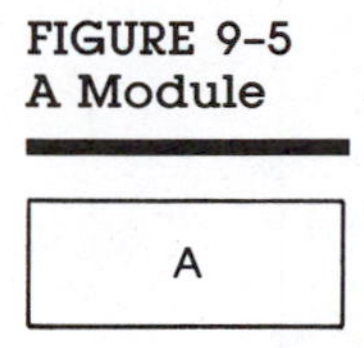

2. The *connection* is represented by a vector linking two modules. It usually means one module has called another module. In Figure 9–6, module A calls module B; it also calls module C.

FIGURE 9–6 Connection—An Example

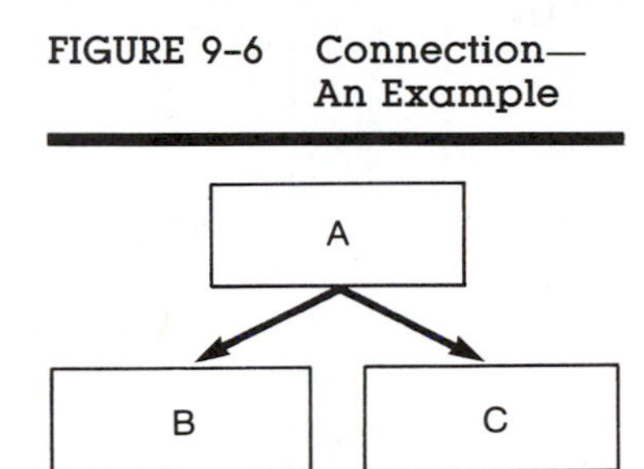

3. The *couple* is represented by an arrow with a circular tail. It represents data items moved from one module to another. In Figure 9–7, O, P, and Q are couples. Module A calls B, passing O downward. Likewise, module A calls C, passing P downward and receiving Q back. More on coupling is described next.

FIGURE 9–7 Coupling—An Example

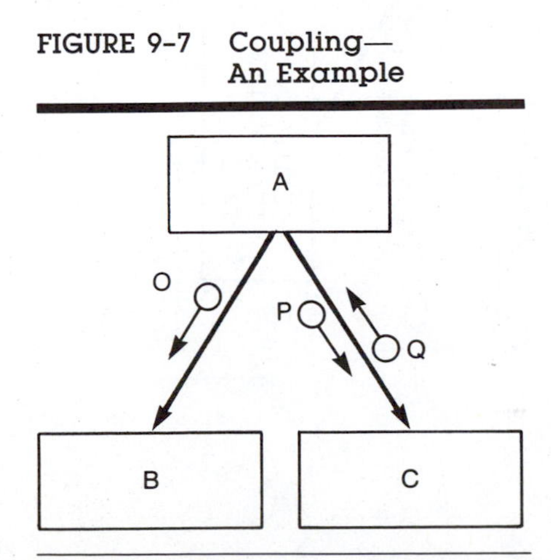

FIGURE 9-8 Structure Chart—An Example

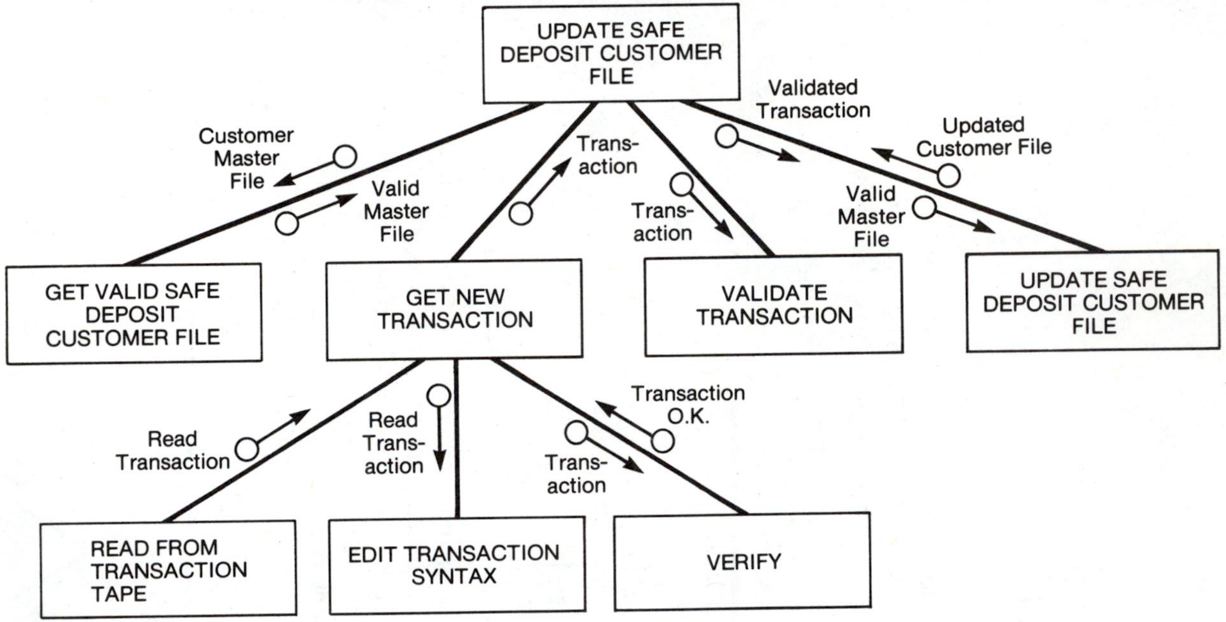

An example of a structure chart, showing the three elements, is shown in Figure 9–8.[1]

In the functional decomposition approach to structured design, software is partitioned into independent modules so that each module is small enough to be manageable. In the evaluation of a program module, two criteria are considered: the module's connections to other modules, called *coupling,* and its intramodule strength, or *cohesion.*[2]

Module coupling refers to the number of connections between a "calling" and a "called" module and the complexity of these connections. There must be at least one connection between a module and a calling module. A design objective for producing an easily understood code is to make the modules as independent as possible. For example, in Figure 9–8, the module GET NEW TRANSACTION is called by UPDATE SAFE DEPOSIT CUSTOMER FILE, which is the calling module. In this case, we have coupling.

Module cohesion refers to the relationship among elements (instructions) within a module. If a module does more than one discrete task, the instructions in that module are said not to be bound together very closely. Modules that perform only one task are said to be more cohesive (and less error-prone) than modules that perform multiple tasks. In Figure 9–8, the detail under the module GET NEW TRANSACTION is made up of strongly cohesive modules. Compare it to the poorly cohesive modules in Figure 9–9 and you can see how important cohesion is.

FIGURE 9–9 Poorly Cohesive Design—An Example

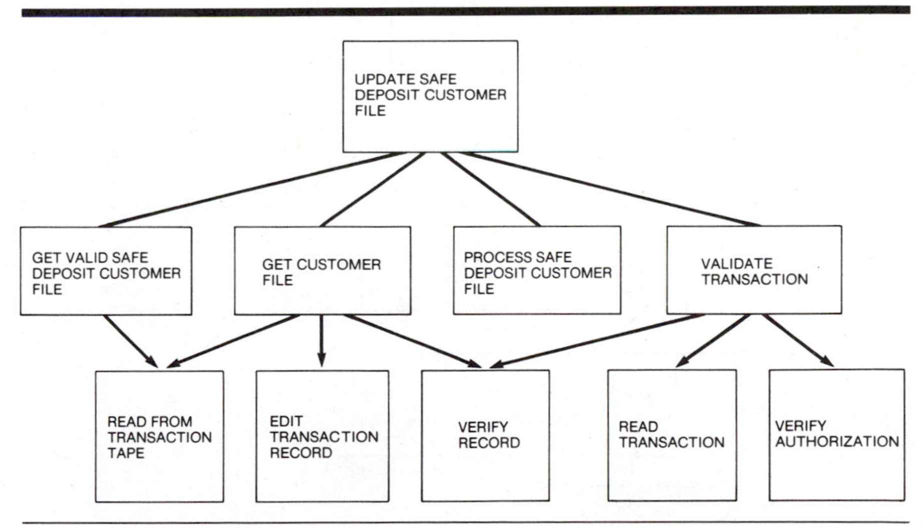

[1] Tom Demarco, *Structured Analysis and Design* (New York: Yourdon Publishing, 1979), p. 307.

[2] See Cyril P. Brosh, Philip J. Grouse, D. Ross Jeffery, and Michael J. Lawence, *Information System Design* (Englewood Cliffs, N.J.: Prentice-Hall, 1982), pp. 168–174.

Forms-Driven Methodology—The IPO Charts

In structured design a hierarchy chart represents a good program design if it meets the criteria of cohesion and coupling. Each module performs a single function (cohesion) and should be independent of the rest of the program (coupling). Each criterion calls for more details than are available. This prompts the analyst to develop input/process/output (IPO) charts for each module in the hierarchy chart.

HIPO and IPO Charts

HIPO is a forms-driven technique in that standard forms are used to document the information. It consists of a hierarchy chart and an associated set of input/process/output charts. HIPO captures the essence of top-down decompostion; it describes the data input and output from processes and defines the data flow composition. It was developed by IBM as a design aid and implementation technique with the following objectives:

1. Provide a structure by which the functions of a system can be understood. A hierarchy structure is shown in Figure 9–10.

FIGURE 9–10 A Hierarchy (Notation) Diagram

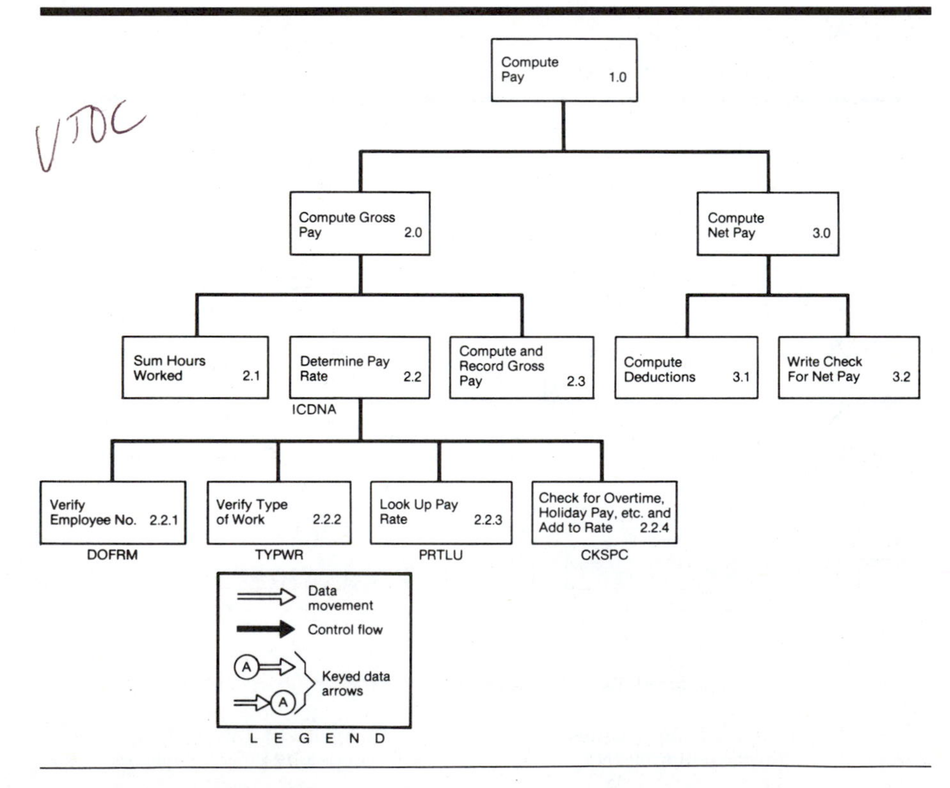

2. State the functions to be performed by the program rather than specifying the program statements to be used to perform the functions.
3. Provide a visual description of input to be used and output to be produced for each level of the diagram. HIPO makes the transformation of input to output data visible. Figure 9–11 is a typical HIPO diagram.

HIPO uses easy-to-draw vector-like symbols between processes that define data communication and data direction. As shown in Figure 9–11, the order of the processes in the process section of the diagram indicates the sequence of execution. The procedure for generating HIPO diagrams is simple:

1. Begin at the highest level of abstraction and define the inputs to the system and the ouputs from it in aggregate terms.
2. Identify the processing steps by those that convert input into output.
3. Document each element using HIPO diagram notation and the associated treelike structure.

FIGURE 9–11 An HIPO Diagram Derived from the Hierarchy Diagram (Figure 9–10)

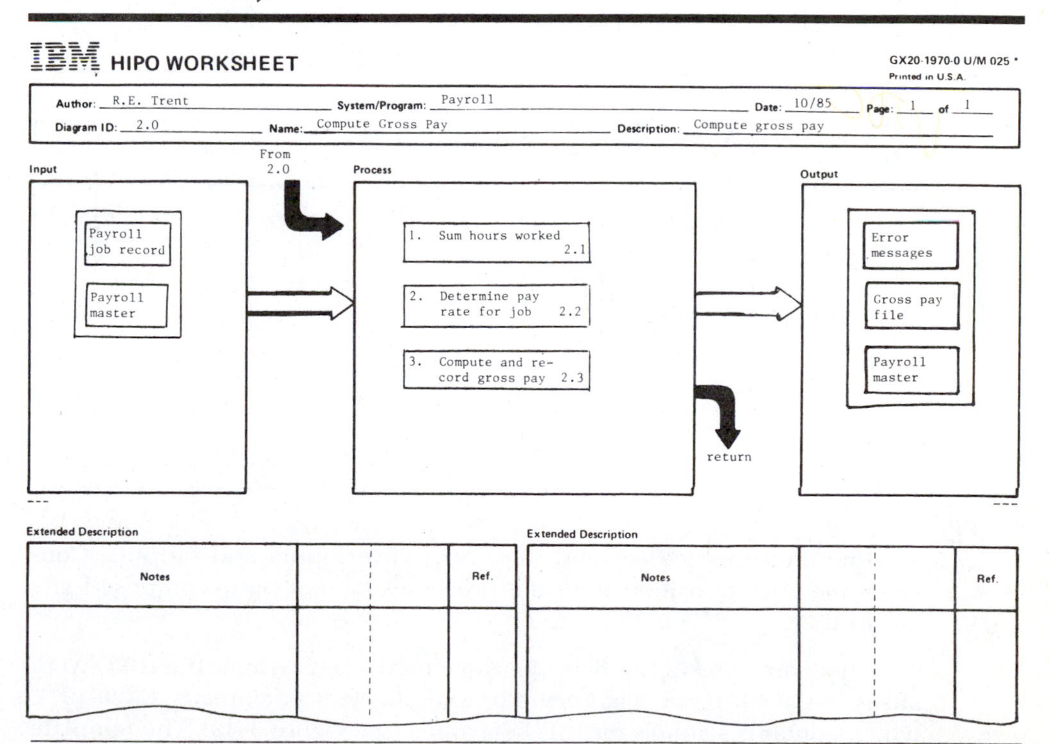
IBM HIPO WORKSHEET
GX20-1970-0 U/M 025 *
Printed in U.S.A.

Author: R.E. Trent System/Program: Payroll Date: 10/85 Page: 1 of 1
Diagram ID: 2.0 Name: Compute Gross Pay Description: Compute gross pay

Input
Payroll job record
Payroll master

From 2.0

Process
1. Sum hours worked 2.1
2. Determine pay rate for job 2.2
3. Compute and record gross pay 2.3

return

Output
Error messages
Gross pay file
Payroll master

Extended Description
Notes | Ref.

Extended Description
Notes | Ref.

FIGURE 9-12 HIPO Work Sheet and a Sample Overview Diagram

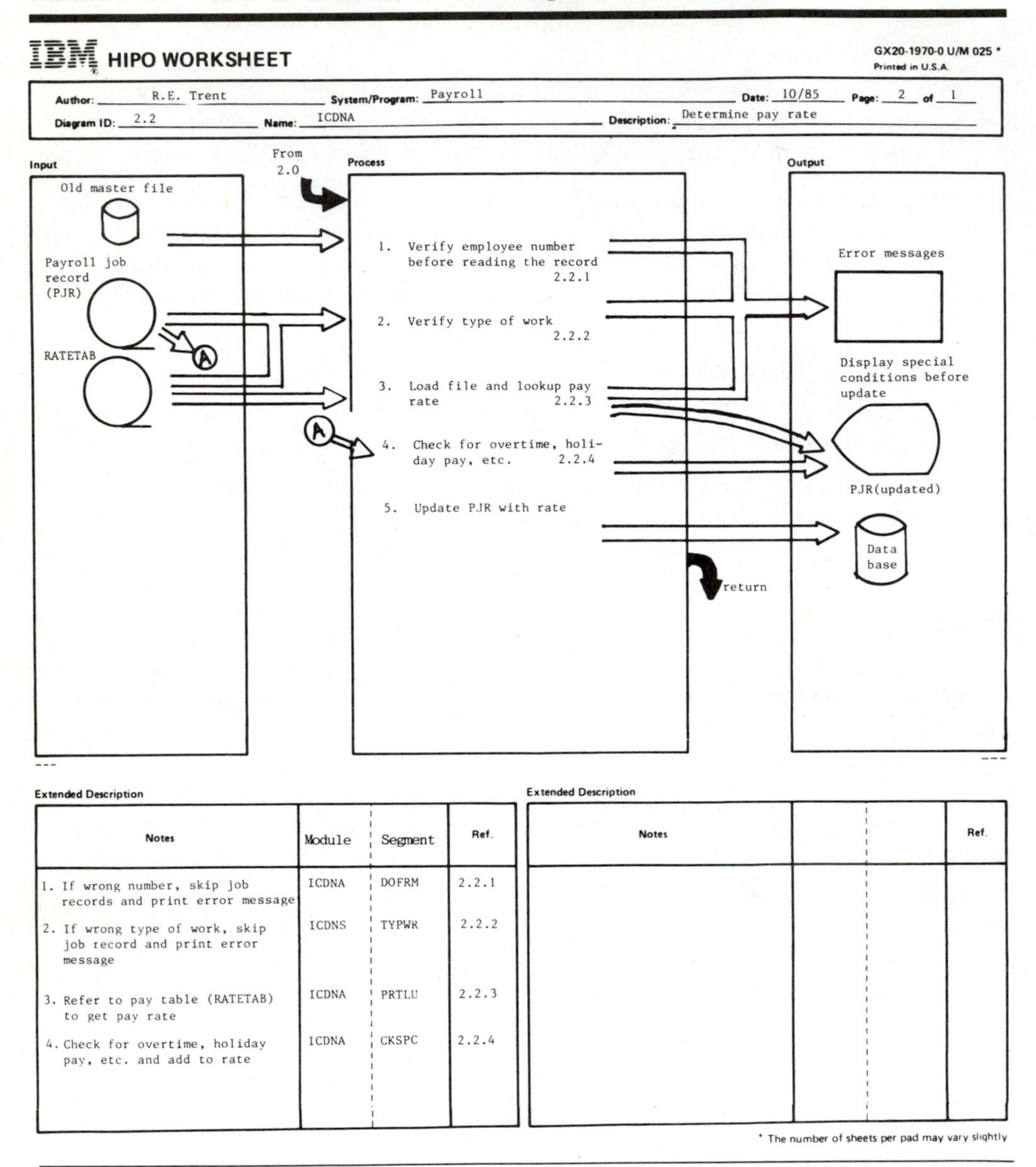

IBM HIPO WORKSHEET

GX20-1970-0 U/M 025 *
Printed in U.S.A.

Author: R.E. Trent System/Program: Payroll Date: 10/85 Page: 2 of 1
Diagram ID: 2.2 Name: ICDNA Description: Determine pay rate

Extended Description

Notes	Module	Segment	Ref.
1. If wrong number, skip job records and print error message	ICDNA	DOFRM	2.2.1
2. If wrong type of work, skip job record and print error message	ICDNS	TYPWR	2.2.2
3. Refer to pay table (RATETAB) to get pay rate	ICDNA	PRTLU	2.2.3
4. Check for overtime, holiday pay, etc. and add to rate	ICDNA	CKSPC	2.2.4

Extended Description

Notes		Ref.

* The number of sheets per pad may vary slightly

4. Identify subprocesses and their respective inputs and outputs. Continue decomposition until the processes cannot be decomposed any further.

There are two aids available for drawing HIPO diagrams: the HIPO work sheet, GX20-1970 (see Figure 9-12), and the HIPO template, GX20-1971, which contains symbols for HIPO diagrams (see Figure 9-13). The template

FIGURE 9–13 HIPO Template and Symbols

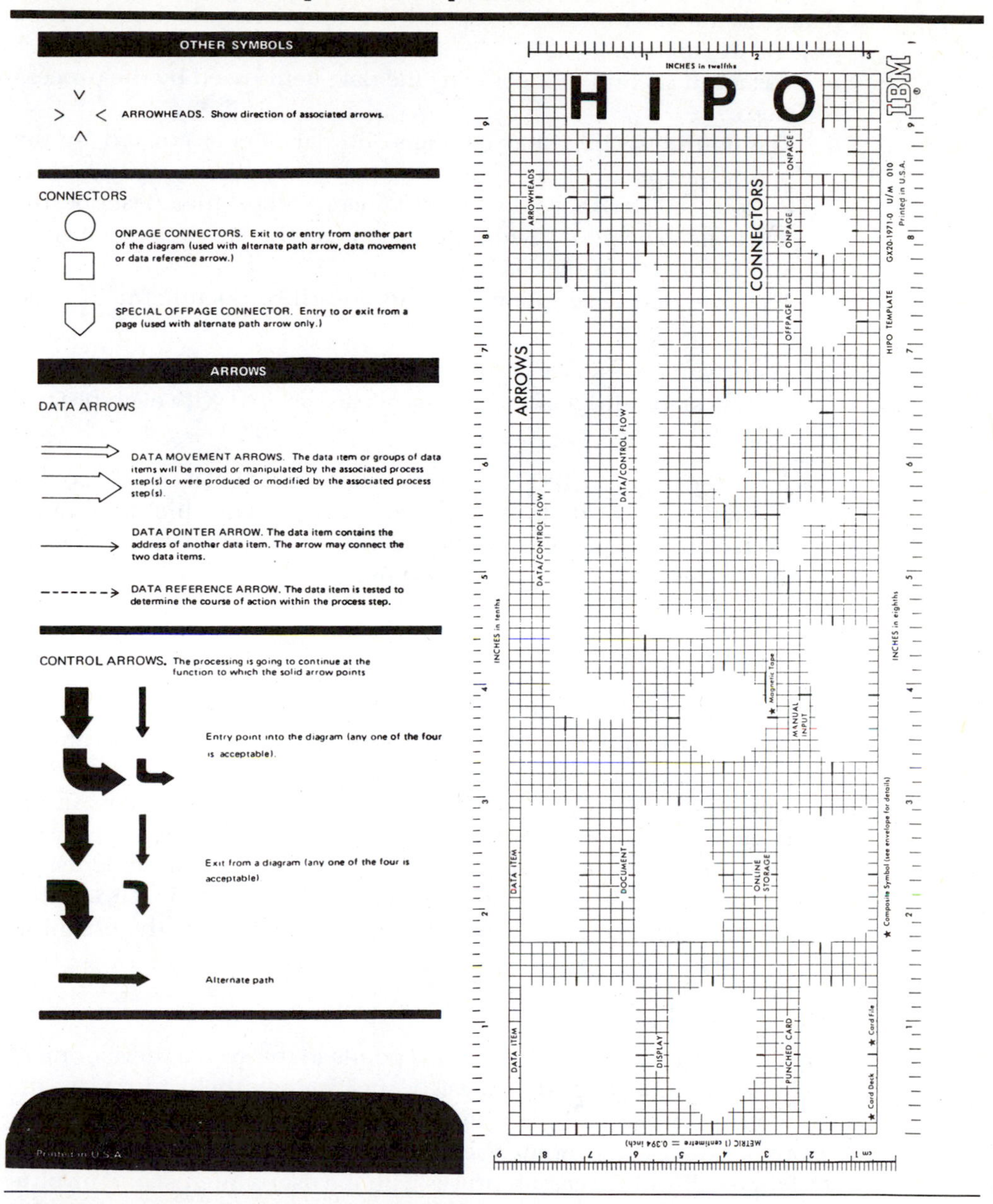

jacket describes the symbols. The HIPO package format consists of the following:

1. *Visual table of contents* shows the structure of the diagram and the relationships of the functions in a hierarchical manner. It also has a legend to show how the symbols are to be used.

2. *Overview diagrams* describe the major functions and reference the detail diagram(s) needed to expand the functions adequately. They provide the following:
   a. The input section that contains the data items used by the process steps.
   b. The output section that contains the data items created by the process steps.
   c Process section that contains numbered steps that describe the functions to be performed. Arrows connect then to the output steps and input/output data items.
   d. The extended description refers to non-HIPO documentation and code.
3. *Detail diagram* contains an extended description section that amplifies the process steps and references the code associated with each process step.

It is important to use HIPO early in the design phase of a project so that designers can document their thoughts concurrently with the design process. Thus, the preparation of HIPO diagrams is a by-product of the thought process of the design rather than an additional chore.

Structured Walkthrough

An activity of all phases of a structured project is the walkthrough. It is an interchange of ideas among peers who review a product presented by its author(s) and agree on the validity of a proposed solution to a problem. In a design walkthrough, the purpose is to anticipate as many problems in the design as possible while they are still "paper tigers." It is cheaper to make changes now than later during conversion. This is a practical implementation of "A stitch in time saves nine." The objective is to come up with a maintainable design that is flexible and adaptable and meets the organizations's standards.

User Involvement

Walkthroughs may be held at various points in the system development life cycle. In addition to system design, they may be held to review the system test plan, program design, and production acceptance. In each case, the people who will be running the system should be consulted.

The probability of success improves with the user's interest and involvement in the design of the system. The user can save a system that might otherwise be of marginal benefit, and likewise he/she can kill a superbly designed system if it is perceived as making only a small contribution to job performance. Thus, promoting a user's contribution in the walkthrough and throughout the design phase can be crucial for successful implementation.

User involvement gives the designer important feedback as the design is being completed. It also provides the user with a basic understanding of what the candidate system will and will not do. User involvement invariably paves the way for acceptance of the system by the user staff. It also bridges the gap between the designer, who as a staff person has an "expert" perspective, and the user, who is more typically "line" with a generalist or managerial view.

MAJOR DEVELOPMENT ACTIVITIES

Several development activities are carried out during structured design. They are data base design, implementation planning, system test preparation, system interface specification, and user documentation (see Figure 9–14). Each of these activities is covered in a separate chapter.

1. *Data base design.* This activity deals with the design of the physical data base. A key is to determine how the access paths are to be implemented. A physical path is derived from a logical path. It may be implemented by pointers, chains, or other mechanisms, as we discuss in Chapter 11.

2. *Program design.* In conjunction with data base design is a decision on the programming language to be used and the flowcharting, coding, and debugging procedure prior to conversion. The operating system limits the programming languages that will run on the system.

When the system design is under way and programming begins, the plans and test cases for implementation are soon required. This means there must be detailed schedules for system testing and training of the user staff. Planned training allows time for selling the candidate system to those who will deal with it on a regular basis. Consequently, user resistance should be minimized.

3 and 4. *System and program test preparation.* Each aspect of the system

FIGURE 9–14 System Design Activities

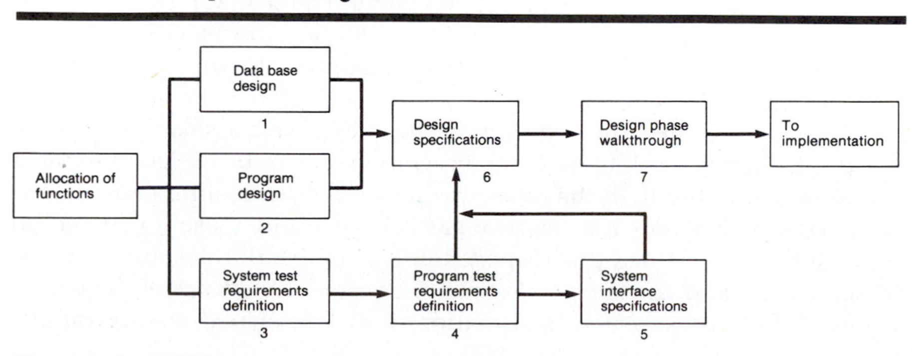

has a separate test requirement. *System testing* is done after all programming and testing are completed. The test cases cover every aspect of the candidate system—actual operations, user interface and so on. System and program test requirements become a part of design specifications—a prerequisite to implementation.

In contrast to system testing is *acceptance testing*, which puts the system through a procedure design to convince the user that the candidate system will meet the stated requirements. Acceptance testing is technically similar to system testing, but politically it is different. In system testing, bugs are found and corrected with no one watching. Acceptance testing is conducted in the presence of the user, audit representatives, or the entire staff.

Since test cases may be shared by both system testing and acceptance testing, system testing may be viewed as a dress rehearsal for the acceptance test. The criteria or plan for acceptance should be available in the structured specification.

5. *System interface specification*. This phase specifies for the user how information should enter and leave the system. The designer offers the user various options. By the end of the design, formats have to be agreed upon so that machine-machine and human-machine protocols are well defined prior to implementation.

Before the system is ready for implementation, user documentation in the form of a user's or operator's manual must be prepared. The manual provides instructions on how to install and operate the system, how to provide input, how to access, update, or retrieve information, how to display or print output, in what format, and so on. Much of this documentation cannot be written until the operation documentation is finalized—a task that usually follows design.

Personnel Allocation

In the past, a medium or large project was handled by a team of programmers with the goal of speeding up implementation. Unfortunately, there was more emphasis on numbers than on talent. The structured approach to design and implementation is useful in facilitating the planning process. Emphasis is on allocating the right programmers to the task within a realistic timetable.

A completed structure chart gives the designer a realistic outline of the programming work to be done. Programmers can be assigned to meet the workload rather than the other way around. A team of programmers is assigned a subsystem that is strongly cohesive and loosely coupled to other subsystems. Once this responsibility is assigned, roles are allocated within each team. The designer, of course, oversees the work of the teams. Figure 9–15 illustrates how a structure chart is assigned among various teams.

FIGURE 9-15 Personnel Allocation Using Structure Chart

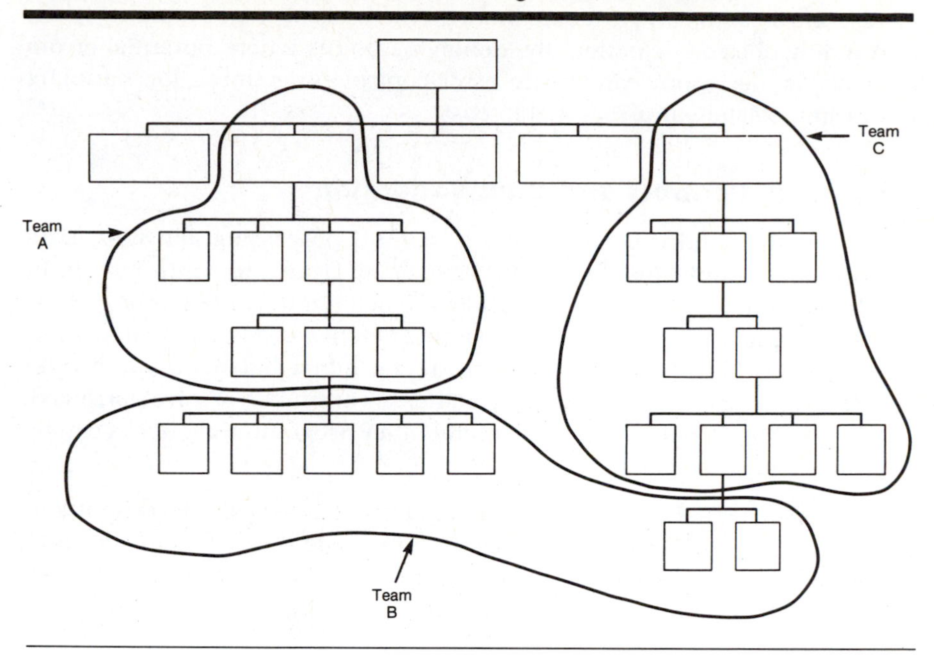

What module(s) in the chart is assigned to what size team can be very important. Modules at the bottom of a structure chart are usually utility and data base access modules. This sector represents the user interface. With today's emphasis on what the Germans call *Benutzefreundlichkeit* or *user-friendliness*, a team with specialized skills should be assigned to the interface subsystem.

AUDIT CONSIDERATIONS

A well-designed system should have controls to ensure proper operation and routine auditing. A candidate system's failure often results from a lack of emphasis on data control. Therefore, standards of accuracy, consistency, and maintainability must be specified to eliminate errors and control for fraud.

A system design introduces new control elements and changes the control procedures. New controls in the form of relational comparisons are designed to detect and check errors that arise from the use of the system. In a manual system, internal control depends on human judgment, personal care, and division of labor. In a computer-based system, the number of persons involved is considerably reduced. A software package is an effective substitute for human judgment in processing routines and error checks.

In designing a new system, the designer should specify the location of error-control points and evaluate them on the basis of error frequency, cost, and timing of error detection. By identifying points where potential errors may occur, designers can create error-control procedures for handling errors immediately at a reasonable cost.

Processing Controls and Data Validation

Several methods have been devised to control processing activities. Data records, for example, may be batched into small groups to control totals. In batch processing, if an error is detected, the batch that contains the error is reviewed without disturbing the remaining batches of the file. If all control totals balance, the batch is accepted. If the batch balances but certain records are rejected, the batch may be held until the errors are corrected.

In addition to batch controls, several other programmed checks can be used for testing the data:

1. *Completeness check* ensures that all fields in a record are present and are read in the proper sequence. In a multiple-record check, the program verifies the self-checking number of the record(s) that make up the transaction. If an error is detected, the entire group of records is rejected.
2. *Consistency check* refers to the relevance of one type of data to another. For example, a retailer may have a policy of extending a 30 percent discount to religious leaders on all merchandise ordered and up to $2,000 credit at no interest. An order received from a customer must be checked against these provisions to ensure consistency with the specified policy.
3. *Reasonableness check* evaluates a transaction against a standard to determine whether it meets the test. For example, if the maximum wage rate is $5 per hour and no overtime is allowed, an employee's weekly gross pay of $210 would be unreasonable, since the limit is $200.
4. *Sequence check* verifies that data records are in sequence prior to processing. A check of duplicate records may also be incorporated in the routine.

Audit Trail and Documentation Control

An important function of system controls is providing for an audit trail. An *audit trail* is a routine designed to allow the analyst, user, or auditor to verify a process or an area in the new system. In a manual system, the audit trail includes journals, ledgers, and other documents that the auditor uses to trace transactions through the system. In a computerized system, record content and format frequently make it difficult to trace a transaction completely. Some reasons are the following:

1. Records stored on a magnetic medium (tape, disk, etc.) can be read only by a computer and with the use of a computer program.
2. Data processing activities are difficult to observe, since they take place within the computer system.
3. The record sequence is difficult to check without the assistance of a computer system.
4. Direct data entry eliminates the physical documentation for an audit program.

One way to overcome these constraints is to keep a file on all transactions as they occur. For example, transactions can be recorded on tape, which can be an input to an audit program. The program pulls selected transactions and makes them available for tracing their status. The systems analyst must be familiar with basic auditing or work closely with an auditor to ensure an effective audit trail during the design phase.

The proper audit of a system also requires documentation. Documentation is the basis for the review of internal controls by internal or independent auditors. It also provides a reference for system maintenance. Preparing documentation occupies much of the analyst's time. When the implementation deadline is tight, documentation is often the first item to be ignored.

Documentation may be internal (in program documentation) or external hard-copy documentation. It must be complete and consistent for all systems prepared according to standards. So a plan to approve a new design should include documentation standards before programming and conversion.

In summary, the primary purpose of auditing is to check that controls built into the design of candidate systems ensure its integrity. Audit considerations must be incorporated at an early stage in the system development so that changes can be made in time. Neglecting this vital step could spell trouble for system implementation.

Summary

1. System design is a transition from a user-oriented document to a document oriented to programmers or data base personnel. It goes through logical and physical design with emphasis on the following:
   a. Preparing input/output specifications.
   b. Preparing security and control specifications.
   c. Specifying the implementation plan.
   d. Preparing a logical design walkthrough before implementation.
2. Structured design is a data-flow-based methodology that identifies inputs and outputs and describes the functional aspects of the system. It partitions a program into a hierarchy of modules organized in a top-down manner with the details at the bottom.

3. The documentation tool is the structure chart. It has three elements—the module, the connection, and the couple. When a module is evaluated, the module's connections to other modules (coupling) and its intramodule strength (cohesion) are considered.
4. One way of developing an input/process/output chart for modules is with the HIPO chart. It consists of the hierarchy chart plus the input/process/output (IPO) chart, thus capturing the essence of top-down decomposition. The aids are the HIPO work sheet and template.
5. A useful activity in all phases of a structured project is the walkthrough, where ideas are interchanged among peers. User involvement is extremely important.
6. The major development activities during structured design are data base design, implementation planning, system test planning, system interface specifications, and user documentation. Much of the documentation is written after conversion, a task that follows design.
7. A well-designed system should provide for controls to eliminate errors, check fraud, and ensure system integrity. Audit trails, documentation, and processing control must be incorporated into the system before it is released to the end user.

Key Words

Audit Trail
Cohesion
Couple
Decomposition
HIPO
IPO Chart
Physical Design
Structure Chart
Structured Walkthrough
System Design
Top-down Design

Review Questions

1. In your own words, describe the process of system design.
2. Distinguish between the following:
   a. Logical and physical design.
   b. HIPO and IPO.
   c. Coupling and cohesion.
3. What design methodologies are used in system design? Be specific.
4. Define structured design. How is it related to a DFD?
5. "Structured design provides the best partitioning of a program into small, independent modules organized in a top-down manner." Do you agree? Illustrate.
6. What are some of the advantages of top-down design? Elaborate.
7. Explain and illustrate the key elements of a structure chart.

8. What is the goal of the functional decomposition approach to structured design?
9. How is a HIPO chart related to structured design? What are its objectives?
10. Explain the components of the HIPO package format. How would one generate a HIPO chart?
11. How is a structured walkthrough conducted? What is the role of the user in this activity? Elaborate.
12. What development activities are carried out during structured analysis? Discuss.
13. How are personnel allocated in system design? Illustrate.
14. What audit considerations are included in system design? Why are they important?

Application Problems

1 Review the partial structure chart below.

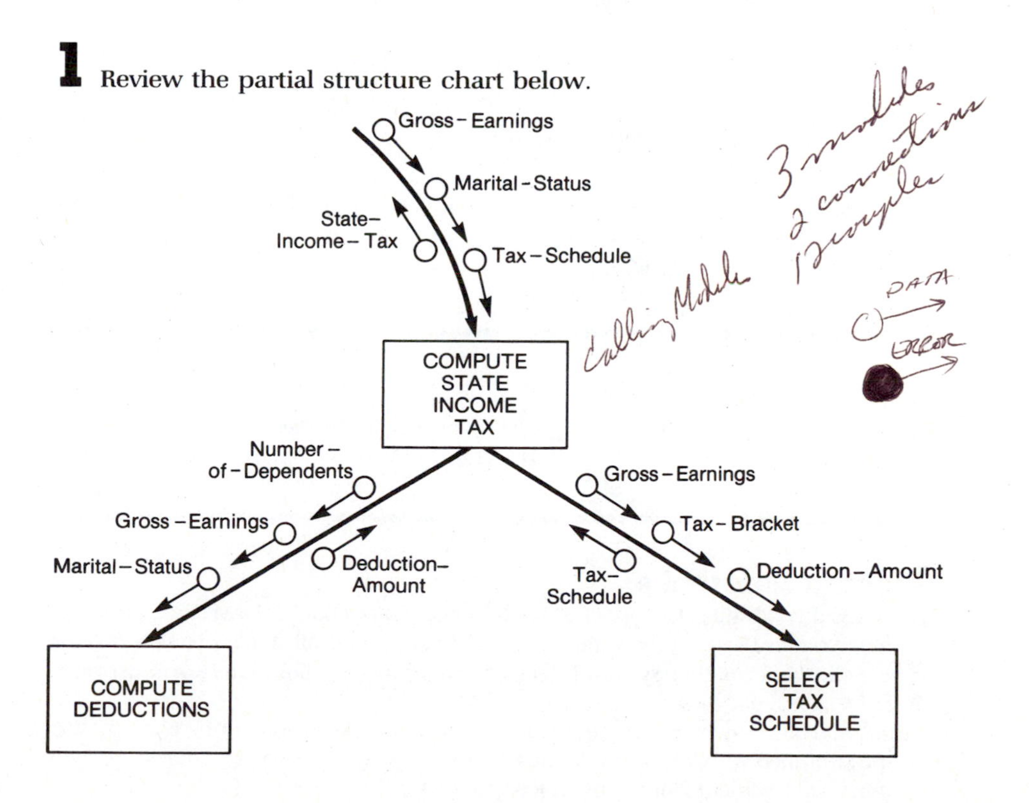

Assignment

a. How many modules, connections, and couples are there in the chart? What do they mean?

b. Which module is the calling module? The called module?

c. How many couples are passed to the calling module?

d. What is the minimum number of call statements inside COMPUTE STATE INCOME TAX?

e. What is the output of SELECT TAX SCHEDULE?

---

2 Design a structure chart using the following information:

a. Calling module: RECORD STUDENT GRADES

b. Called modules: GET ACADEMIC RECORD
GET VALID GRADES
ADD NEW GRADES
REPORT ERRORS
CHECK FOR PROBATION
CHECK FOR DEAN'S LIST

c. Include the required input and output couples, showing the direction and meaning.

d. In the same chart, show CHECK FOR PROBATION as a calling module and factor a called module called CALCULATE GPA. Show input and output couples.

---

3 Use the data flow diagrams in Exhibit 9–1 to derive a structure chart. Start with the calling module HANDLE CASH WITHDRAWAL INQUIRY.

Selected References

Brosh, Cyril P., Philip J. Grouse, J. Ross Jeffery, and Michael J. Lawrence. *Information System Design.* Englewood Cliffs, N.J.: Prentice-Hall, 1982, pp. 168–74.

Bryce, Milt. "Information System Design Methodologies." *Computerworld,* November 7, 1983, p. 51ff.

Demarco, Tom. *Structured Analysis and Design.* New York: Yourdon, 1979, p. 307.

Dimino, Stephen A. "Corporate Politics and the System's Process." *Journal of Systems Management,* September 1983, pp. 6–9.

EXHIBIT 9–1 Response to Cash Withdrawal Inquiry

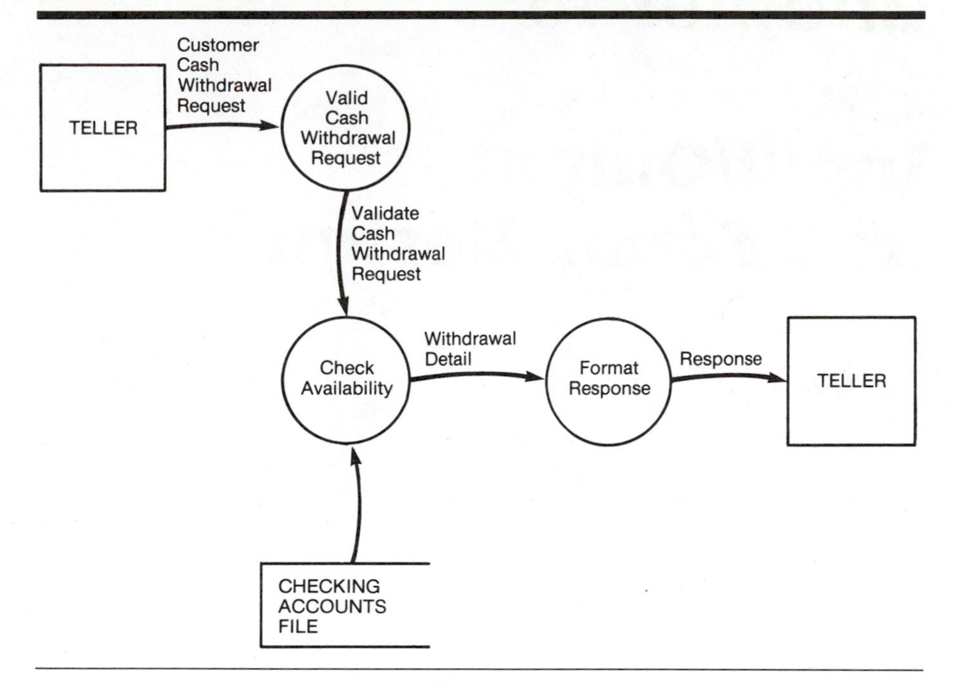

Hershauer, James C. "What's Wrong with Systems Design Methods? It's Our Assumption!" *Journal of Systems Management*, April 1978, pp. 25–28.

Lapointe, Joel R. "Userware: The Merging of System Design and Human Needs." *Data Management*, February 1982, pp. 29–33.

Patrick, Rober L. "A Checklist for System Design." *Datamation*, January 1980, pp. 147–48ff.

Peters, Lawrence, and Leonard Tripp. "Comparing Software Design Methodologies." *Datamation*, November 1977, pp. 89–94.

Sadek, Konrad E., and Edward Tomeski. "Different Approaches to Information Systems." *Journal of Systems Management*, April 1981, pp. 17–27.

Sides, S. R. "Walkthroughs and Reviews." *Computerworld (In Depth)*, August 10, 1981, p. 15ff.

Tsichritzis, Dionysios. "The Architects of System Design." *Datamation*, September 1980, pp. 201–202.

Yourdon, Edward. *How to Manage Structured Programming*. New York: Yourdon, 1976, pp. 31–50, 100–112, 137–50.

Chapter 10

Input/Output and Forms Design

At a Glance

The first step in systems design is to design output and input within predefined guidelines. In input design, user-originated inputs are converted to a computer-based format. It also includes determining the record media, method of input, speed of capture, and entry into the system. Online data entry accepts commands and data through a keyboard. The data (input or output) are displayed on a cathode-ray tube (CRT) screen for verification. The major approaches to input design are the menu, the formatted, and the prompt design. In each alternative, the user's options are predefined.

In output design, the emphasis is on producing a hard copy of the information requested or displaying the output on a CRT screen in a predefined format. Form design elaborates on the way output is presented and the layouts available for capturing information.

By the end of this chapter, you should know:

1. How source documents are designed.
2. The various data capture media and devices.
3. How data are entered into the computer.
4. What can be displayed on a CRT screen.
5. The classes of forms and how they are designed.

INTRODUCTION

In Chapter 9, we defined systems design as the process of developing specifications for a candidate system that meet the criteria established in systems analysis. A major step in design is the preparation of input and the design of output reports in a form acceptable to the user. This chapter reviews input and output design and the basics of forms design. As we shall see, these steps are necessary for successful implementation.

INPUT DESIGN

Inaccurate input data are the most common cause of errors in data processing. Errors entered by data entry operators can be controlled by input design. *Input design* is the process of converting user-originated inputs to a computer-based format. In the system design phase, the expanded data flow diagram identifies logical data flows, data stores, sources, and destinations. A systems flowchart specifies master files (data base), transaction files, and computer programs. Input data are collected and organized into groups of similar data. Once identified, appropriate input media are selected for processing.

Input Data

The goal of designing input data is to make data entry as easy, logical, and free from errors as possible. In entering data, operators need to know the following:

1. The allocated space for each field.
2. Field sequence, which must match that in the source document.
3. The format in which data fields are entered; for example, filling out the date field is required through the edited format mm/dd/yy.

When we approach input data design, we design the source documents that capture the data and then select the media used to enter them into the computer. Let us elaborate on each step.

Source Documents

Source data are captured initially on original paper or a source document. For example, a check written against an account is a source document. When it reaches the bank, it is encoded with special magnetic ink character recognition (MICR) so that it can be processed by a reader that is part of the information system of the bank. Therefore, source documents initiate a processing cycle as soon as they are entered into the system.

Source documents may be entered into the system from punch cards, from diskettes, or even directly through the keyboard. A source document may or may not be retained in the candidate system. Thus, each source document may be evaluated in terms of (1) its continued use in the candidate system, (2) the extent of modification for the candidate system, and (3) replacement by an alternative source document.

A source document should be logical and easy to understand. Each area in the form should be clearly identified and should specify for the user what to write and where to write it. For example, a field as simple as date of birth may be written in four different ways:

1. 19 September 1935
2. Sept. 19, 1935
3. 9/19/35
4. 19/9/35 (European style)

Unless it is clear in a source document that two digits are allowed for the month, day, and year (MM/DD/YY), we could expect such combinations of responses.

In source documents where the user chooses from a list of options, it is more efficient to direct the person to check the appropriate box than to enter a character. Figure 10–1 illustrates this point.

Input Media and Devices

Source data are input into the system in a variety of ways. The following media and devices are suitable for operation:

1. *Punch cards* are either 80 or 96 columns wide. Data are arranged in a sequential and logical order. Operators use a keypunch to copy data from source documents onto cards. This means that the source document and card design must be considered simultaneously.

2. *Key-to-diskette* is modeled after the keypunch process. A diskette replaces the card and stores up to 325,000 characters of data—equivalent to the data stored in 4,050 cards. Like cards, data on diskettes are stored in sequence and in batches. The approach to source document and diskette design is similar to that of the punch card. Data must be in sequence and logically cohesive.

3. *MICR* translates the special fonts printed in magnetic ink on checks into direct computer input.

FIGURE 10–1 Partial Source Document with Check-off Options

Recommended	Inefficient
Shirt size (check one) ☐ Small ☐ Medium ☐ Large ☐ X large	Shirts (enter size) ☐

4. *Mark-sensing* readers automatically convert pencil marks in predetermined locations on a card to punched holes on the same card.

5. *Optical character recognition (OCR)* readers are similar to MICR readers, except that they recognize pencil, ink, or characters by their configuration (shape) rather than their magnetic pattern. They are often used in remote locations as free-standing input preparation devices or direct input media to the system.

6. *Optical bar code readers* detect combination of marks that represent data (see Figure 10–2). The most widely known system is the Universal Product Code (UPC), which codes retail items in stores. Automatic tag reading is a major breakthrough in speeding up customer service and eliminating costly data input errors at the point of sale. It is virtually impossible for the sales clerk to enter incorrect merchandise information such as department and class type data. Automatic tag reading is the ideal way to collect unit inventory information fast, accurately, and economically.

7. *Cathode-ray tube (CRT) screens* are used for online data entry. CRT

FIGURE 10–2 Bar Code Data Entry—An Example

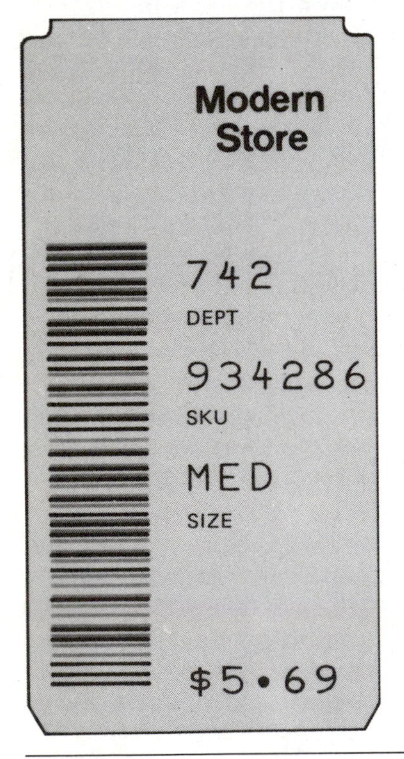

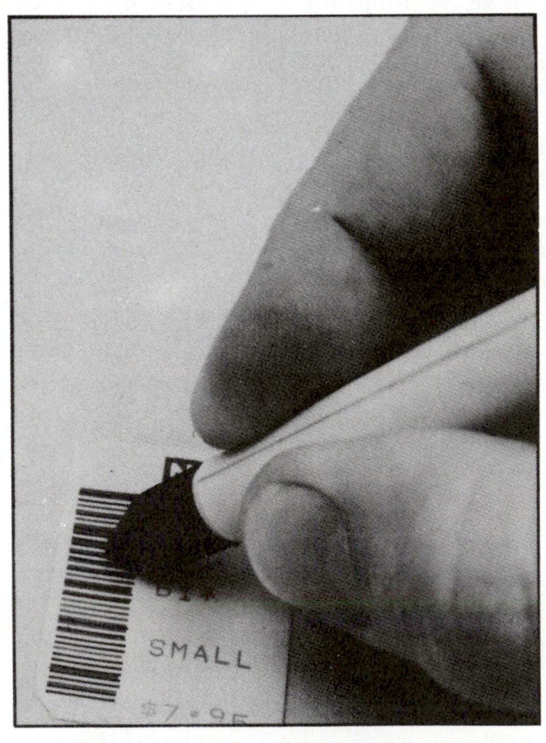

Courtesy National Cash Register.

screens display 20, 40, or 80 characters simultaneously on a television-like screen. They show as many as 24 lines of data.

In addition to determining record media, the analyst must decide on the method of input and the speed of capture and entry into the system. Processing may be batched (a group of records handled as a unit), online (records processed directly), sequential (sorted records), or random (unsorted). For example, magnetic tape may be suitable for batch sequential processing, whereas diskettes are ideal for online processing and random inquiries. Figure 10–3 summarizes selected methods of data capture and entry.

Online Data Entry

We live in the age of the microcomputer and at a time where more and more CRT screens are used for online data entry. As terminal prices decline and microcomputers become more popular, entering data directly through the keyboard will become a way of life. The number of applications that rely on direct data entry is too long to list. Two examples are automated teller machines (ATMs) in banking and point of sale (POS) in retailing.

Online data entry makes use of a processor that accepts commands and data from the operator through a keyboard or a device such as a touch-sensitive screen or voice input. The input received is analyzed by the processor. It is then accepted or rejected, or further input is requested. The request for input is in the form of a message displayed on the screen or by audio output.

Most keyboards have keys for alphabetic, numeric, as well as special functions. CRT screens display 24, 40, or 80 characters simultaneously or one line at a time, depending on the application and options offered by the vendor. Care must be taken that the hardware facilitates easy data entry into the system.

There are three major approaches for entering data into the computer: menus, formatted forms, and prompts.

The Menu. A menu is a selection list that simplifies computer data access or entry. Instead of remembering what to enter, the user chooses

FIGURE 10–3 Selected Data Capture Media/Devices

Input Device/Media	Type of Processing
Key punch/punch card	Batch, sequential
Key-to-diskette/diskette	Batch, sequential (or random)
MICR reader	Batch, sequential (or random)
Mark-sensing reader	Batch, sequential (or random)
OCR reader	Batch, sequential (or random)
Optical bar code	Batch, sequential (or random)
Online data entry	Online, sequential, or random

from a list of options and types the option letter associated with it. For example, Figure 10–4 shows a menu for entering, adding, or deleting box types in our safe deposit billing system. A curser blinking in the space reserved for ( ) YOUR CHOICE requests the user to type the letter that represents the option wanted. Other menus require the user to position the curser in front of a menu choice to make a selection.

A menu limits a user's choice of responses but reduces the chances for error in data entry.

The Formatted Form. A formatted form is a preprinted form or a template that requests the user to enter data in appropriate locations. It is a fill-in-the-blank type form. The form is flashed on the screen as a unit. The curser is usually positioned at the first blank. After the user responds by filling in the appropriate information, the curser automatically moves to the next line, and so on until the form is completed. During this routine, the user may move the curser up, down, right, or left to various locations for making changes in the response. Figure 10–5 is a safe deposit customer set-up form. The system requests information about the customer's name, address, renewal date, and so on. The user fills out the information on the dotted lines.

FIGURE 10–4 The Menu Approach to Data Entry—An Example

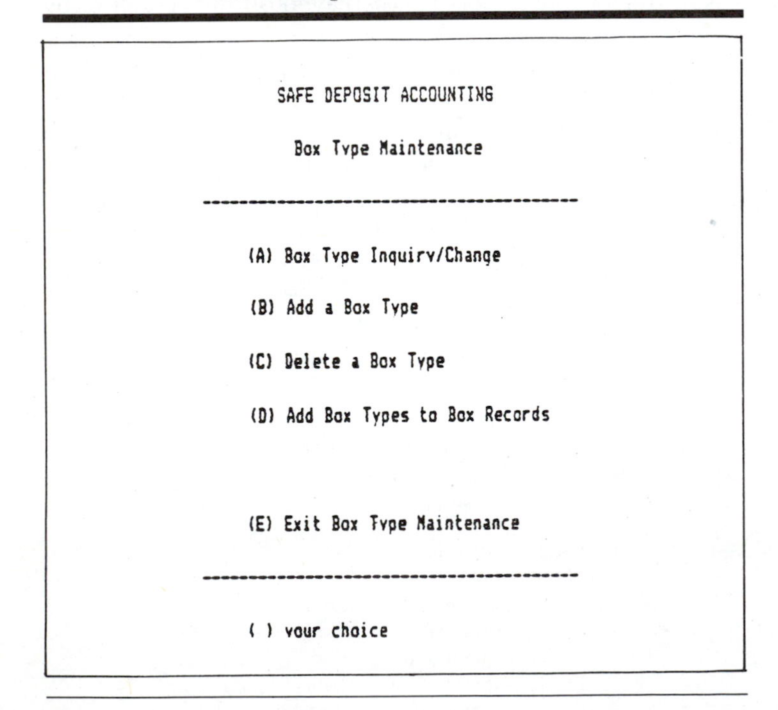
SAFE DEPOSIT ACCOUNTING

Box Type Maintenance

(A) Box Type Inquiry/Change

(B) Add a Box Type

(C) Delete a Box Type

(D) Add Box Types to Box Records

(E) Exit Box Type Maintenance

( ) your choice

Courtesy Aurora Systems, Inc.

FIGURE 10-5 The Formatted Form—An Example

```
                    CUSTOMER SET UP

Office #
Box #
-----------------------------------------
(0) Open Date   07/11/83
(1) Last Pmt    __/__/__
(2) BOX TYPE __  (3) EXCEPTION CODE  0
(4) Customer
       Last Name          -------------------
       First Name         -------------------

(5) Address 1             -------------------
            2             --------------
            3             -----------

(6) Renewal Month      --
(7) Mo. btw. bills     --        Mo. btw. bills means months between
(8) Account Type       _         bills.  Allowable inputs are:
(9) Account Number     ---------  1, 2, 3, 4, 6, 12
```

The Prompt (conversational mode). In prompt the system displays one inquiry at a time, asking the user for a response. For example, the following dialogue represents a prompt approach to data entry:

System: ENTER PASSWORD
User: A1260
System: ENTER FILENAME
User: Inventory
System: INPUT DATA NOW? Y/N
User: Y

Most systems edit the data entered by the user. For example, if the password exceeds a maximum number of digits or if the password is illegal, the system responds with a message like UNAUTHORIZED ENTRY or ILLEGAL NUMBER. The user has three chances to enter the correct code, after which the system "locks up." In banks' automated teller machines (ATMs), if the customer entered his/her code wrong three times, the system retains the card and displays a message on the screen that the user should check with an officer during banking hours.

The prompt method also allows the user to key questions that determine the next response of the system. In our example INPUT DATA NOW? Y/N, if the response is Y, the system might display a record format for entering data. Otherwise, it might automatically go back to the menu for a different set of options. This method along with the menu and template is designed to improve the efficiency and accuracy of online data entry.

In each of the alternative approaches to data entry, the user's options are predefined. An inexperienced user is guided through complex functions by a formatted procedure. The main limitation with many of the available menus or prompts is that they require only one item to be entered at a time rather than a string of data items simultaneously.

CRT Screen Design

Many online data entry devices are CRT screens that provide instant visual verification of input data and a means of prompting the operator. The operator can make any changes desired before the data go to the system for processing. A CRT screen is actually a display station that has a buffer (memory) for storing data. A common size display is 24 rows of 80 characters each or 1,920 character.

There are two approaches to designing data on CRT screens: manual and software utility methods. The manual method uses a work sheet much like a print layout chart. The menu or data to be displayed are blocked out in the areas reserved on the chart and then they are incorporated into the system to formalize data entry. For example, in Figure 10–6, we use dBASE II software commands (explained in Chapter 11) to display a menu on the screen. The first command in the partial program is interpreted by the system as follows: "Go to row 10 and column 10 on the screen and display (SAY) the statement typed between quotes." The same applies to the next three commands. The command "WAIT TO A" tells the system to keep the menu on the screen until the operator types the option next to the word "WAITING."

The main objective of screen display design is simplicity for accurate and quick data capture or entry. Other guidelines are:

1. Use the same format throughout the project.
2. Allow ample space for the data. Overcrowding causes eye strain and may tax the interest of the user.
3. Use easy-to-learn and consistent terms, such as "add," "delete," and "create."
4. Provide help or tutorial for technical terms or procedures.

The second approach to designing screen layouts is through software utility, usually provided by the CRT vendor. For example, IBM provides a

FIGURE 10-6 CRT Screen Design

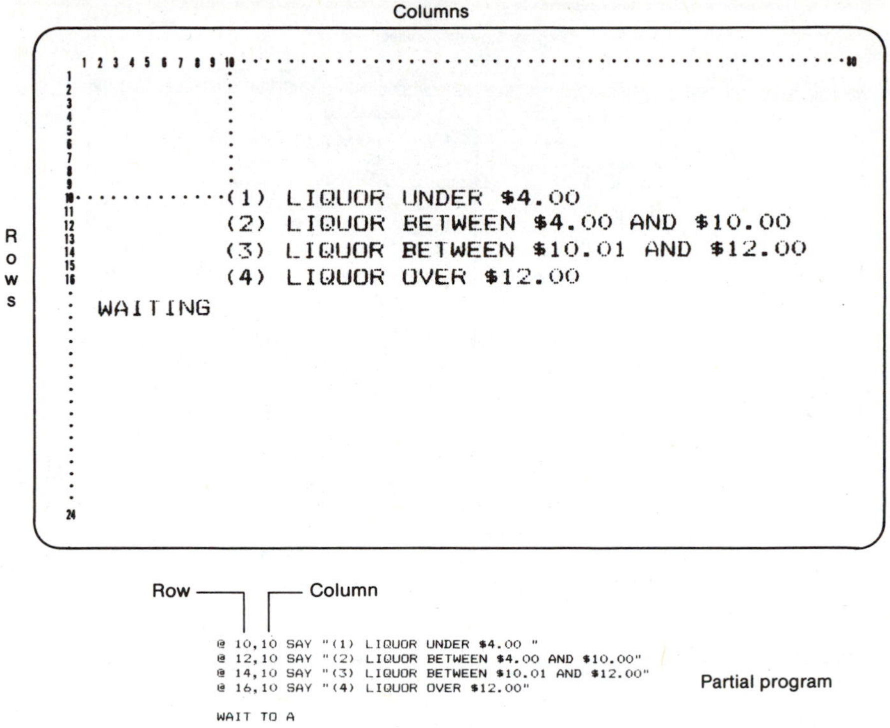

Columns

Rows

```
(1) LIQUOR UNDER $4.00
(2) LIQUOR BETWEEN $4.00 AND $10.00
(3) LIQUOR BETWEEN $10.01 AND $12.00
(4) LIQUOR OVER $12.00
WAITING
```

Row — Column

```
@ 10,10 SAY "(1) LIQUOR UNDER $4.00 "
@ 12,10 SAY "(2) LIQUOR BETWEEN $4.00 AND $10.00"
@ 14,10 SAY "(3) LIQUOR BETWEEN $10.01 AND $12.00"
@ 16,10 SAY "(4) LIQUOR OVER $12.00"

WAIT TO A
```

Partial program

Screen Design Aid (SDA) package that allows the designer (at the terminal) to modify the display components (see Figure 10-7).

OUTPUT DESIGN

Computer output is the most important and direct source of information to the user. Efficient, intelligible output design should improve the system's relationships with the user and help in decision making. A major form of output is a hard copy from the printer. Printouts should be designed around the output requirements of the user. The output devices to consider depend on factors such as compatibility of the device with the system, response time requirements, expected print quality, and number of copies needed. The following media devices are available for providing computer-based output:

1. MICR readers.
2. Line, matrix, and daisy wheel printers.
3. Computer output microfilm (COM).

FIGURE 10-7 Manual (*above*) and Computer-Generated Screen Display Design

Display Layout Sheet

COLUMN

1-10 11-20 21-30 31-40 41-50 51-60 61-70 71-80

1 2 3 4 5 6 7 8 9 0 1 2 3 4 5 6 7 8 9 0 1 2 3 4 5 6 7 8 9 0 1 2 3 4 5 6 7 8 9 0 1 2 3 4 5 6 7 8 9 0 1 2 3 4 5 6 7 8 9 0 1 2 3 4 5 6 7 8 9 0 1 2 3 4 5 6 7 8 9 0

ROW

01
02
03
04
05
06
07
08
09
10
11
12
13
14
15
16
17
18
19
20
21
22
23
24

1-10 11-20 21-30 31-40 41-50 51-60 61-70 71-80

1 2 3 4 5 6 7 8 9 0 1 2 3 4 5 6 7 8 9 0 1 2 3 4 5 6 7 8 9 0 1 2 3 4 5 6 7 8 9 0 1 2 3 4 5 6 7 8 9 0 1 2 3 4 5 6 7 8 9 0 1 2 3 4 5 6 7 8 9 0 1 2 3 4 5 6 7 8 9 0

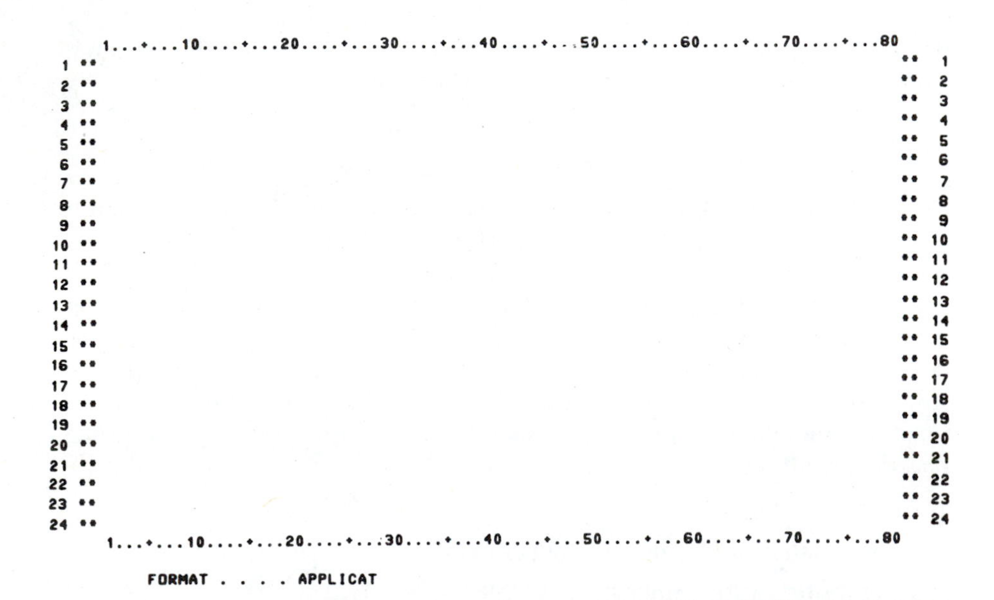

4. CRT screen display.
5. Graph plotters.
6. Audio response.

In addition to deciding on the output device, the systems analyst must consider the print format and the editing for the final printout. Editing ranges from suppressing unwanted zeros to merging selected records to produce new figures. In either case, the task of output preparation is critical, requiring skill and ability to align user requirements with the capabilities of the system in operation.

The standards for printed output suggest the following:

1. Give each output a specific name or title.
2. Provide a sample of the output layout, including areas where printing may appear and the location of each field.
3. State whether each output field is to include significant zeros, spaces between fields, and alphabetic or any other data.
4. Specify the procedure for proving the accuracy of output data.

In online applications, information is displayed on the screen. The layout sheet for displayed output is similar to the layout chart used for designing input. Areas for displaying the information are blocked out, leaving the rest of the screen blank or for system status information. Allowing the user to review sample screens can be extremely important because the user is the ultimate judge of the quality of the output and, in turn, the success (or failure) of the system. For example, the following shows editing output for a student birthdate:

DISPLAY: DATE OF BIRTH (mm/dd/yy) 23/19/80

RESPONSE: MONTH EXCEEDS 12

SUGGESTS A RETRY: DATE OF BIRTH (mm/dd/yy)

FORMS DESIGN

We have learned that data provide the basis for information systems.[1] Without data there is no system, but data must be provided in the right form for input and the information produced must be in a format acceptable to the user. In either case, it is still data—the basic element of a printed form.

What Is a Form?

People read from forms, write on forms, and spend billions of hours handling forms and filing forms. The data the forms carry come from people,

[1] The terms *data* and *information* are used here interchangeably, since forms design encompasses both input and output forms. Thus, *data* may be related to input forms design and *information* may be used with regard to output forms design.

and the informational output of the system goes to people. So the form is a tool with a message; *it is the physical carrier of data—of information*. It also can constitute authority for action. For example, a purchase order says BUY, a customer's order says SHIP, and a paycheck says PAY TO THE ORDER OF. Each form is a request for action. It provides information for making decisions and improving operations.

With this in mind, it is hard to imagine a business operating without using forms. They are the vehicles for most communications and the blueprint for many activities. As important as a printed form is, however, the majority of forms are designed by poorly trained people. People are puzzled by confusing forms; they ask for directions on how to read them and how to fill them out. When a form is poorly designed, it is a poor (and costly) administrative tool.

Classification of Forms

A printed form is generally classified by what it does in the system. There are three primary classifications: action, memory, and report forms. An *action* form requests the user to do something—get action. (Examples are purchase orders and shop orders.) A *memory* form is a record of historical data that remains in a file, is used for reference, and serves as control on key details. (Examples are inventory records, purchase records, and bond registers.) A *report* form guides supervisors and other administrators in their activities. It provides data on a project or a job. (Examples are profit and loss statements and sales analysis reports.) Figure 10-8 is a summary of the characteristics and examples of these forms.

FIGURE 10-8 Three Classes of Forms—A Summary

Class	Characteristics	Examples
Action	1. Orders, instructs, authorizes 2. Achieves results 3. Goes from one place (person) to another	Application form Purchase order Sales slip Shop order Time card
Memory	1. Represents historical data 2. Data generally used for reference 3. Stationary and remains in one place, usually in a file 4. Serves as control on certain details	Bond register Inventory record Journal sheet Purchase record Stock ledger
Report	1. Summary picture of a project 2. Provides information about job or details that need attention 3. Used by a manager with authority to effect change 4. Used as a basis for decision making	Balance sheet Operating statement Profit and loss statement Sales analysis Trial balance

Requirements of Forms Design

Forms design follows analyzing forms, evaluating present documents, and creating new or improved forms. Bear in mind that detailed analysis occurs only after the problem definition stage and the beginning of designing the candidate system. Since the purpose of a form is to communicate effectively through forms design, there are several major requirements:

1. *Identification and wording.* The form title must clearly identify its purpose. Columns and rows should be labeled to avoid confusion. The form should also be identified by firm name or code number to make it easy to reorder.

2. *Maximum readability and use.* The form must be easy to use and fill out. It should be legible, intelligible, and uncomplicated. Ample writing space must be provided for inserting data. This means analyzing for adequate space and balancing the overall forms layout, administration, and use.

3. *Physical factors.* The form's composition, color, layout (margins, space, etc.), and paper stock should lend themselves to easy reading. Pages should be numbered when multipage reports are being generated for the user.

4. *Order of data items.* The data requested should reflect a logical sequence. Related data should be in adjacent positions. Data copied from source documents should be in the same sequence on both forms (see Figure 10–9). Much of this design takes place in the forms analysis phase.

5. *Ease of data entry.* If used for data entry, the form should have field positions indicated under each column of data (see Figure 10–9) and should have some indication of where decimal points are (use broken vertical lines).

6. *Size and arrangement.* The form must be easily stored and filed. It should provide for signatures. Important items must be in a prominent location on the form.

7. *Use of instructions.* The instructions that accompany a form should clearly show how it is used and handled.

8. *Efficiency considerations.* The form must be cost effective. This means eliminating unnecessary data and facilitating reading lines across the form. To illustrate, if a poorly designed form causes 10 supervisors to waste 30 seconds each, then 5 minutes are lost because of the form. If the firm uses 10,000 of these forms per year, then 833 hours of lost time could have been saved by a better forms design.

9. *Type of report.* Forms design should also consider whether the content is executive summary, intermediate managerial information, or supporting-data. The user requirements for each type often determine the final form design.

FIGURE 10-9 Item Transfer from Source Document to Record

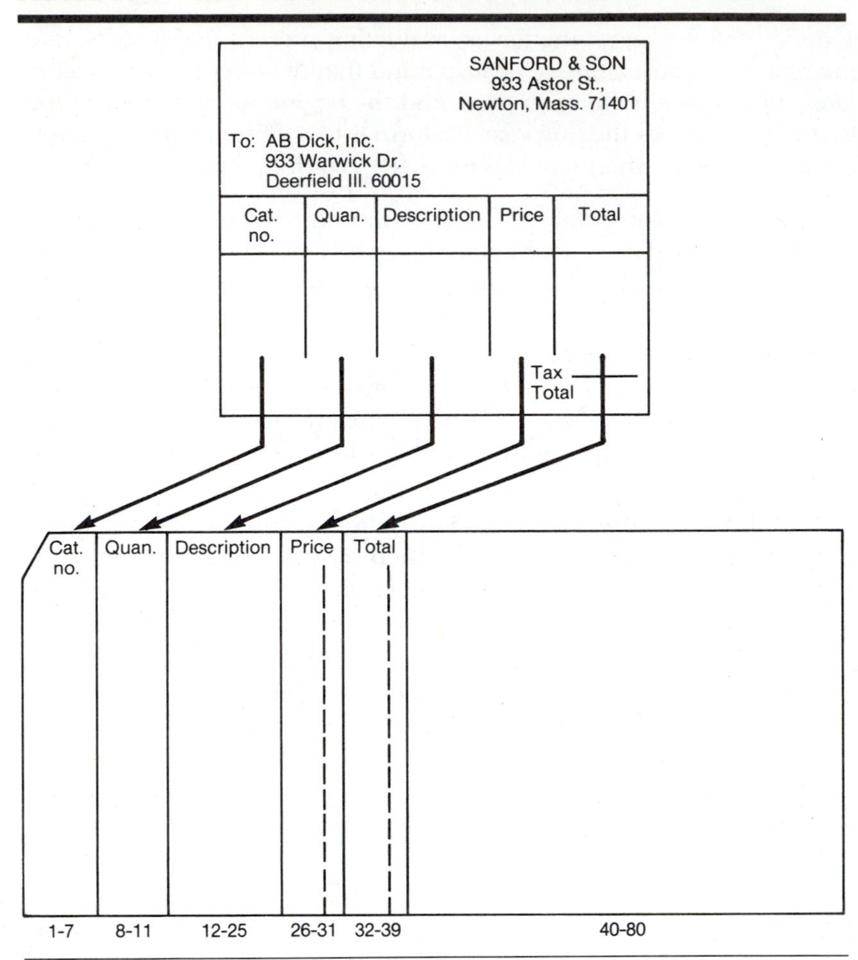

Carbon Paper as a Form Copier

There are many ways of duplicating forms but none is as handy and versatile as carbon paper. There are two primary types of carbon, classified by the action they encounter:

1. *Glide action* carbon inserted between a set of forms allows the glide action of the pencil (or ballpoint) to transfer dye to the surface of the sheet beneath.
2. *Hammer action* carbon is used in typewriters and line printers of computers. The hammer action of the key(s) transfers the carbon coating to the sheet beneath.

Various methods of transferring impressions between copies are also used:

1. *One-time carbon.* Made of inexpensive Kraftex paper, it is interleaved between two sheets in the form, used once, and then thrown away. It is the most cost effective (also the messiest) method for multipart forms.
2. *Carbon-backed paper.* The back of each form copy is coated with carbon, which transfers data to the copy beneath.
3. *NCR (no carbon required) paper.* The top sheet is chemically treated with invisible dye, which allows impressions to be tranferred to the next lower copy. It is the cleanest method of copying but also the costliest. Erasing removes the coating permanently.

The readability of carbon copies depends on color and outline. Regardless of the carbon quality, copies beyond the third or fourth become progressively poorer, so that by the tenth copy all one gets is a smudge of color—with hardly any sharpness of outline. In multicopy forms, the copies below the original should be lighter-weight paper for easier transfer of the carbon. The bottom copy should be heavier paper. Because carbon smears, carbon copies should not be used for data entry copy.

Selected data are deleted from printing on certain copies by using split carbons or short carbons or by printing a random design in the area where the data will be printed. The random design blurs out the readibility of the printed data (see Figure 10–10). Generally, one-time carbon is preferred when a small number of copies is required. If carbon is unacceptable, NCR paper is recommended. When selected data are restricted to specific copies, split carbons are used.

FIGURE 10–10 Blocked-out (Split) Carbon Paper *(left)* Short Carbon Paper

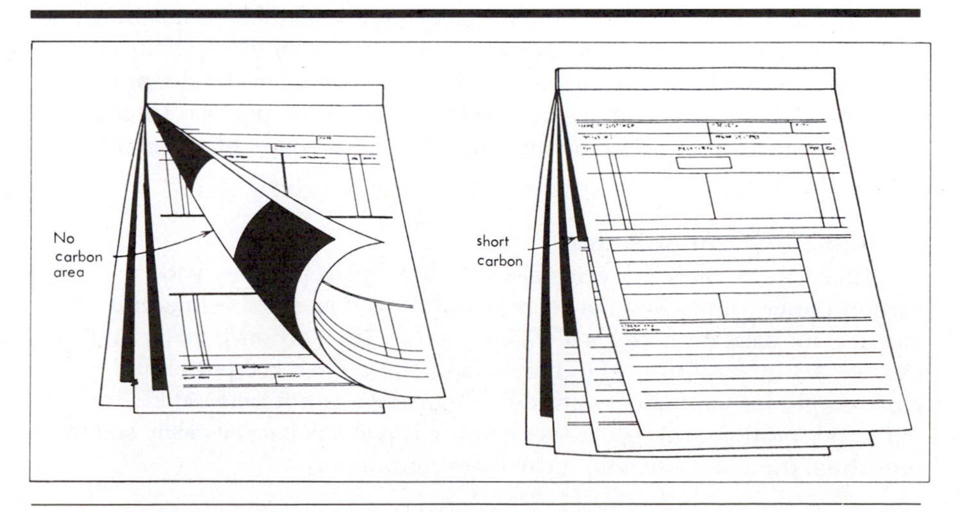

FIGURE 10-11 Flat Form—An Example

Types of Forms

Forms are classified into several categories: flat forms, unit-set/snapout forms, continuous strip/fanfold forms, NCR paper, and preprinted forms. These types are described briefly.

Flat Forms

A flat form is single-copy form prepared manually or by a machine and printed on any grade of paper. For additional copies of the original, carbon paper is inserted between copies. It is the easiest form to design, print, and reproduce; it has a low-volume use; and it is the least expensive. Often a pad of the flat forms is printed identical to the original copy of a unit set. (see Figure 10–11).

Unit-Set/Snapout Forms

These forms have an original copy and several copies with one-time carbon paper interleaved between them. The set is glued into a unit (thus, *unit set)* for easy handling (see Figure 10–12). The carbon paper is approximately 3/8 inch shorter than the copies. The copies are perforated at the glue margin for tearing out, although the carbon is not perforated. Because of the perforation and the shorter carbon, the forms can be easily snapped out (thus, the name *snapout* form) after completion.

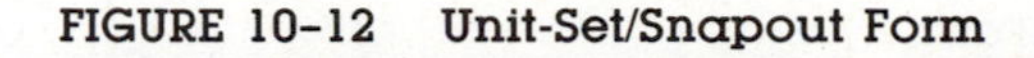

FIGURE 10–12 Unit-Set/Snapout Form

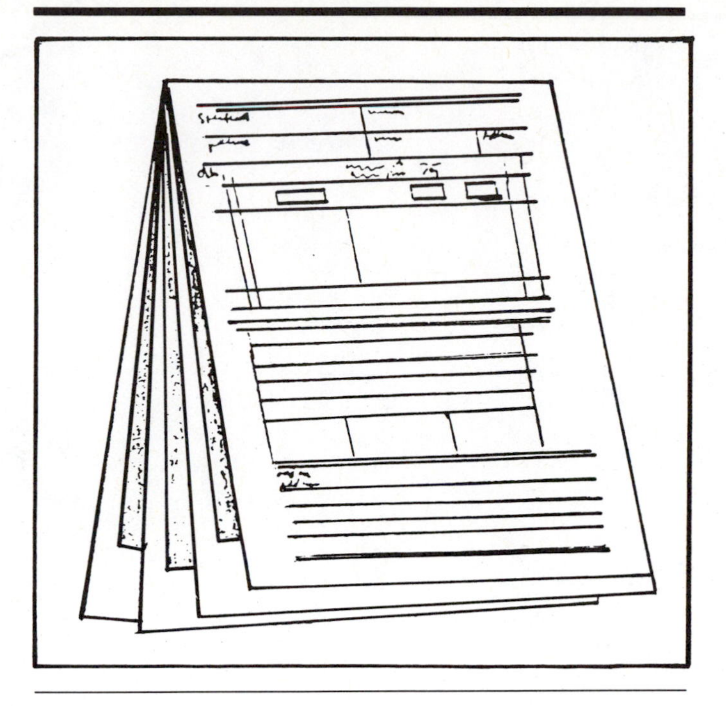

Continuous Strip/Fanfold Forms

These are multiple-unit forms joined together in a continuous strip with perforations between each pair of forms. One-time carbon is interleaved between copies, which are stacked in a fanfold arrangement. In Figure 10–13, note the pin-feed holes punched in both margins for mounting the forms onto the sprocket wheels of the printer. The device eliminates individual insertion of forms. The fanfold is the least expensive construction for large-volume use. Computer printouts are invariably produced on them; they are virtually part of systems design.

NCR (no carbon required) Paper

Several copies can be made by pressing a chemical undercoating on the top sheet into a claylike coating on the top of the second sheet. The writing (or printing) pressure forms an image by the coating material. The same process applies to the back of the second sheet for producing a carbon copy on the face of the succeeding sheet, and so on (see Figure 10–14).

NCR paper has many applications in salesbooks, checkbooks, inventory tickets, and deposit slips. It offers cleaner, clearer, and longer-lasting copies than carbon-interleaved forms. No carbon means no smears or smudges.

FIGURE 10-13 Continuous Strip/Fanfold Form

```
AUDIT REPORT FOR  EDEMO          10-JUN-83   09:03
  DEMONSTRATION EVENT
       TICKETMASTER
       TICKET SERVICE
       (312) 559-1950
  10 S. RIVERSIDE PLAZA
       CHICAGO IL 60606

LEVEL       1          2          3         4        5        6        7     TOTAL        SALE
ADULT     10.00       8.00       6.00
J-TYPE     5.00       4.00       3.00
G-TYPE     9.00       7.00       5.00
K-TYPE     8.50       6.50       4.50
L-TYPE     8.00       6.00       4.00
M-TYPE     7.50       5.50       3.50
V-TYPE     6.00       5.00       4.00
NOTICE:  PRICES HAVE BEEN CHANGED
--------------------------- TODAY'S SALES ---------------------------
                            OUTLET SALES
                            AGENCY SALES
                          BOX OFFICE SALES
                     SECONDARY BOX OFFICE SALES

                            TOTAL SALES
------------------------- TOTAL SALES FOR EVENT -----------------------
                            OUTLET SALES
ADULT                        1                                        1        6.00

TOTAL                        1                                        1        6.00
                            AGENCY SALES
                          BOX OFFICE SALES
ADULT         26        64       199                                289     1966.00
J-TYPE         2         2                                            4       18.00
G-TYPE         2         2                                            4       32.00
K-TYPE         3         2                                            5       38.50
L-TYPE         4                                                      4       32.00
M-TYPE         1                                                      1        7.50
V-TYPE         4        10                                           14       38.50

TOTAL         42        80       199                                321     2132.50
                     SECONDARY BOX OFFICE SALES

                            TOTAL SALES
ADULT         26        64       200                                290     1972.00
J-TYPE         2         2                                            4       18.00
G-TYPE         2         2                                            4       32.00
K-TYPE         3         2                                            5       38.50
L-TYPE         4                                                      4       32.00
M-TYPE         1                                                      1        7.50
V-TYPE         4        10                                           14       38.50

TOTAL         42        80       200                                322     2138.50

%SALES   84.00%  100.00%  100.00%                                 97.55%     96.35%

******************************************************************************
      THE FOLLOWING DOLLAR AMOUNTS REPRESENT POTENTIAL SALES ONLY

------------------------- UNSOLD TICKETS -------------------------
OPEN           6                                                      6       60.00
HOLD           2                                                      2       20.00

UNSOLD         8                                                      8       80.00

SEATING
CAP           50        80       200                                330     2218.50
```

The fanfolded form

One problem is the sensitivity of the chemical: It shows every unintended scratch. Other disadvantages are difficulty with erasures and high cost. NCR paper costs as much as 25 percent more than the carbon-interleaved forms. Considering the labor savings of the NCR process, however, cost may be well justified in the long run.

FIGURE 10-14 The NCR Process

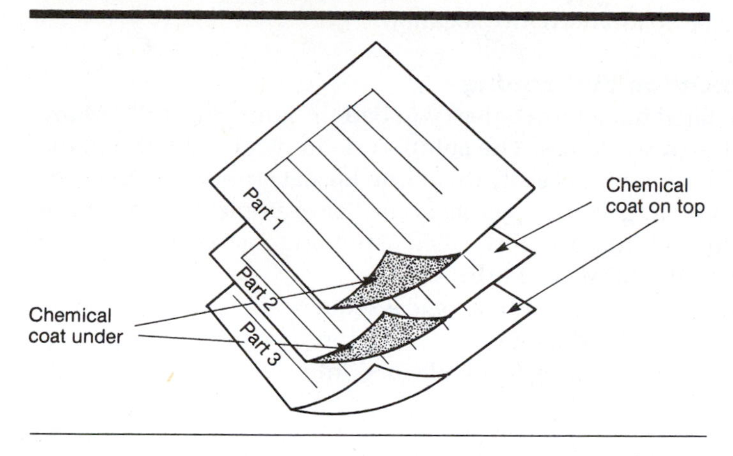

Layout Considerations

When a form is designed, a list is prepared of all the items to be included on the form and the maximum space to be reserved. The list should be checked by the form user to make sure it has the required details.

Form Title and Number

The first consideration in forms design is a brief, descriptive title that tells what the form is and what it does. Since we read from left to right and from top to bottom, the upper left corner of the form is an appropriate place for a title. On forms that go outside the organization, the title is placed in the center at the top of the form.

Long titles with vague words should be avoided. Good titles often are no longer than two words. Consider the following form titles:

Form Title	Comment
INVOICE	Too general
SALES INVOICE	Better
FURNITURE SALES INVOICE	Overmodified
INVOICE FOR SALE OF FURNITURE	Too long, unwieldy

A form number identifies the form, serves as a reference, and provides a control check for the supply room. Since the title and number are parts of the form identification, they should be placed together—say, in the upper left corner of the form. For an oversized form, the title is positioned in the center in bold type. In other situations where the upper left corner is used, the lower left corner is an alternative location.

Data Classification and Zoning

Have you ever filled out a form where you had to "jump around"? Many forms have this design weakness. The solution is simple: List all the items that must go on the form and classify them into logical groupings. All other items are listed under a "general" classification. For example, in a purchase order, several items fall into a group called "shipping instructions." They include SHIP TO, SHIP FROM, and SHIP VIA.

After the items are classified into a logical sequence by group, the next consideration is placing the data groups in appropriate areas (zones). Figure 10–15 shows two interlocking forms divided into seven zones each. To summarize:

1. A form is divided into zones; each zone represents a similar group of information.
2. The zones are organized in a logical flow based on the upper left corner (ULC) method. As shown in Figure 10–16, a form should be designed to be read or filled out in the same way we read or write in English (left to right and top to bottom).
3. When more than one form is involved, the sequence of data in related forms should follow the same flow.

Rules and Captions

In designing forms, use rules (lines) to *guide* the human eye to read and write data groups. In this respect, printed rules aid in zoning a form. A

FIGURE 10–15 A Zoned Form

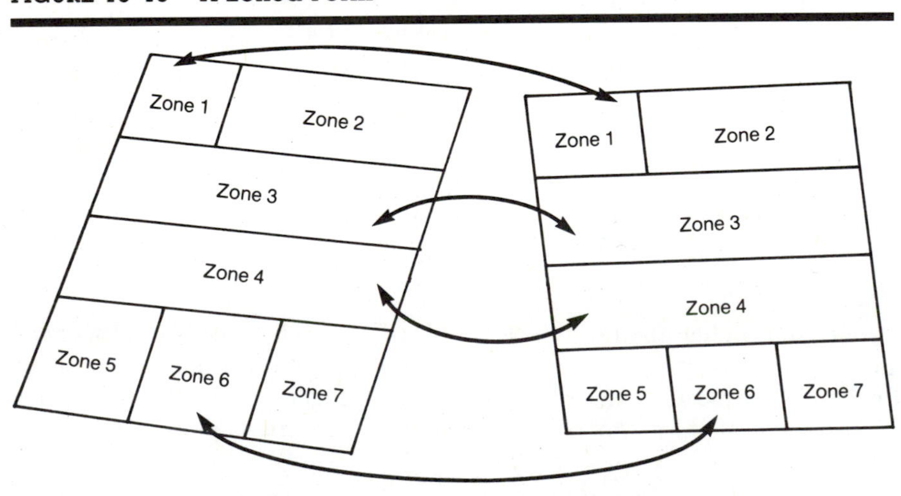

FIGURE 10-16 ULC Method of Data Flow

Start
Move down and left
Continue
Move down and left
Continue
Etc.

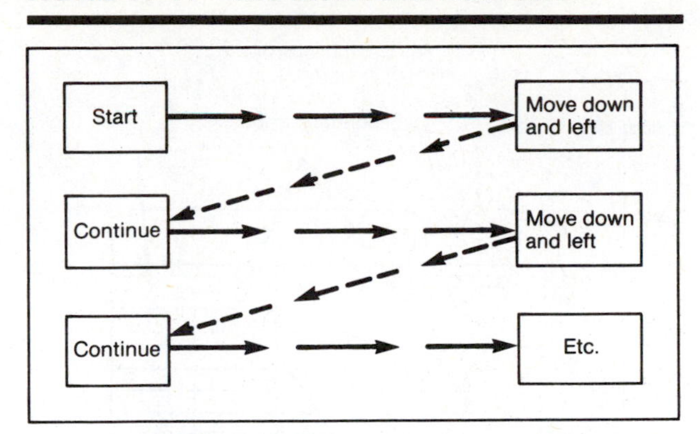

caption is similar to a column heading. It specifies what information to write in the space provided. Rules and captions go together: *Rules guide and separate, whereas captions guide and instruct*.

Since a caption is used to guide, one or two different sizes of captions are usually used. A caption should not be set in such bold type that it "barks" at the form user. As shown in Figure 10-17, the information filled in rather than the caption should stand out.

Light, single "hairline" rules should be used unless there is a definite need for separating parts of the form; in that case, heavier printed rules can be used. Printed rules and captions are shown in Figure 10-18.

A *column heading* is a caption used to refer to more than one rule or box

FIGURE 10-17 Acceptable *(bottom)* and Unacceptable Captions

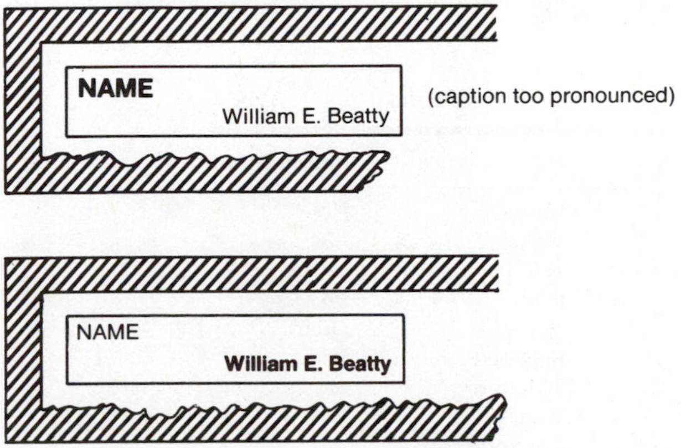

NAME
William E. Beatty
(caption too pronounced)
NAME
William E. Beatty

FIGURE 10-18 Variations of Printed Rules and Captions

Level of emphasis	*Caption*		*Rule*	
Negative	NAME	4 pt. caps	----	Broken
Primary	NAME	6 pt. caps	——	Hairline
Secondary	NAME	8 pt. caps	——	1/2 pt.
Third	NAME	10 pt. caps	——	1 pt.

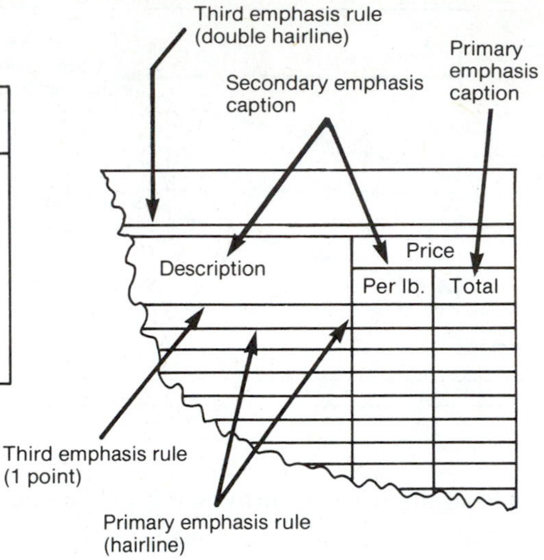

Third emphasis rule (double hairline)
Secondary emphasis caption
Primary emphasis caption
Description
Price
Per lb.
Total
Third emphasis rule (1 point)
Primary emphasis rule (hairline)

on a form. In Figure 10-18, the word *Price* is a major caption, consisting of two secondary captions: *Per lb.* and *Total*. Acceptable abbreviations and those that are not misleading may be used. Sample captions and their abbreviations are listed in Figure 10-19.

To summarize, a form is designed with a combination of rules and captions. Rules can also be used to make boxes in which the user places data. The caption tells the user *what* information goes in a particular position. Line-caption positions are shown in Figure 10-20.

Box Design

Whenever possible, it is advisable to design the form using the box-style rule, with captions in the upper left corner. The box design gets the

FIGURE 10-19 Form Caption Abbreviations

Caption Word	Abbreviation	Caption Word	Abbreviation
Account	Acct	Hours	Hrs
Amount	Amt	Manager	Mgr
Average	Avg	Merchandise	Mdse
Balance	Bal	Paid	Pd
Check	Ck	Quantity	Quan
Department	Dept	Received	Rec'd
Discount	Disc	Signature	Sig
Each	Ea	Weight	Wt
Freight	Frt	Zip code	Zip

FIGURE 10-20 Common Line-Caption Positions

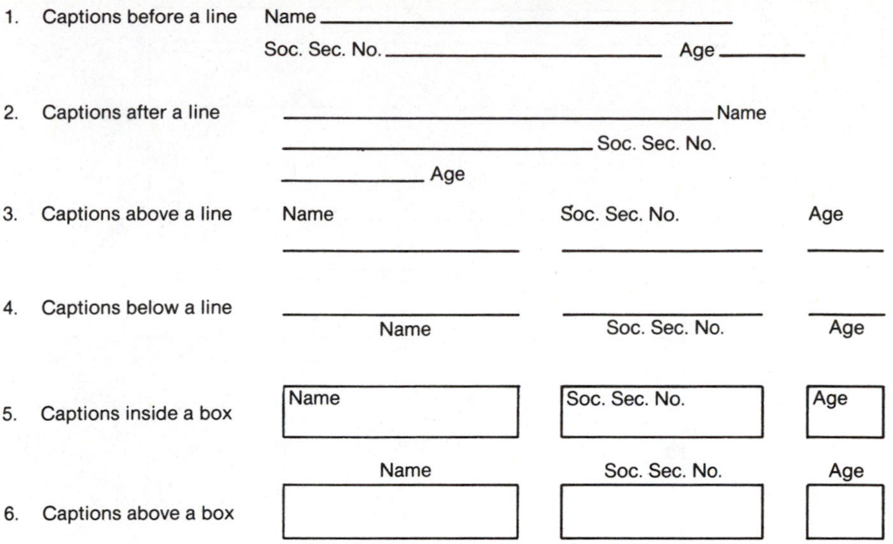

1. Captions before a line — Name ______ Soc. Sec. No. ______ Age ______
2. Captions after a line — ______ Name ______ Soc. Sec. No. ______ Age
3. Captions above a line — Name / Soc. Sec. No. / Age (above lines)
4. Captions below a line — (lines) Name / Soc. Sec. No. / Age
5. Captions inside a box — [Name] [Soc. Sec. No.] [Age]
6. Captions above a box — Name / Soc. Sec. No. / Age (above boxes)

captions up out of the way and reduces the form size by 25 to 40 percent. It also makes the data entry uninterrupted from left to right. Figure 10–21 contrasts the traditional style used on many forms with the box design. The traditional design is acceptable for a handwritten form, but it can be tricky to position the captions on the typewriter.

Spacing Requirements

If you pick 20 printed forms at random, you will find a great variety of spacing on them. Most forms seem to be spaced haphazardly. The method of preparing a form tells whether to allow for handwritten, printed entries,

FIGURE 10-21 Traditional and Box-Style Line and Caption Arrangements

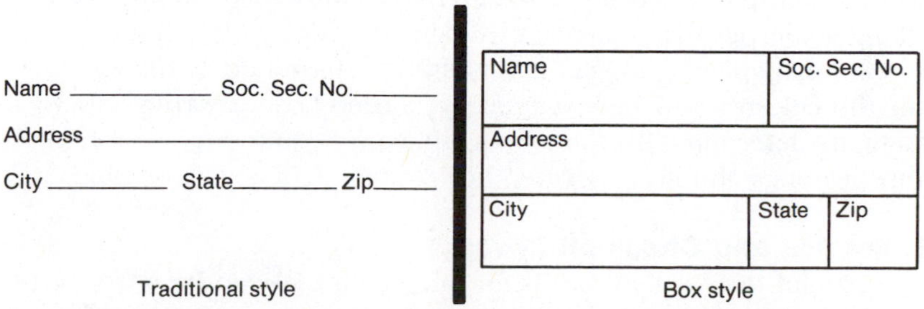

Name ______ Soc. Sec. No. ______
Address ______
City ______ State ______ Zip ______

Traditional style

Name		Soc. Sec. No.
Address		
City	State	Zip

Box style

FIGURE 10–22 The 3/5 Spacing Method

Name
Soc. Sec. No.
Address
City
State
Zip

Cat. no.	Quan.	DESCRIPTION	Unit price	Total

3
5
1234567890 Pica (10 cpi)
012345678901 Elite (12 cpi)
1 2 3 4 5 Hand spacing for workers (5 cpi)
12345678 Hand spacing for clerks (8 cpi)
1 inch of space
Space 3 lines per inch vertically
Space 5 characters per inch horizontally
cpi = characters per inch

or both. In either case, there must be sufficient space to allow for data capture. A standard is needed.

A commonly used standard, called 3/5 spacing, is illustrated in Figure 10–22. The 3 refers to the number of lines per vertical inch, while the 5 refers to the number of characters that fit in one horizontal inch. This approach is related to spacing for clerks (8 characters per inch or cpi), workers (5 cpi), and printers (10/12 cpi).

There are times when a certain amount of space must have a minimum number of lines. One way of determining lines is with the *diagonal spacing* method. As illustrated in Figure 10–23, a 4-inch space is divided into nine writing lines by placing a ruler diagonally across the space so that the $4\frac{1}{2}$-inch mark, (equaling nine half-inch spaces) spans the 4-inch area. In this case, $\frac{1}{2}$-inch multiples are used. A one-inch multiple can be obtained simply by making a sharper diagonal slant of the ruler.

In columnar spacing, the column width is determined by the amount of data in the column and how data are recorded. The 3/5 rule should be adequate to determine the columnar spacing required on most forms. Column headings should be written horizontally whenever possible.

Ballot Box and Check-off Designs

Using ballot or check-off boxes for questions that can be answered by yes or no can greatly reduce the amount of required writing. The user

FIGURE 10–23 The Diagonal Method of Spacing

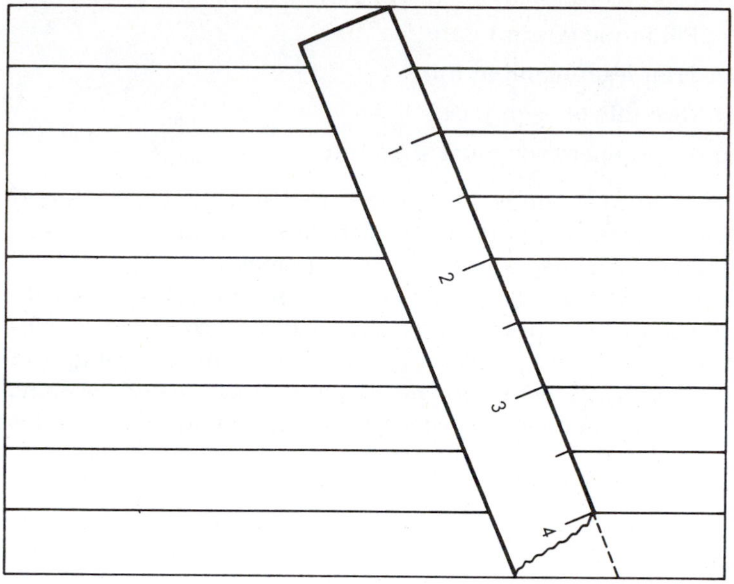

indicates a preference simply by checking off the desired box or placing an X in the appropriate space Figure 10–24 illustrates both designs.

Form Instructions

A well-designed form with clearly stated captions should be self-instructing. In a recent consulting job, an eight-page procedure included two

FIGURE 10–24 Ballot Box and Check-off Designs—Examples

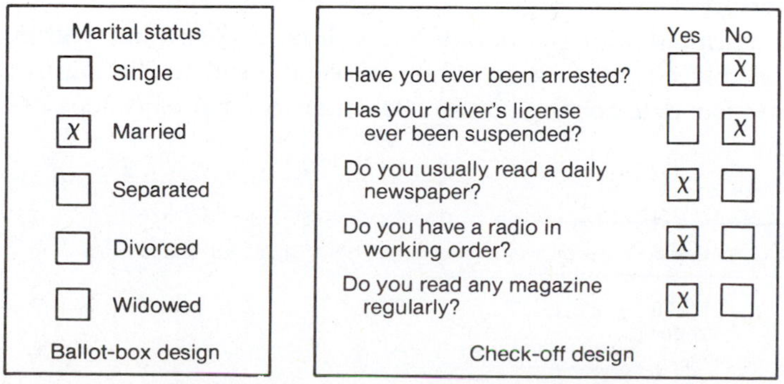

Marital status

- [ ] Single
- [X] Married
- [ ] Separated
- [ ] Divorced
- [ ] Widowed

Ballot-box design

	Yes	No
Have you ever been arrested?	☐	☒
Has your driver's license ever been suspended?	☐	☒
Do you usually read a daily newspaper?	☒	☐
Do you have a radio in working order?	☒	☐
Do you read any magazine regularly?	☒	☐

Check-off design

pages telling how to fill out the printed forms. A sample of the instructions is as follows:

Date: Fill in the current date.
Name: Print legal name in full.
Description: Give title of each part.
Signature: Your supervisor must sign here.

The first form had 29 captions. The procedure listed the captions and explained the information required under each. Much of this work would have been eliminated if the captions were self-explanatory.

Forms such as application blanks that are filled out once should be self-instructing. Other forms (e.g., purchase orders) that are processed repeatedly by several people should be designed for easy writing, typing, and sorting. A form becomes self-instructing by means of clearly stated captions and brief, but specific, procedural instructions. The following examples illustrate these points.

Purpose	Unclear Caption	Self-Instructing Caption
Date of mailing the form	Date	Today's date
Number of copies required	Purchase copies	Number of purchase copies required
Quantity to be shipped	Quantity	Quantity shipped
Inspector of unit shipped	Inspector	Inspected by

The procedural instructions on the form should be self-explanatory—for example, "Enclose your stub with your payment."

Instructions are placed on the cover of padded forms or in the stub area of snap-out forms. Some forms have instructions on the back. A notation on the front upper left corner of the page tells the user where to find the instructions.

Paper Selection

Forms may be printed on paper of different colors, grades, and weights. Colored paper or colored printing on white paper is used to distinguish among copies and facilitate sorting copies. Common color preferences are listed in the table.

Order of Copy	Color
First	White
Second	Yellow
Third	Pink
Fourth	Blue
Fifth	Buff (or goldenrod)

Paper weight is based on a ream of 500 sheets that are 17 by 22 inches; it ranges from 4 pounds to 220 pounds (see Figure 10–25). Cutting the 500 sheets into quarters results in the standard size of a typewriter page—8½ by 11 inches.

There are three major factors to consider in paper selection: appearance, longevity, and handling. The form designer needs to know (1) the number of times the form will be handled (daily, weekly, etc.), (2) the amount of folding it will receive, and (3) the extent of exposure to the environment.

Paper is generally classified as onionskin, bond, duplicator, ledger, index, and card stock. Its thickness is expressed in pound weight. *Onionskin* paper (9-pound weight) is used for inner copies of multiple-part sets. *Bond* paper is usually rag paper that has the best feel and quality, depending on the rag content (25, 50, or 100 percent). *Duplicator* paper (16–20-pound weight) is used for duplicating and xeroxing machines. *Ledger* paper (28–32-pound weight) is used for checks, accounting records, and permanent ledger cards. *Index* paper (more than 72-pound weight) is strictly for printing cards. *Card stock* is the heaviest paper, although it has a lower grade than the other types mentioned. It comes in various weights, ranging from 90 to 140 pounds. The 90-pound weight is used for durable records. The 140-pound weight is strictly heavyweight card stock.

Cost Considerations

Various cost factors go into the final decision to produce a form. Costs consist of both one-time (flat) and running costs. Flat charges center around the preparation of the system used to create the first copy. Charges such as the cost of paper, ink, machine, and labor are all running charges. One way of reducing costs is to order "two-up" or side-by-side forms attached by a perforated line. Other cost-reducing alternatives are:

1. Using standard size and weight paper.

FIGURE 10–25 Bond Paper Weight

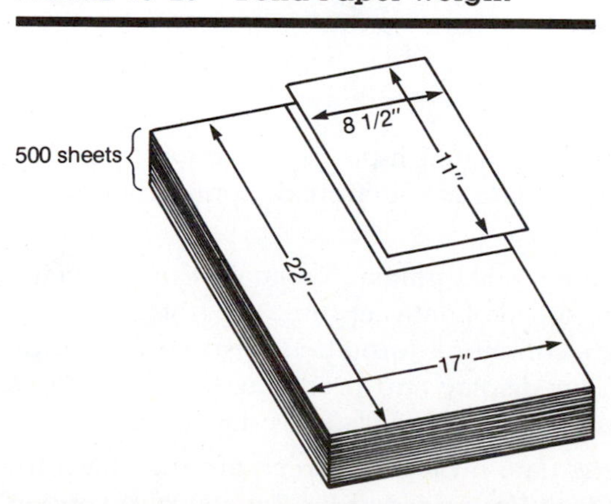

2. Ordering in larger quantities at a discount.
3. Discouraging the use of unnecessary color.
4. Using standard locations for key captions and form title.

Forms Control

The first step in forms control is to determine whether a form is necessary. Managing the hundreds of forms in a typical organization requires a forms control program. Forms control is a procedure for (1) providing improved and effective forms, (2) reducing printing costs, and (3) securing adequate stock at all times.

The first step in a procedure for forms control is to collect, group, index, stock, and control the forms of the organization. Each form is identified and classified by the function it performs and whether it is a flat form, a snap-out form, or something else. Once classified, a form is evaluated by the data it requires, where it is used, and how much it overlaps with other forms. The object is to get rid of unnecessary forms and improve those forms that are necessary.

Before launching a forms control program, the designer needs to consider several questions:

1. Who will be responsible for improving forms design?
2. Should forms be produced in house or assigned to an outside printer?
3. What quantity should be printed? What is the break-even point on printing forms?
4. How much lead time is required to replenish forms?
5. How will one handle reorders? Who will initiate them? In what way?
6. How will obsolete forms be handled?
7. What should be the life of the form?
8. Where and how should the form be stored?

If questions of this nature are not addressed in advance, the organization is probably not ready to launch a forms control program.

Summary

1. Input design shows how user-originated input is converted to a computer-based format. After input data are identified, appropriate input media are selected for processing.
2. In addition to determining the record media, the analyst must decide how input is entered and the speed of data capture. The fastest method is online data entry through a menu, a formatted form, or a prompt. This requires a CRT screen for display and predefined user's options that standardize data capture and provide visual verification.
3. Two approaches to designing data on a CRT screen are manual print layout charts and a software utility provided by the software vendor.

The latter method is more versatile, allowing for instant modifications directly on the screen.

4. Standards for printed output suggest giving each output a name or title, providing a sample of the output layout, and specifying the procedure for proving the accuracy of the output data. The output devices to consider depend on the compatibility of the device with the system, response time requirements, and the printed quality required.
5. A form is the physical carrier of data. It can carry authority for action. It is classified by what it does in the system. There are actions, memory, and report forms. In any case, a form is a tool with a message.
6. In designing a form, attention is given to proper identification and wording, readability and use, composition and layout, order of data items, and clarity of instructions. For form duplication, one-time carbon or carbon-backed paper is chosen, depending on how the form will be used.
7. Forms are classified as flat (single copy), snapout, fanfold, NCR, and preprinted. Fanfold forms are ideal for computer output. NCR is cleaner but costs more than carbon-interleaved forms.
8. A well-designed form with clearly stated captions should be self-instructing. If instructions are needed, they are placed on the cover of padded forms, in the stub area of snapout forms, or in some cases on the back of the form.
9. Forms can be printed on paper of different colors, grades, and weights. Color distinguishes among copies. In deciding on the kind of paper to select, the forms designer must evaluate appearance, longevity, and handling. These factors are considered against cost.
10. An organization's forms must be centrally controlled for efficient handling. Some planning is required prior to implementation. If forms are handled successfully, unauthorized forms should be minimized.

Key Words

Ballot Box Design
Caption
Computer Output Microfilm (COM)
Fanfold Form
Form
Forms Design
Magnetic Ink Character Recognition (MICR)
Menu
NCR Paper
Prompt
Rule
Snapout Form
3/5 Space

Review Questions

1. What is the goal of input design? Output design?
2. If you were asked to adopt a method for tagging merchandise for a retail store, what input medium would you choose? Why?

3. What is unique about online data entry? What role does a CRT terminal play for input and output?
4. Explain briefly three approaches for data entry.
5. How would one design data outputs on a CRT screen? Illustrate.
6. What is a form? Summarize the characteristics of action, memory, and report forms.
7. In what respect is carbon paper a form copier? How is it classified?
8. Distinguish between the following:
   a. Snapout and fanfold forms.
   b. Rule and caption.
   c. Ballot box and check-off design.
9. What are the abbreviations of the following captions:
   a. Accumulate.
   b. Insurance.
   c. Manufacturing.
   d. Purchase order.
   e. Reference.
10. Illustrate how a 5-inch space can be divided into 13 writing lines.
11. How much instruction does a form need? Are written instructions required on a printed form? Explain.
12. What factors determine paper selection? Explain.
13. If you were asked to develop a forms control program for your firm, how would you proceed? How would you control for unauthorized forms?

Application Problems

1 Review the form shown in Exhibit 10–1 and complete the following assignment:

a. What flaws are there in the form? Be specific.

b. Develop an updated version of the form and explain in detail the basis for the change.

2 The manager of the purchasing department asked the firm's analyst to design a new purchase requisition form. As a first step, the analyst reviewed the existing form and ones used by other firms with which she had previously consulted. She developed a rough draft showing the

EXHIBIT 10-1 Application Form—An Example

UNIVERSITY OF VIRGINIA
McINTIRE SCHOOL OF COMMERCE

SPECIAL STUDENT APPLICATION

Name ______________________________ Soc. Sec. No. ____________
(Last) *(First)* *(Middle)*

1. I hereby apply for admission to the McIntire School of Commerce as a Special Student for the semester beginning ____________________ .

2. Permanent address ______________________ Phone __________

3. Present address ______________________ until __________

4. Date of birth __________ Place of Birth __________ Sex* __________

5. Are you a veteran? ________ SAT scores: Verbal ________ Math ________

6. What is your academic average? ____________________

7. List below all accounting and business-related courses you have completed and where they were taken:

____

____

____

____

____

____

8. List below all colleges you have attended, with the dates of attendance at each and degree(s) awarded. (Attach transcript if other than U. Va.)

College and Location Degree Dates attended

____

____

____

____

* The McIntire School of Commerce does not discriminate with regard to sex; information requested is for reports the university makes and provides to federal agencies collecting data to assure equal opportunity.

EXHIBIT 10–1 *(concluded)*

9. Professional goals (a brief statement of your purpose for taking courses at the McIntire School of Commerce).

____

____

____

____

____

____

____

10. Extracurricular and other activities participated in (intramurals, employment, professional affiliations, etc.)

____

____

____

____

____

Attach here a recent small photograph of yourself, preferably full face. Write your name on the back of the photo. Attach along upper edge only, using paste or staples.

11. Signature ______________________

12. Date ______________________

13. Potential area of concentration:
(Circle one)

Accounting Marketing Finance

Management Information Systems

Organizational Management

Please check to see that this form has been filled in completely. Mail to Associate Dean, McIntire School of Commerce, 111 Monroe Hall, University of Virginia, Charlottesville, Virginia 22903, no later than March 1 for September admission.

items and their locations. The draft was then shown to employees in charge of purchase requisition and those in accounting who process purchase orders.

The analyst was surprised to find that each person had a different view of what should go in the form and how it should be laid out. After much debate, she had everyone agree on the overall content. Using her experience and the feedback, she went through a second draft and laid out the items so that the form was easy to read and fill out. Certain areas were set in boldface type for emphasis. A boxed design was introduced to make it easy to find where the answer should be entered. Filing information was placed close to the top of the form.

The form's width and length conformed to standard paper and cabinet size for filing. The spacing also conformed to the company's typewriters, which were pica (12 characters per inch). The typewriters were standard electric units about nine years old.

The form consisted of an original and three copies attached, with one-time carbon in between. Since the form carried essentially the same information, no specific instructions were printed on it. As a final touch, a border was added to give it an official look.

The analyst produced a final draft and showed it to an experienced forms sales person and a systems analyst who works for a competing firm. An order was placed for 10,000 copies of the form. A few weeks after the form was in use, it turned out to be a disaster. The analyst did a thorough job, but some things were apparently missing.

Assignment

a. What major steps were overlooked in the overall design of the form? Expain.

b. If you were the analyst, how would you have handled the project? Elaborate.

---

3 Examine the form shown in Exhibit 10–2 on p. 318. List the items that need to be changed. How would you change them? Why?

EXHIBIT 10-2 Invoice Form—An Example

U.Va. AO-001 June/81

UNIVERSITY OF VIRGINIA

INTER-DEPARTMENTAL TRANSFER INVOICE

Supplied to: CHARGE	Supplied by: CREDIT
Department:	Department:
Messenger Address:	Messenger Address:
	Preparation Date:

DATE OF DELIVERY OR SERVICE	ARTICLES OR SERVICES AND DESCRIPTION	QUANTITY	UNIT	UNIT PRICE	AMOUNT
	DEPARTMENTAL APPROVAL →				

1-11 CHARGE CODE	50-59 AMOUNT	69-79 CREDIT CODE	80-88 I.D. NO.	92-99 THIRD REF.

DO NOT WRITE BELOW THIS LINE — ACCOUNTING OPERATIONS USE ONLY

61-68 VOUCHER NO.	22-25 DATE	
		Signature of Approving Officer

White, Green & Canary - Voucher   Pink & Goldenrod - Departmental Use

Selected References

Filkins, James M. "Forms Analysis—More Than Just Design." *Information and Records Management*, August 1978, p. 18ff.

Kirk, Nathan. "Legal Restraints in Forms Design." *Journal of Systems Management*, June 1979, pp. 38–42.

Koeneke, W. "Forms Control—Fortune or Flop?" *Journal of Systems Management*, January 1981, pp. 11–14.

Mathies, Leslie H., and Gibbs Myers. "Good Forms Design Is Vital to Management." *The Office*, September 1981, p. 76ff.

Myers, Gibbs. "Forms Management—Part 2: How to Design Business Forms." *Journal of Systems Management,* October 1976, pp. 15–19.

Nussbeck, Dan. "Forms Control at Hallmark: A Philosophy as Well as a System." *Information and Records Management,* August 1978, pp. 16–17.

Seager, Dave. "The Ten Commandments of Forms Design." *Canadian Datasystems,* October 1977, pp. 40–45.

Stubbs, Jack. "Forms Are Main User Connection." *Data Management,* June 1977, pp. 12–18.

Chapter 11

File Organization and Data Base Design

At a Glance

After designing the input and output, the designer begins to concentrate on file design or how data should be organized around user requirements. How data are organized depends on the data and response requirements that determine hardware configurations. File organization may be sequential, indexed-sequential, inverted list, or random. Each method has its pros and cons.

An integrated approach to file design is the data base. The general theme is to handle information as an integrated whole, with a minimum of redundancy and improved performance. Software languages are used to manipulate, describe, and manage data. Regardless of the type of data structure used, the objectives of the data base are accuracy and integrity, successful recovery from failure, privacy and security of data, and good overall performance.

By the end of this chapter, you should know:

1. Alternative methods of file organization.
2. Objectives of a data base.
3. Types of data structures.
4. The difference between schemas and subschemas.
5. How to normalize files.
6. The role of the data base administrator.

INTRODUCTION

Perhaps the most important aspect of building candidate systems is file design. After the input, output, and various forms are designed, files and the data they store must be organized according to user requirements and the constraints of the hardware and operating system. Some applications are processed daily and affect each record in the file. Such records are handled in sequential order when tape is the appropriate medium. In contrast, records that are updated, accessed, or retrieved at random are stored on disks or diskettes. How a file is organized has a lot to do with the nature of the data and the response requirements that determine hardware configurations. This chapter reviews the ways files are organized and data base design for implementation success.

FILE STRUCTURE

To learn about files, we need to understand basic terms used to describe the file hierarchy. The terms we shall cover are *byte, data item, record, file,* and *data base* (see Figure 11–1).

Byte. A byte is an arbitrary set of eight bits that represent a character. It is the smallest addressable unit in today's computers.

Data item (element). One or more bytes are combined into a data item to describe an attribute of an object. For example, if the object is an employee,

FIGURE 11–1 Hierarchy of Files

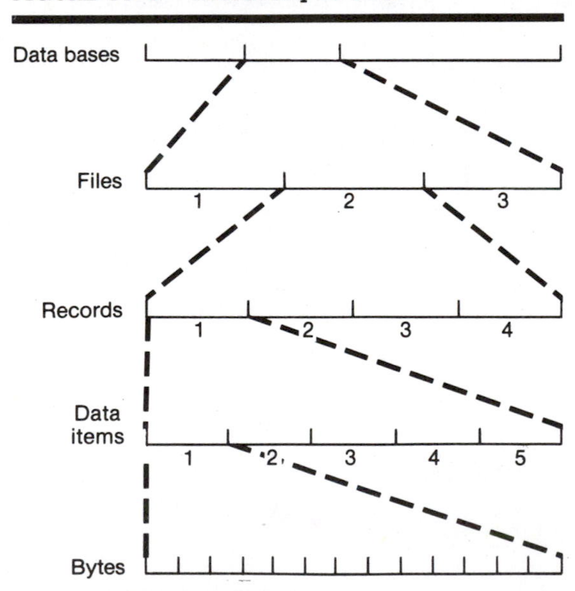

one attribute may be sex, name, age, or social security number. A data item is sometimes referred to as a *field*. A field is actually a physical space on tape or disk, whereas a data item is the data stored in the field.

Record. The data items related to an object are combined into a record. A hospital patient (object) has a record with his/her name, address, health insurance policy, and next of kin. Each record has a unique key or ID number. The patient's tag number, insurance policy number, or a unique number could be used as an identifier for processing the record.

In record design, we distinguish between logical and physical records. A *logical* record maintains a logical relationship among all the data items in the record. It is the way the program or user sees the data. In contrast, a *physical* record is the way data are recorded on a storage medium. To illustrate, Figure 11–2 shows a programmer who requires a five-record file in a particular sequence (A, D, C, B, S). The programmer does not know about the physical "map" on the disk. The software presents the logical records in the required sequence. This capability is unique to data base design.

File. A collection of related records makes up a file. The size of a file is limited by the size of memory or the storage medium. Two characteristics determine how files are organized: activity and volatility. *File activity* specifies the percentage of actual records processed in a single run. If a small percentage of records is accessed at any given time, the file should be organized on disk for direct access. In contrast, if a fair percentage of records is affected regularly, then storing the file on tape would be more efficient and less costly. *File volatility* addresses the properties of record changes. File records with substantial changes are highly volatile, meaning that disk design would be more efficient than tape. Think of the airline reservation system and the high volatility through cancellations, additions, and other transactions compared to the traditional payroll, which is relatively dormant. The higher the volatility, the more attractive is disk design.

Data base. The highest level in the hierarchy is the data base. It is a set of interrelated files for real-time processing. It contains the necessary data for problem solving and can be used by several users accessing data concurrently. Data bases are covered later in the chapter.

FILE ORGANIZATION

A file is organized to ensure that records are available for processing. As mentioned earlier, it should be designed in line with the activity and volatility of the information and the nature of the storage media and devices. Other considerations are (1) cost of file media (highest for disk, lowest for tape), (2) inquiry requirements (real-time versus batch processing), and (3) file privacy, integrity, security, and confidentiality.

There are four methods of organizing files: sequential, indexed-sequential, inverted list, and direct access. Each method is explained.

FIGURE 11-2 Physical and Logical Records—A Contrast

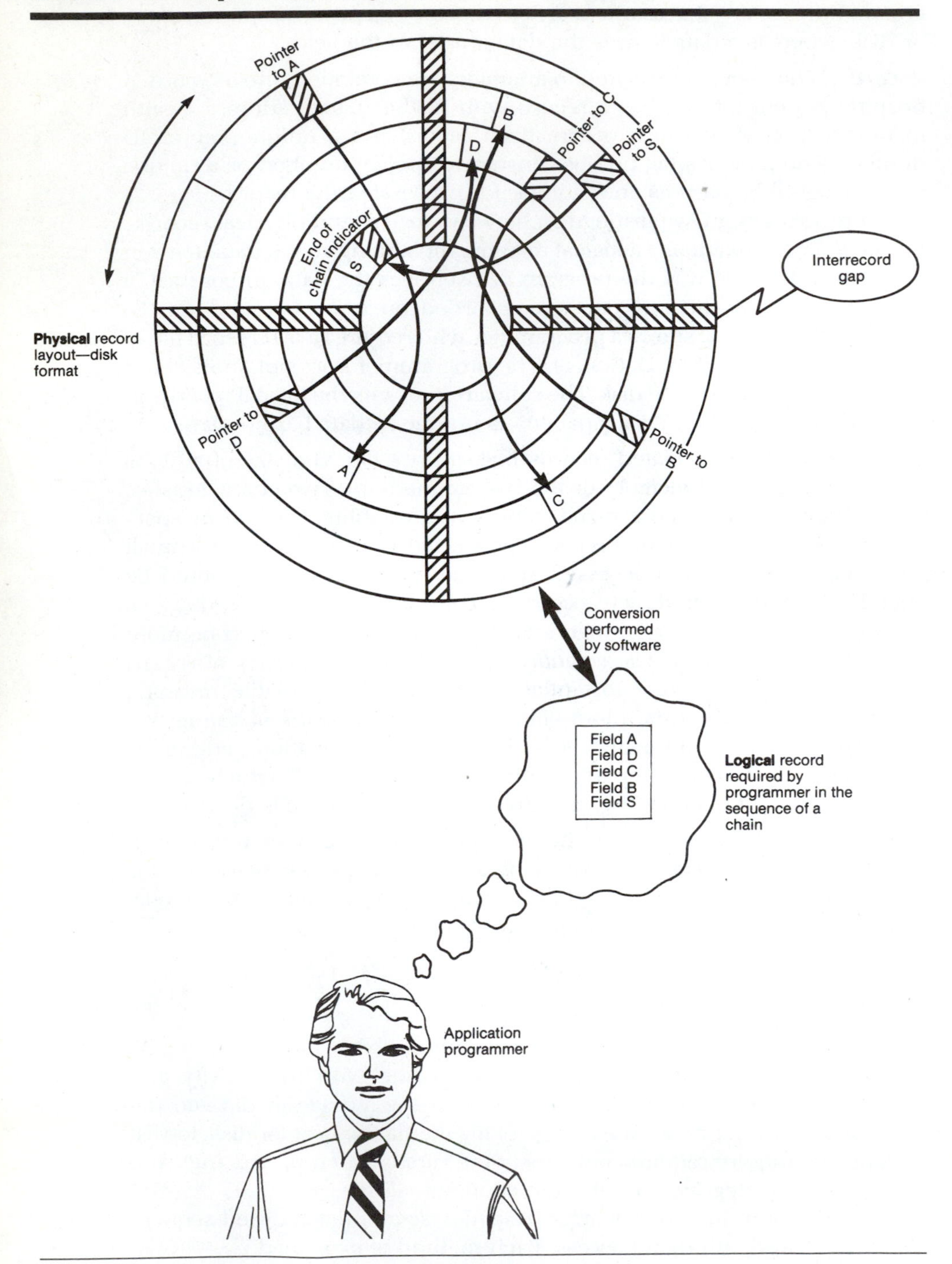

Sequential Organization

Sequential organization simply means storing and sorting in physical, contiguous blocks within files on tape or disk. Records are also in sequence within each block. To access a record, previous records within the block are scanned. Thus sequential record design is best suited for "get next" activities, reading one record after another without a search delay (see Figure 11–3).

In a sequential organization, records can be added only at the end of the file. It is not possible to insert a record in the middle of the file without rewriting the file. In a data base system, however, a record may be inserted anywhere in the file, which would automatically resequence the records following the inserted record. Another approach is to add all new records at the end of the file and later sort the file on a key (name, number, etc.). Obviously, in a 60,000-record file it is less time-consuming to insert the few records directly than to sort the entire file.

In a sequential file update, transaction records are in the same sequence as in the master file. Records from both files are matched, one record at a time, resulting in an updated master file. As shown in Figure 11–4, the system changes the customer's city of residence as specified in the transaction file (on floppy disk) and corrects it in the master file. A "C" in the record number specifies "replace"; an "A," "add"; and a "D," "delete."

In a personal computer with two disk drives, the master file is loaded on a diskette into drive A (left), while the transaction file is loaded on another diskette into drive B. Updating the master file transfers data from drive B to A, controlled by the software in memory.

Indexed-Sequential Organization

Like sequential organization, keyed sequential organization stores data in physically contiguous blocks. The difference is in the use of indexes to locate records. To understand this method, we need to distinguish among three areas in disk storage: prime area, overflow area, and index area. The *prime* area contains file records stored by key or ID numbers. All records are initially stored in the prime area. The *overflow* area contains records added to the file that cannot be placed in logical sequence in the prime area. The *index* area is more like a data dictionary. It contains keys of records and their locations on the disk. A pointer associated with each key is an address that tells the system where to find a record.

Figure 11–5 illustrates an airline reservation file. The index area contains pointers to the Chicago and Houston flights. The Chicago flight points to the Chicago flight information stored in the prime area. The Houston flight points to the Houston flight information in the prime area. Lack of space to store the Huntsville flight in sequential order made it necessary to load it in the overflow area. The overflow pointer places it logically in

FIGURE 11-3 Sequential Organization—An Example

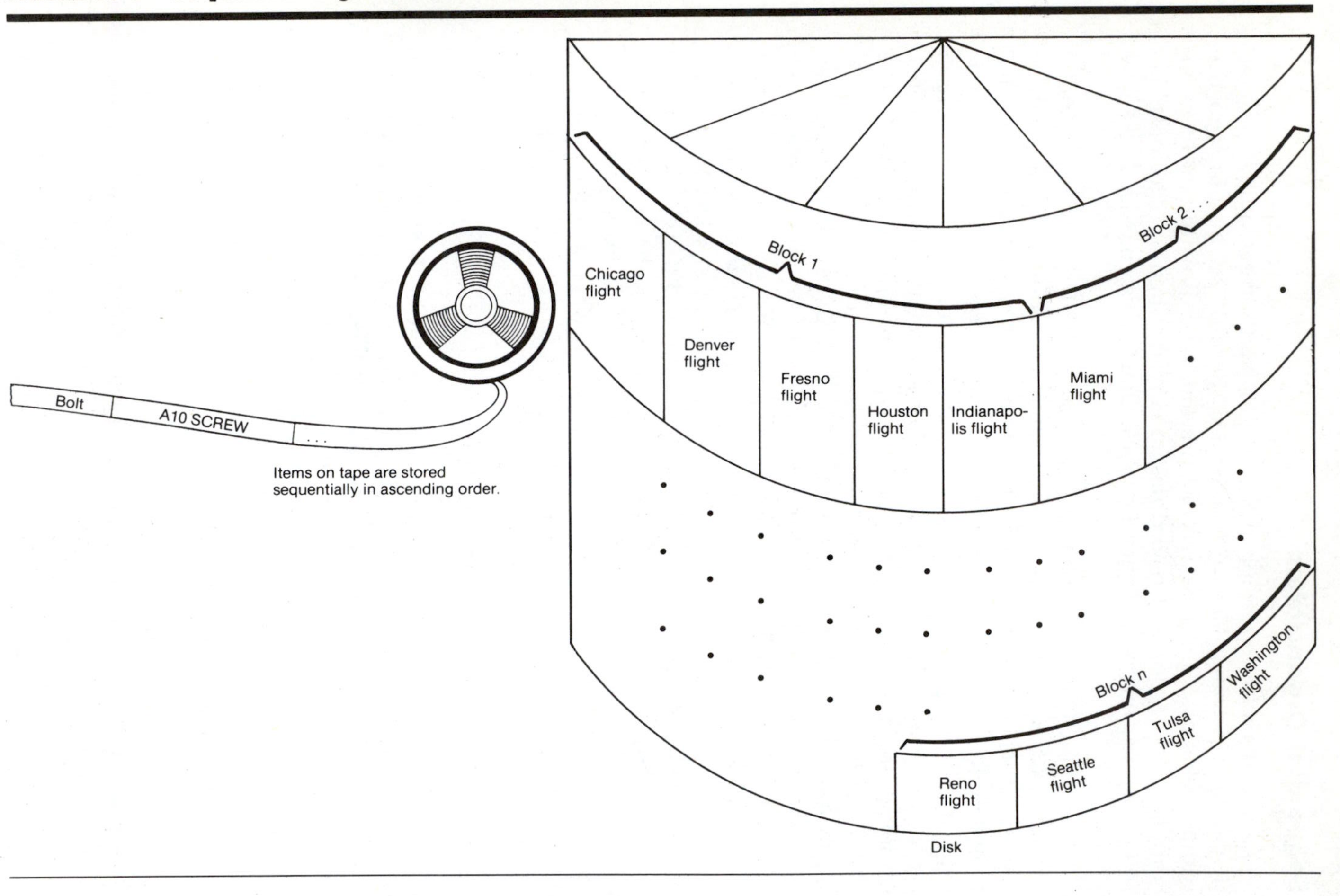

FIGURE 11-4 Updating Sequential File

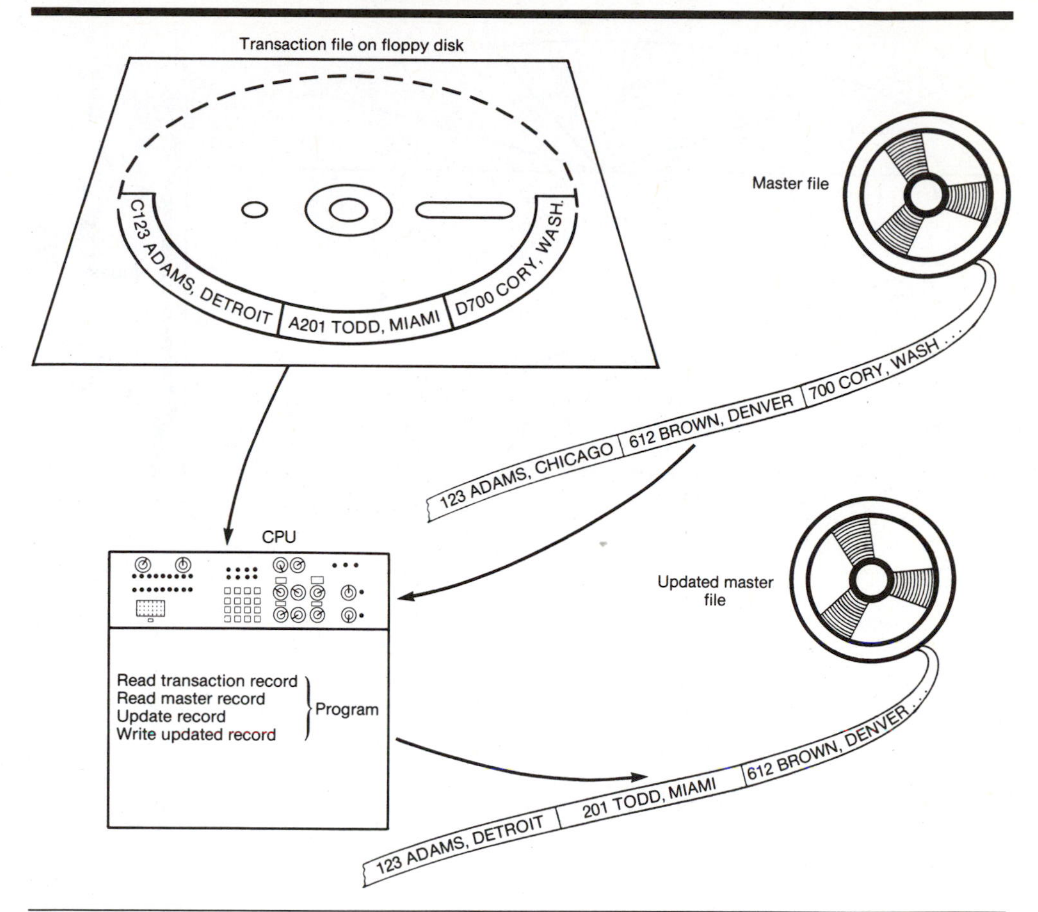

sequential order in the prime area. The same arrangement applies to the Louisville flight.

Indexed-sequential organization reduces the magnitude of the sequential search and provides quick access for sequential and direct processing. The primary drawback is the extra storage space required for the index. It also takes longer to search the index for data access or retrieval.

Chaining

File organization requires that relationships be established among data items. It must show how characters from fields, fields form files, and files relate to one another. Establishing relationships is done through chaining or the use of pointers. Figure 11-5 showed how pointers link one record to another. Figure 11-6 is a list of auto parts—mufflers, windshields, fenders—stored on disk. The file is a indexed-sequential access file sequenced by key

FIGURE 11-5 Indexed-Sequential Organization

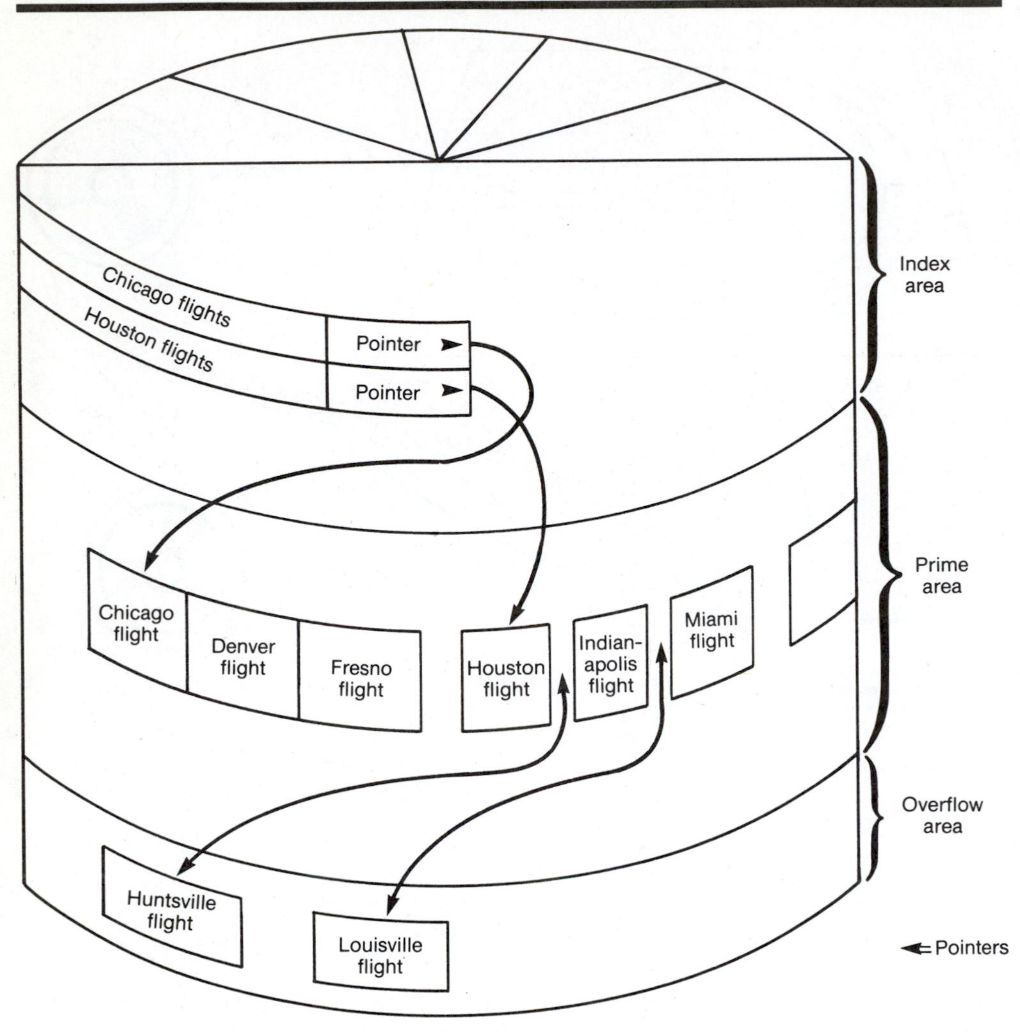

1 that reports part numbers. A record is retrieved by part number. To retrieve all the fenders sold, we can scan the file and determine the total. However, a better way is to chain the records by linking a pointer to each. The pointer gives the address of the next part type of the same class. In the case of fenders, the first fender has a pointer that indicates 092, the address (part number) of the next fender in the second table. Fender 092, in turn, has a pointer that indicates 172, the address of the third fender in the fourth table. Fender 172 has a pointer that indicates END, signaling the last fender available. The search method applies similarly to other parts in the file. A graphic method of linking or chaining records is shown by the connecting lines in the figure.

FIGURE 11-6 Use of Pointers in Chaining Records

Address	Key 1: Part #	Key 2: Part Type	Pointer
101	014	muffler	076
102	021	windshield	081
103	022	fender	092
104	073	tire	130
105	076	muffler	112
106	080	radiator	116
107	081	windshield	090
201	090	windshield	199
202	092	fender	172
203	097	water pump	150
204	112	muffler	168
205	116	radiator	END
301	130	tire	165
302	149	horn	173
303	150	water pump	177
304	165	tire	END
401	168	muffler	END
402	172	fender	END
403	173	horn	198
404	177	water pump	END
405	198	horn	END
406	199	windshield	END

Inverted List Organization

Like the indexed-sequential storage method, the inverted list organization maintains an index. The two methods differ, however, in the index level and record storage. The indexed-sequential method has a multiple index for a given key, whereas the inverted list method has a single index for each key type. In an inverted list, records are not necessarily stored in a particular sequence. They are placed in the data storage area, but indexes are updated for the record keys and location.

Data for our flight reservation system are shown in Figure 11-7. The flight number, flight description, and flight departure time are all defined as keys, and a separate index is maintained for each. Note that in the data location area, flight information is in no particular sequence. Assume that a passenger needs information about the Houston flight. The agent requests the record with the flight description "Houston flight." The data base management system (DBMS) then reads the single-level index sequentially until it finds the key value for the Houston flight. This value has two records associated with it: R3 and R6. The DBMS essentially tells the agent that flight #170 is departing at 10:10 A.M. (R3) and flight #169 is departing at 8:15 A.M. (R6).

FIGURE 11-7 Flight Reservation System—Partial View

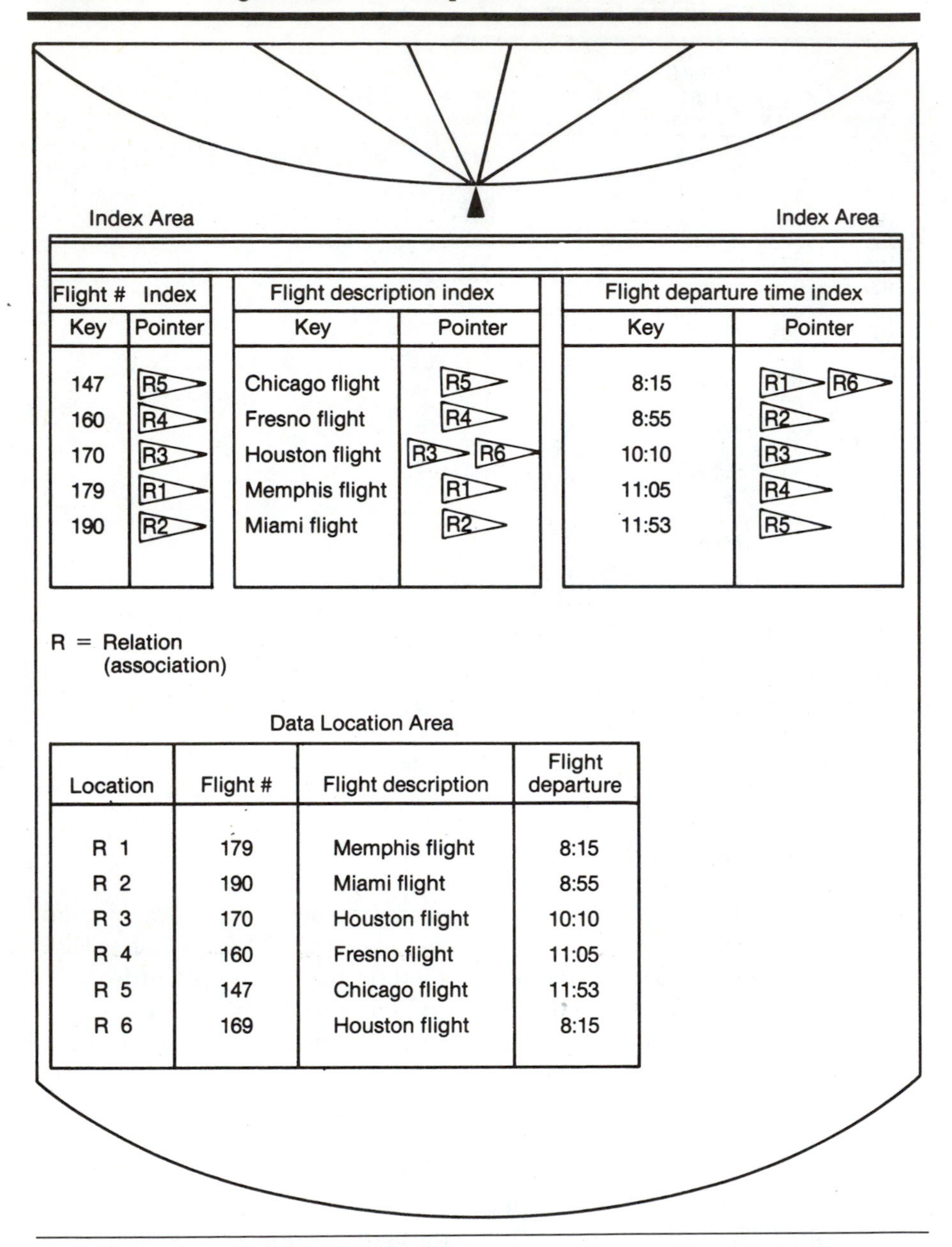

Looking at inverted-list organization differently, suppose the passenger requests information on a Houston flight that departs at 8:15. The DBMS first searches the flight description index for the value of the "Houston flight." It finds R3 and R6. Next it searches the flight departure index for these values. It finds that the R3 value departs at 10:10, but the R6 value departs at 8:15. The record at location R6 in the data location area is displayed for follow-up.

It can be seen that inverted lists are best for applications that request specific data on multiple keys. They are ideal for static files because additions and deletions cause expensive pointer updating.

Direct-Access Organization

In direct-access file organization, records are placed randomly throughout the file. Records need not be in sequence because they are updated directly and rewritten back in the same location. New records are added at the end of the file or inserted in specific locations based on software commands.

Records are accessed by addresses that specify their disk locations. An address is required for locating a record, for linking records, or for establishing relationships. Addresses are of two types: absolute and relative. An *absolute* address represents the physical location of the record. It is usually stated in the format of sector/track/record number. For example, 3/14/6 means go to sector 3, track 14 of that sector, and the sixth record of the track. One problem with absolute addresses is that they become invalid when the file that contains the records is relocated on the disk. One way around this is to use pointers for the updated records.

A *relative* address gives a record location relative to the beginning of the file. There must be fixed-length records for reference. Another way of locating a record is by the number of bytes it is from the beginning of the file (see Figure 11–8). Unlike relative addressing, if the file is moved, pointers need not be updated, because the relative location of the record remains the same regardless of the file location.

Each file organization method has advantages and limitations; a summary is given in Figure 11–9. Many applications by their nature are best done sequentially. Payroll is a good example. The system goes through the employee list, extracts the information, and prepares pay slips. There are no lengthy random-access seeks. In contrast, real-time applications where response requirements are measured in seconds are candidates for random-access design. Systems for answering inquiries, booking airlines or stadium seats, updating checking or savings accounts in a bank, or interacting with a terminal are examples for random-access design.

DATA BASE DESIGN

A decade ago, *data base* was unique to large corporations with mainframes. Today it is recognized as a standard of MIS and is available for virtually every size of computer. Before the data base concept became operational, users had programs that handled their own data independent of other users. It was a conventional file environment with no data integration or sharing of common data across applications. In a data base environment, common data are available and used by several users. Instead of each program (or user) managing its own data, data across applications are

FIGURE 11-8 Absolute and Relative Addressing—An Example

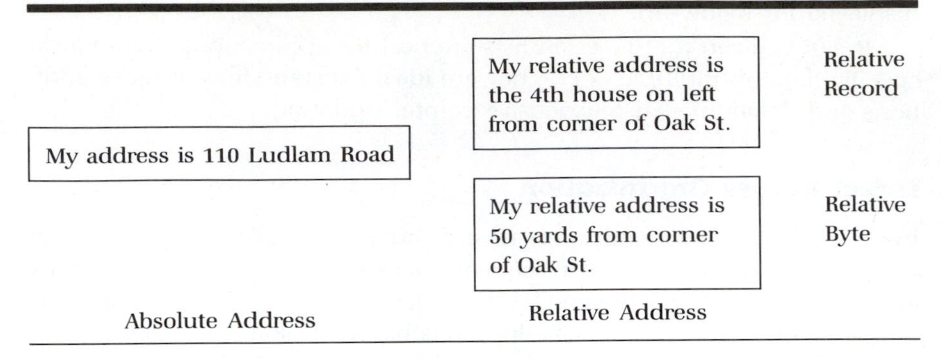

shared by authorized users with the data base software managing the data as an entity. A program now requests data through the data base management system (DBMS), which determines data sharing (see Figure 11–10).

Objectives of Data Base

The general theme behind a data base is to handle information as an integrated whole. There is none of the artificiality that is normally embedded in separate files or applications. A data base is a collection of *interrelated* data stored with *minimum redundancy* to serve many users quickly and efficiently. The general objective is to make information access easy, quick, inexpensive, and flexible for the user. In data base design, several specific objectives are considered:

FIGURE 11-9 File Organization Methods—A Summary

Method	Advantages	Disadvantages
Sequential	Simple to design Easy to program Variable length and blocked records available Best use of storage space	Records cannot be added to middle of file
Indexed-sequential	Records can be inserted or updated in middle of file Processing may be carried out sequentially or randomly	Unique keys required Processing occasionally slow Periodic reorganization of file required
Inverted list	Used in applications requesting specific data on multiple keys	
Random	Records can be inserted or updated in middle of file Better control over record allocation	Calculating address required for processing Variable-length records nearly impossible to process

FIGURE 11-10 Conventional and DBMS Environments

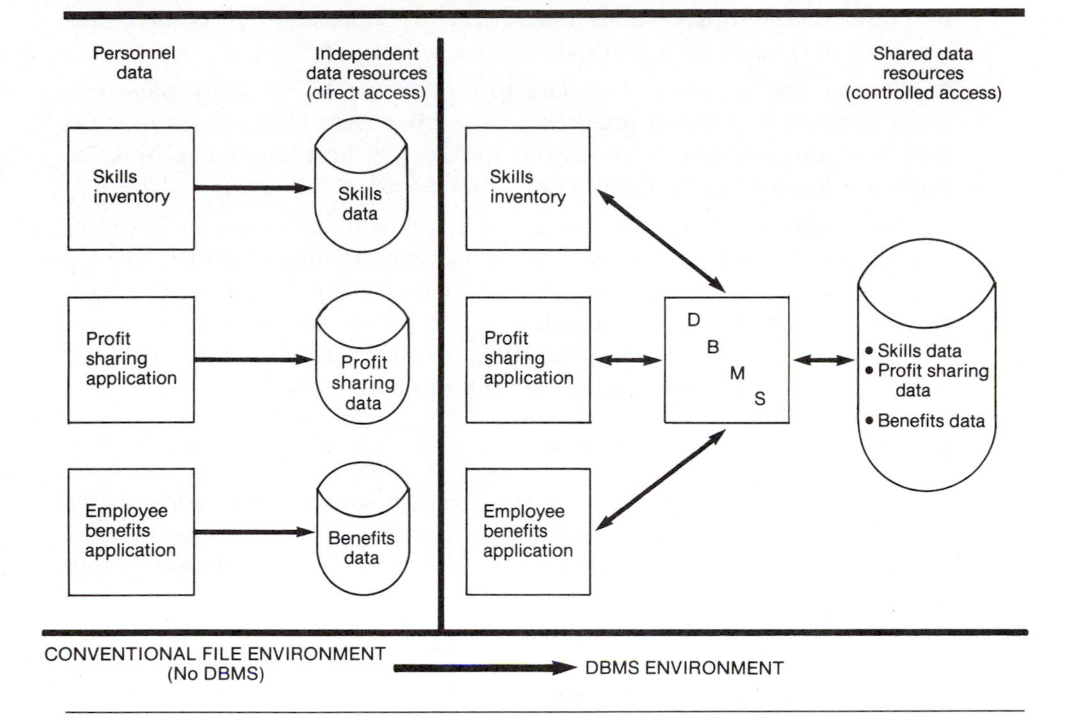

1. *Controlled redundancy.* Redundant data occupies space and, therefore, is wasteful. If versions of the same data are in different phases of updating, the system often gives conflicting information. A unique aspect of data base design is storing data only once, which controls redundancy and improves system performance.

2. *Ease of learning and use.* A major feature of a user-friendly data base package is how easy it is to learn and use. Related to this point is that a data base can be modified without interfering with established ways of using the data.

3. *Data independence.* An important data base objective is changing hardware and storage procedures or adding new data without having to rewrite application programs. The data base should be "tunable" to improve performance without rewriting programs.

4. *More information at low cost.* Using, storing, and modifying data at low cost are important. Although hardware prices are falling, software and programming costs are on the rise. This means that programming and software enhancements should be kept simple and easy to update.

5. *Accuracy and integrity.* The accuracy of a data base ensures that data quality and content remain constant. Integrity controls detect data inaccuracies where they occur.

6. *Recovery from failure.* With multiuser access to a data base, the system must recover quickly after it is down with no loss of transactions. This objective also helps maintain data accuracy and integrity.

7. *Privacy and security.* For data to remain private, security measures must be taken to prevent unauthorized access. Data base security means that data are protected from various forms of destruction; users must be positively identified and their actions monitored.

8. *Performance.* This objective emphasizes response time to inquiries suitable to the use of the data. How satisfactory the response time is depends on the nature of the user–data base dialogue. For example, inquiries regarding airline seat availability should be handled in a few seconds. On the other extreme, inquiries regarding the total sale of a product over the past two weeks may be handled satisfactorily in 50 seconds.

Key Terms

In data base design, we need to be familiar with several terms. Suppose we have a *sales status* system designed to give the sales activities of each salesperson. Using the basic model in Figure 11–11, we run into four terms:

FIGURE 11–11 Conventional File Environment

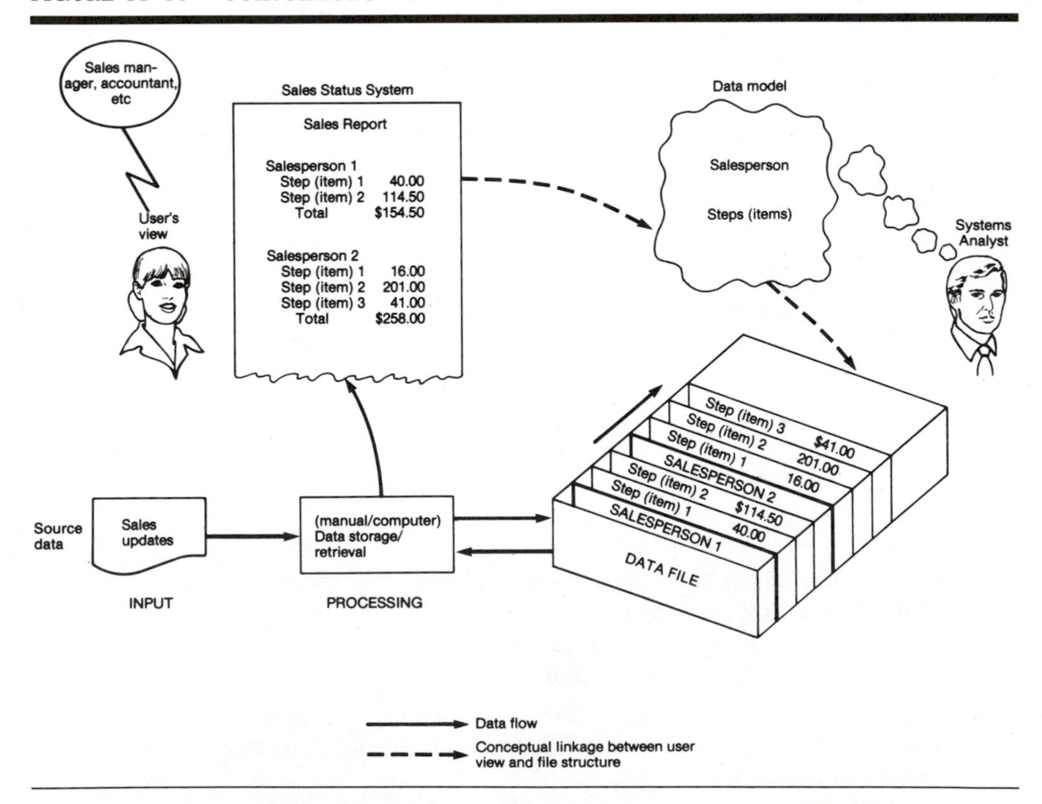

1. *User's view* is a profile that the user expects to see on a report. In our example, the user's view is a sales report.
2. *Processing* refers to the changes made to produce the sales report.
3. *Data model* is a framework or a mental image of the user's view. In Figure 11–11, the data model consists of two steps or entities: the salesperson and the items sold. The steps are determined by the analyst as a framework for producing the user's view—the report.
4. *Data file* is the area where the actual files are stored. The storage sequence is determined by the data model. In our example, the data are organized by salesperson and the items under each salesperson.

In a data base environment, the DBMS is the software that provides the interface between the data file on disk and the program that requests processing. As shown in Figure 11–12, the DBMS stores and *manages* data. The procedure is as follows:

1. The user requests a sales report through the application program. The application program uses a data manipulation language (DML) to tell the DBMS what is required.
2. The DBMS refers to the data model, which describes the view in a language called the data definition language (DDL). The DBMS uses DDL to determine *how* data must be structured to produce the user's view.
3. The DBMS requests the input/output control system (IOCS) to retrieve the information from physical storage as specified by the application program. The output is the sales report.

To summarize,

1. DML *manipulates* data; it specifies *what* is required.
2. DDL *describes how* data are structured.
3. DBMS *manages* data according to DML requests and DDL descriptions.

As you can tell, DBMS performs several important functions:

1. Storing, retrieving, and updating data.
2. Creating program and data independence. Either one can be altered independently of the other.
3. Enforcing procedures for data integrity. Data are immune from deliberate alteration because the programmer has no direct method of altering physical data bases.
4. Reducing data redundancy. Data are stored and maintained only once.
5. Providing security facilities for defining users and enforcing authorizations. Access is limited to authorized users by passwords or similar schemes.
6. Reducing physical storage requirements by separating the logical and physical aspects of the data base.

FIGURE 11-12 Data Base File Environment

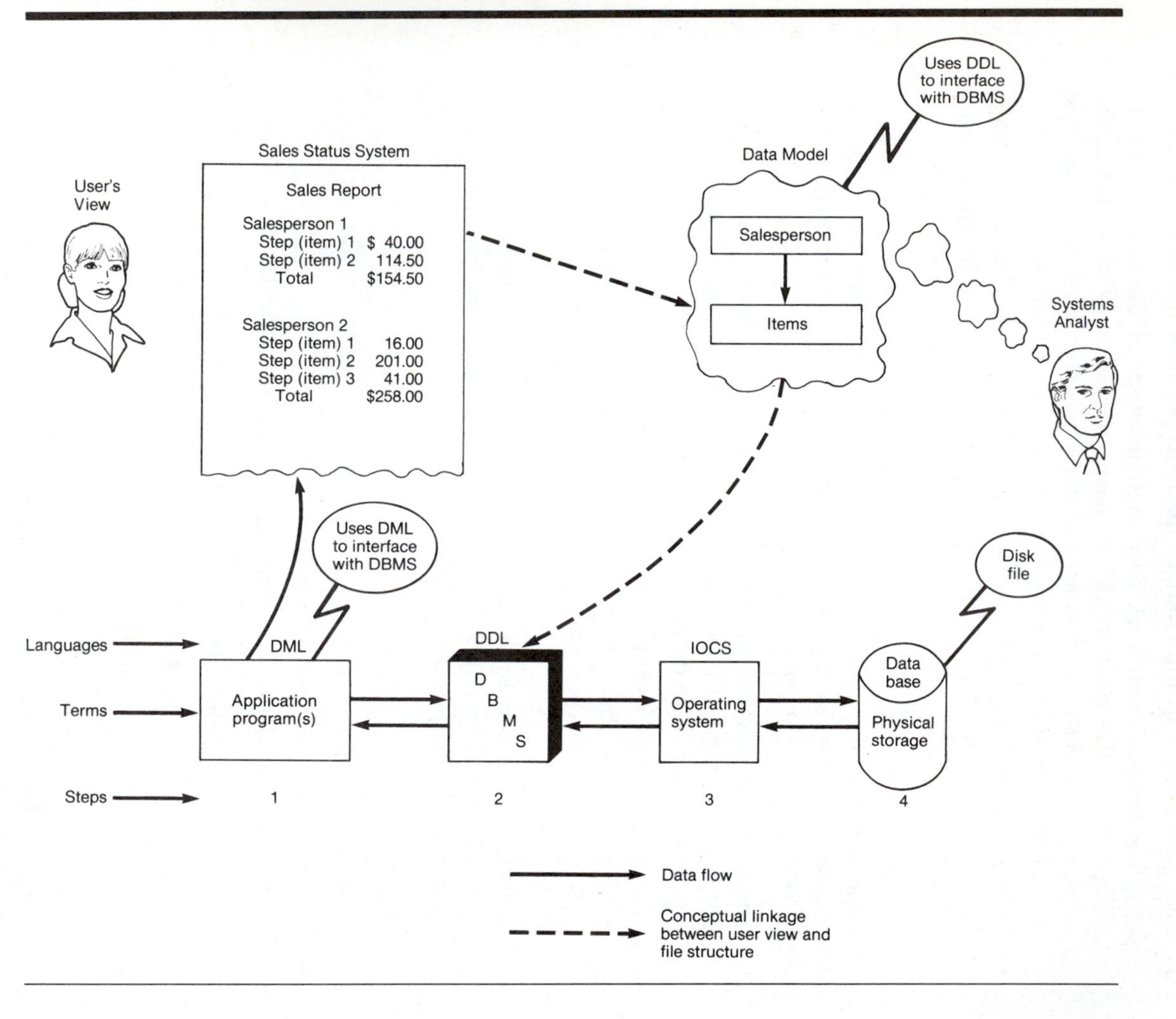

Logical and Physical Views of Data

In data base design, several views of data must be considered along with the persons who use them. In addition to data structuring, where relationships are reflected between and within entities, we need to identify the application program's logical views of data within an overall logical data structure. The *logical* view is what the data look like, regardless of how they are stored. The *physical* view is the way data exist in physical storage. It deals with how data are stored, accessed, or related to other data in storage (see Figure 11–2). Figure 11–13 shows four views of data: three logical and one physical. The logical views are the user's view, the programmer's view, and the overall logical view, called a *schema*.

Schemas and Subschemas

The schema is the view that helps the DBMS decide what data in storage it should act upon as requested by the application program. An example of a schema is the arrival and departure display at an airport. Scheduled flights and flight numbers (schema) remain the same, but the actual departure and arrival times may vary. The user's view might be a particular flight arriving or departing at a scheduled time. How the flight actually takes off or lands is of little concern to the user. The latter view is of *subschema*. It is a programmer's (pilot's) view. Many subschemas can be derived from one schema, just as different pilots visualize different views of a landing approach, although all (it is hoped) arrive at the sheduled time indicated on the CRT screen display (schema).

Different application programmers visualize different subschemas. The

FIGURE 11–13 Four Views of Data

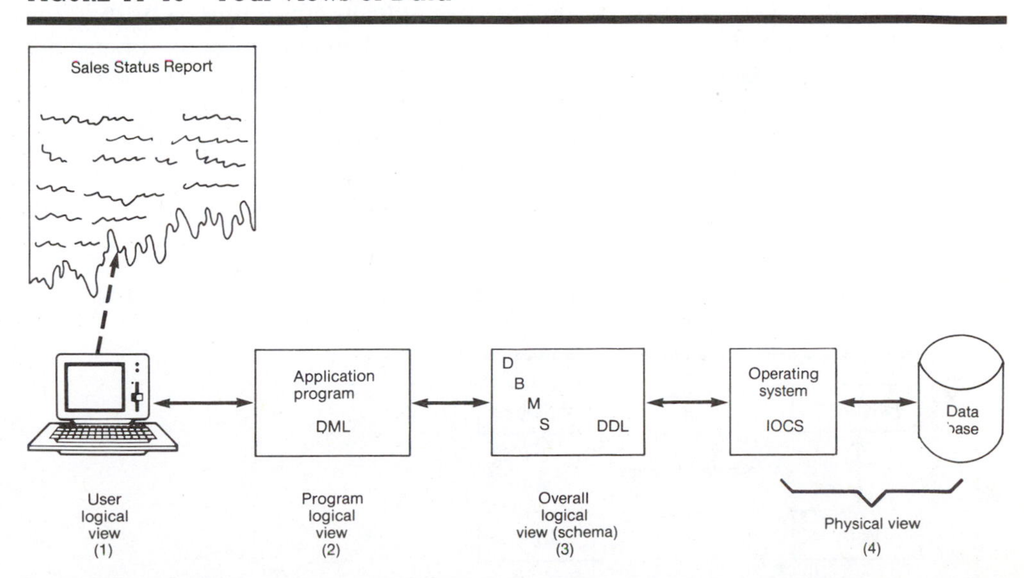

FIGURE 11–14 Entities and Relationships

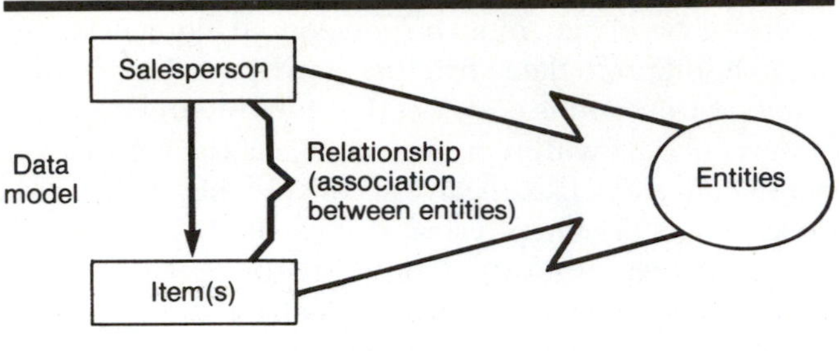

relationships among the schema, subschema, and physical structure are provided by the software.

Data Structure

Data are structured according to the data model. In our example (see Figure 11–12), sales items are linked to the salesperson who sold them. The salesperson is called an entity and the item sold is also an entity. An *entity* is a conceptual representation of an object. Relationships between entities make up a *data structure*. A data model represents a data structure that is described to the DBMS in DDL (see Figure 11–14).

Types of Relationships

Three types of relationships exist among entities: one-to-one, one-to-many, and many-to-many relationships. A *one-to-one* (1:1) relationship is an association between two entities. For example, in our culture, a husband is allowed one wife (at a time) and vice versa, and an employee has one social security number (see Figure 11–15).

A *one-to-many* (1:M) relationship describes an entity that may have two or more entities related to it. For example, a father may have many children, and an employee may have many skills (see Figure 11–16).

A *many-to-many* (M:M) relationship describes entities that may have

FIGURE 11–15 1:1 Relationship—Examples

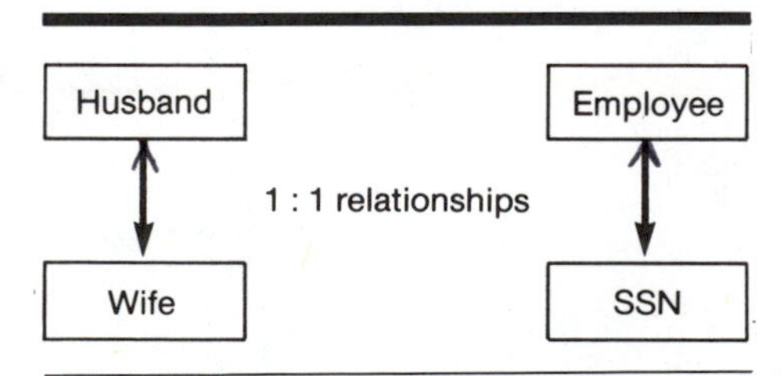

FIGURE 11-16 1:M Relationship—Examples

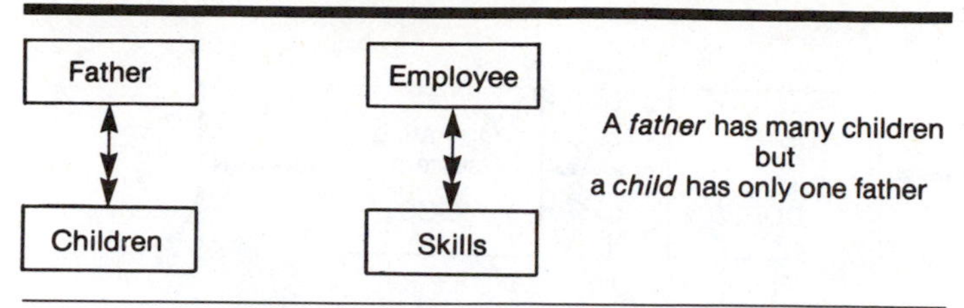

many relationships in both directions. For example, children may have many toys, and students may have many courses (see Figure 11-17).

Types of Data Structure

Data structuring determines whether the system can create 1:1, 1:M, or M:M relationships among entities. Although all DBMSs have a common approach to data management, they differ in the way they structure data. There are three types of data structure: hierarchical, network, and relational (see Figure 11-18).

Hierarchical Structuring. Hierarchical (also called *tree*) structuring specifies that an entity can have no more than one owning entity; that is, we can establish a 1:1 or a 1:M relationship. The owning entity is called the *parent;* the owned entity, the *child.* A parent with no owners is called the *root.* There is only one root in a hierarchical model. In Figure 11-19, a customer record has a name, address, and date—each with three subfields. The customer record is the parent or the root; it has six children (2-7). As parents, 3, 4, and 5 each has three children (8-10, 11-13, and 14-16, respectively), giving the profile of a tree structure.

A parent can have many children (1:M), whereas a child can have only one parent. Elements at the ends of the branches with no children (2, 5, and 7-16) are called *leaves.* Trees are normally drawn upside down, with the root at the top and the leaves at the bottom.

The hierarchical model is easy to design and understand. Some ap-

FIGURE 11-17 M:M Relationship—Examples

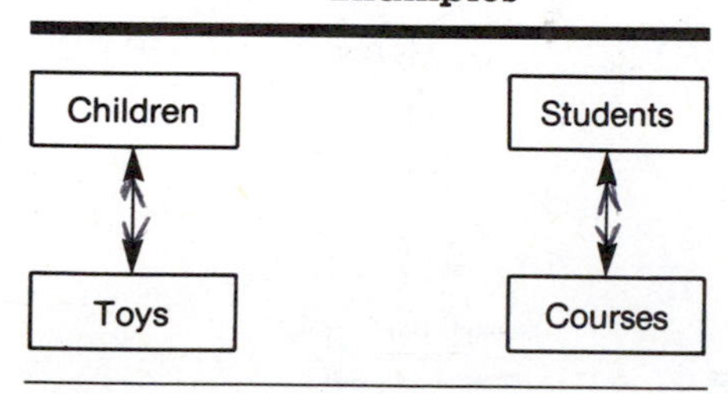

FIGURE 11-18 Types of Data Structures

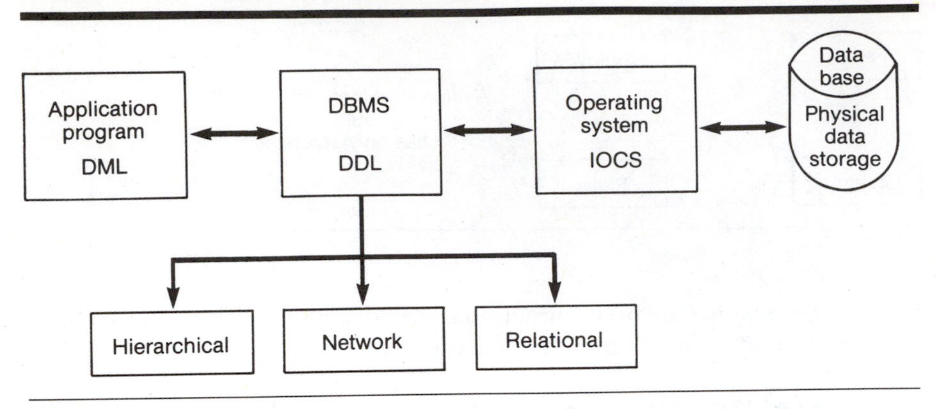

plications, however, do not conform to such a scheme. In Figure 11-19, if we allowed more than one person in a joint account, we would have a nonhierarchical structure, which complicates programming or the DBMS description. The problem is sometimes resolved by using a network structure.

Network Structuring. A network structure allows 1:1, 1:M, or M:M relationships among entities. For example, an auto parts shop may have dealings with more than one automaker (parent). In Figure 11-20, spare parts come from Ford and GM, so they are owned by both entities—a

FIGURE 11-19 Tree Structure of Customer Checking Account

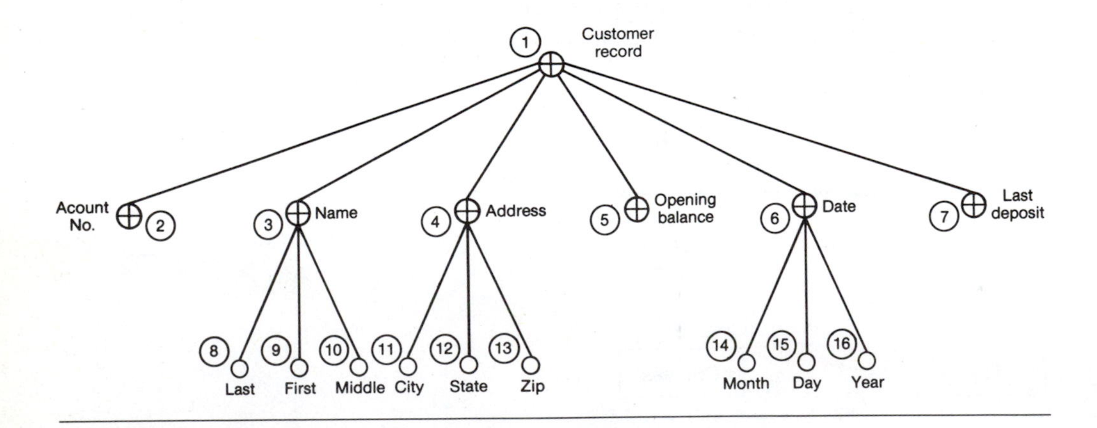

FIGURE 11–20 Network Structuring—An Example

structure that can best be supported by a network. Now consider the automaker and the auto parts shops it deals with. If the automaker sold spare parts to only one shop (say, a new car dealer), then there is a 1:1 relationship. If it supplied many other dealers, then there is a 1:M relationship. The 1:1 and 1:M relationships can be supplied by a hierarchy. When auto parts dealers are supplied by many automakers, however, there is an M:M relationship, which is a network structure.

A network structure reflects the real world, although a program structure can become complex. The solution is to separate the network into several hierarchies with duplicates. This simplifies the relationship to no more complex than 1:M. A hierarchy, then, becomes a subview of the network structure. The simplified network structure is shown in Figure 11–21.

Relational Structuring. In relational structuring, all data and relationships are represented in a flat, two-dimensional table called a *relation*. A

FIGURE 11–21 Simplified Networking

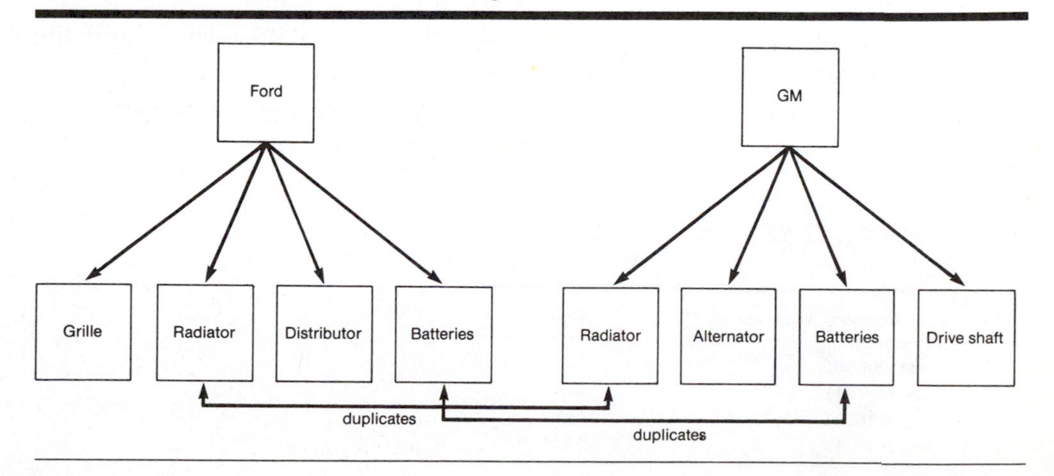

FIGURE 11-22 Employee Relation

Employee Number	Name	Years with the Firm
211306801	Arnold, Jim	1
301421011	Zmud, Bill	3
419846204	Belohlov, Jim	2
612047216	Boynton, Tom	6

relation is equivalent to a file, where each line represents a record. Figure 11-22 is a relation that describes the entity EMPLOYEE by social security number, name, and years with the firm. Note that all the entries in each column are of the same kind. Furthermore, each column has a unique name. Finally, no two rows in the table are identical. A row is referred to as a *tuple* (rhymes with "couple").

A relational DBMS has several features:

1. It allows the user to update (add, modify, or delete) the table's contents. Any position can be changed.
2. It provides inquiry capabilities against a label. Using our example, an inquiry might be: "How many years has Boynton been with the firm?" The response is "6."
3. Two or more tables can be merged to form one relation. Unlike hierarchical or network structuring where all relationships are predefined, a relational DBMS develops new relations on user commands.

To illustrate, suppose a relational DBMS maintains two relations: the EMPLOYEE relation (Figure 11-22) and the EMPLOYEE EDUCATION relation (Figure 11-23). A query about employees with more than two years in the firm and an MBA puts the relational DBMS through the following routine:

1. A relationship is implied between the EMPLOYEE relation and the EMPLOYEE EDUCATION relation.
2. A temporary table of employees who have been in the firm more than two years is obtained from the EMPLOYEE relation and placed in the

FIGURE 11-23 Employee Education Relation

Employee Number	Degree
211306801	High school
301421011	MBA
419846204	MA
612047216	MBA

file. This is called a *temporary relation* and will be deleted after the user has been satisfied.

3. The information in the temporary table is taken along with the EMPLOYEE EDUCATION relation to determine which employee who has been in the firm more than two years has an MBA. This results in a second temporary relation. In our example, Zmud and Boynton are the only employees. The temporary relation is also deleted after the user query has been satisfied.

A relational structure is simpler to construct than a hierarchical or a network structure. It may be inefficient, though, since a relational DBMS responds to queries by an exhaustive review of the relations involved.

Entities and Attributes

An entity is something of interest to the user about which to collect or store data. It is also called a *data aggregate* because it represents a number of data elements. In our sales status system, the "sales" entity contains data elements such as the salesperson's number, name, and date of employment, and the sales period covered by the report. The "item" entity has data elements such as item number, item description, and the sale price of each item.

Data entities are explained by the use of several terms: *attribute, value key,* and *instance of an entity*. For example, a salesperson (entity) is described by attributes such as number, name, sex, age, and height. So *attributes* describe an entity. They are physically stored in fields or data elements (see Figure 11–24).

Each attribute takes on a unique *value*. For example, "211306801" is a unique value of the attribute "salesperson number." An attribute, then, takes on a value for a specific occurrence (or *instance*) of an entity. In Figure 11–24, one instance of the entity "salesperson" is "Name." The value is Jim Arnold. Other values that describe him are male, 34 years old, and 5 feet 10 inches tall. So the instance of the entity is a 34-year-old, 5 feet 10 inches tall, male salesman with social security number 211306801.

A *key* is a unique identifier of the entity. In our example, the key

FIGURE 11–24 Attributes of an Entity

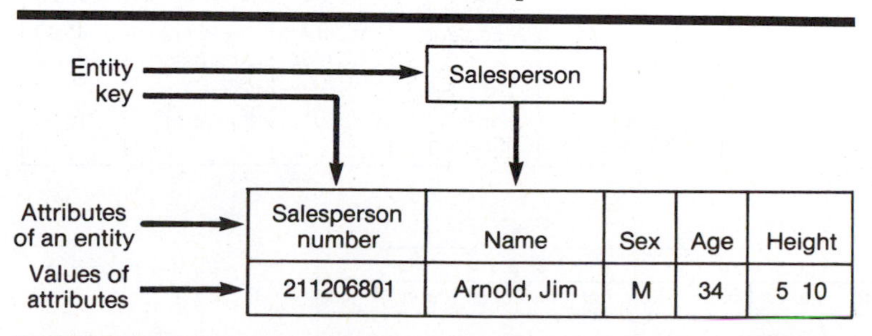

211306801 is a unique identifier of Jim Arnold. Sex, age, and height are not identifiers, since they are not unique. They are nonkey identifiers.

Normalization

Data structuring is refined through a process called *normalization*. Data are grouped in the simplest way possible so that later changes can be made with a minimum of impact on the data structure. When too many attributes are grouped together to form entities, some attributes are found to be entities themselves. Further normalization of these entities into attributes linked by common data elements to form relationships improves the effectiveness of the DBMS.

Steps in Normalization

1. Isolate repeating groups from an entity because they are easier to process separate from the rest of an entity. Figure 11-25 is an unnormalized file structure. The first four attributes (employee number, name, branch, and department) are virtually constant. The remaining three attributes (item number, item description, and price) contain data that change and are repeated with different salespersons. Therefore, the repeating groups should be separated from the entity "salesperson."

The reorganized file is shown in Figure 11-26. It consists of two files: the

FIGURE 11-25 Unnormalized File

Salesperson				Sales		
Employee number	Employee name	Store branch	Department	Item number	Item description	Sale price
211306801	Arnold, Jim	Downtown	Hardware	TR100 SA 10 PT 65 AB165	Router Saw Drill Lawnmower	$ 35.00 19.00 21.00 245.00
301421011	Zmud, Bill	Dadeland	Home appliances	TT 14 DS104	Humidifier Dishwasher	114.00 262.00
419846204	Belohlov, Jim	Cutler Ridge	Auto parts	MC164 AC1462 BB1000	Snow tire Alternator Battery	85.00 65.00 49.50
612047216	Boynton, Tom	Fashion Square	Men's clothing	HS101	3-pc. suit	215.00

Salesperson				Sales		
Number	Name	Branch	Dept.	Item no.	Item description	Price

FIGURE 11-26 First Normalization

* =Key

* Employee number	Employee name	Store branch	Department
211306801	Arnold, Jim	Downtown	Hardware
301421011	Zmud, Bill	Dadeland	Home appliances
419846204	Belohlov, Jim	Cutler Ridge	Auto parts
612047216	Boynton, Tom	Fashion Square	Men's clothing

Salesperson Data file

* Employee number	* Item number	Item description	Sale price
211306801	TR 100	Router	$ 35.00
211306801	SA 10	Saw	19.00
211306801	PT 65	Drill	21.00
211306801	AB 165	Lawnmower	245.00
301421011	TT 14	Humidifier	114.00
301421011	DS 104	Dishwasher	262.00
419846204	MC 164	Snow tire	85.00
419846204	AC1462	Alternator	65.00
419846204	BB1000	Battery	49.50
612047216	HS 101	3-pc. suit	215.00

Salesperson Item File

salesperson data file with employee number as the primary key and the *salesperson item file* with employee number and item number as new attributes. They are added to relate the records in the file to the salesperson data file. The two attributes are used together for accessing data. Two keys used together are called a *concatenated key.*

2. After isolating repeating groups from the rest of an entity, try to simplify the relation further. The second normalization makes sure that each nonkey attribute depends on a key attribute or concatenated key. Nonkey attributes that do not meet this condition are split into simpler entities. In Figure 11–26, each attribute in the salesperson data file depends on the primary key "employee number." In the salesperson item file, the attribute "sales price" depends on a concatenated key ("employee number" and "item number"). The way the file is set up, the sales price is strictly related to the salesperson number and the item number of the sale. Alternatively, the attribute "item description" tags to "item number," which is part of the concatenated key. This causes several concerns. Sales information is availabe only by salesperson number, which is unwieldy. Worse yet, an employee transfer would make it difficult to maintain records because they would be dropped when the salesperson leaves the department.

To solve the problem, we create new independent entities for "item description" and "sales price." In one file, we create the item description attribute with item number keys from the salesperson item file. The remaining attributes (employee number, item number, and sales price) become the second normalized form (see Figure 11–27.)

The second normalization offers several benefits. Sales items can be added without being tagged to a specific salesperson. If the item changes, we need to change only the item file. If a salesperson leaves the department, it would have no direct effect on the status of the items sold.

3. Looking at the second normalization, we find further room for improvement. In the salesperson data file, the attribute "store branch" is tagged to the primary key "employee number," while the attribute "department" is related to "store branch," which is a nonkey attribute. Making "store branch," a key attribute requires isolating "department" along with "store branch" in a new entity as shown in Figure 11–28.

With the third normalization, we can have store branch information independent of the salespersons in the branch. We can also make changes in the "department" without having to update the record of the employees in it. In this respect, normalization simplifies relationships and provides logical links between files without losing information.

One inherent problem with normalization is data redundancy. For example, to produce a sales report by store branch, the system goes through three steps:

FIGURE 11–27 Second Normalization

* Employee #	Employee name	Store branch	Department
211306801	Arnold, Jim	Downtown	Hardware
301421011	Zmud, Bill	Dadeland	Home appliances
419846204	Belohlov, Jim	Cutler Ridge	Auto parts
612047216	Boynton, Tom	Fashion Square	Men's clothing

Salesperson Data File

1

2

* Employee #	* Item #	Sale price
211306801	TR 100	$ 35.00
211306801	SA 10	19.00
211306801	PT 65	21.00
211306801	AB 165	245.00
301421011	TT 14	114.00
301421011	DS 104	262.00
419846204	MC 164	85.00
419846204	AC1462	65.00
419846204	BB1000	49.50
612047216	HS 101	215.00

Salesperson Item File

* Item #	Item description
TR 100	Router
SA 10	Saw
PT 65	Drill
AB 165	Lawnmower
TT 14	Humidifier
DS 104	Dishwasher
MC 164	Snow tire
AC1462	Alternator
BB1000	Battery
HS 101	3-pc. suit

Item File

FIGURE 11–28 Third Normalization

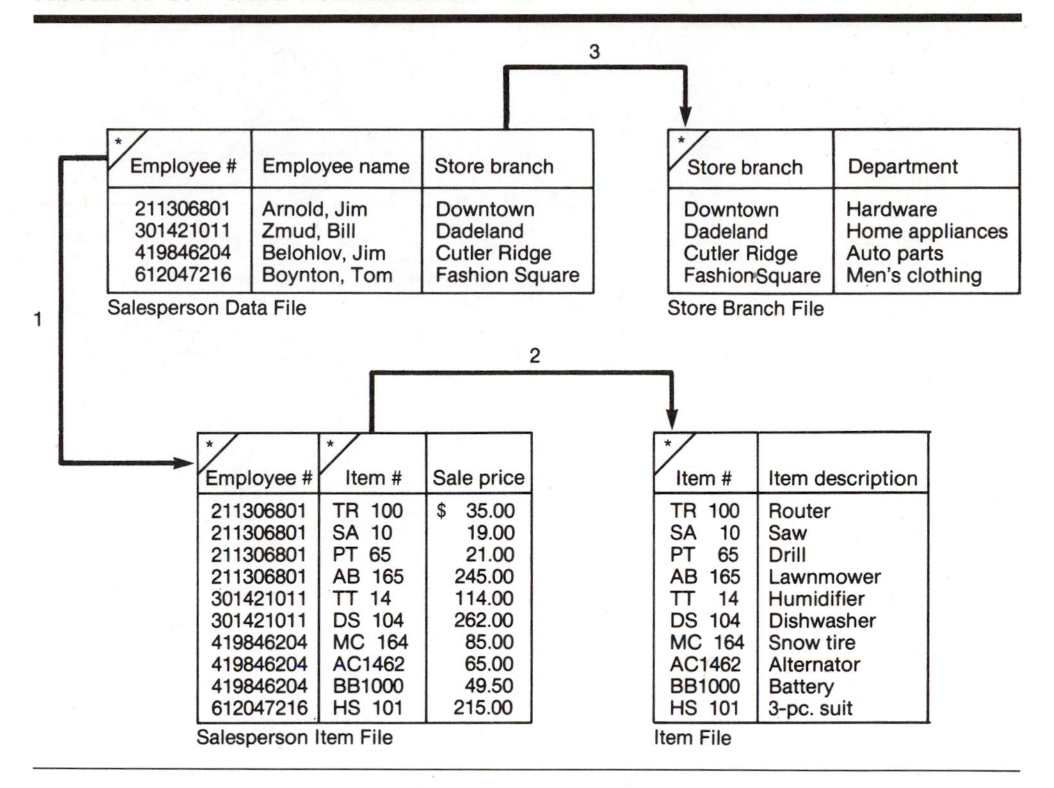

* Employee #	Employee name	Store branch
211306801	Arnold, Jim	Downtown
301421011	Zmud, Bill	Dadeland
419846204	Belohlov, Jim	Cutler Ridge
612047216	Boynton, Tom	Fashion Square

Salesperson Data File

* Store branch	Department
Downtown	Hardware
Dadeland	Home appliances
Cutler Ridge	Auto parts
Fashion Square	Men's clothing

Store Branch File

* Employee #	* Item #	Sale price
211306801	TR 100	$ 35.00
211306801	SA 10	19.00
211306801	PT 65	21.00
211306801	AB 165	245.00
301421011	TT 14	114.00
301421011	DS 104	262.00
419846204	MC 164	85.00
419846204	AC1462	65.00
419846204	BB1000	49.50
612047216	HS 101	215.00

Salesperson Item File

* Item #	Item description
TR 100	Router
SA 10	Saw
PT 65	Drill
AB 165	Lawnmower
TT 14	Humidifier
DS 104	Dishwasher
MC 164	Snow tire
AC1462	Alternator
BB1000	Battery
HS 101	3-pc. suit

Item File

1. Computes total sales for each salesperson from the salesperson item file.
2. Goes to the employee data file to look up the store branch to which the salesperson is assigned.
3. Accumulates each salesperson's sales in a special field in the store branch file.

This procedure is repeated for each salesperson in the file. Figure 11–29 illustrates the processing cycle for salesperson 211306801.

A normalized data base must serve the application needs of the organization. Normalization simplifies the data structure and makes it easy to use and modify. It also establishes an easy relationship between attributes based on common attributes. These benefits should be weighed against redundancies that often result in processing inefficiency.

THE ROLE OF THE DATA BASE ADMINISTRATOR

A data base is a shared resource. When two or more users are tied to a common data base, certain difficulties in sharing are likely to occur. Perceptions regarding data ownership, priority of access, and the like become

FIGURE 11-29 Processing Cycle for #211306801

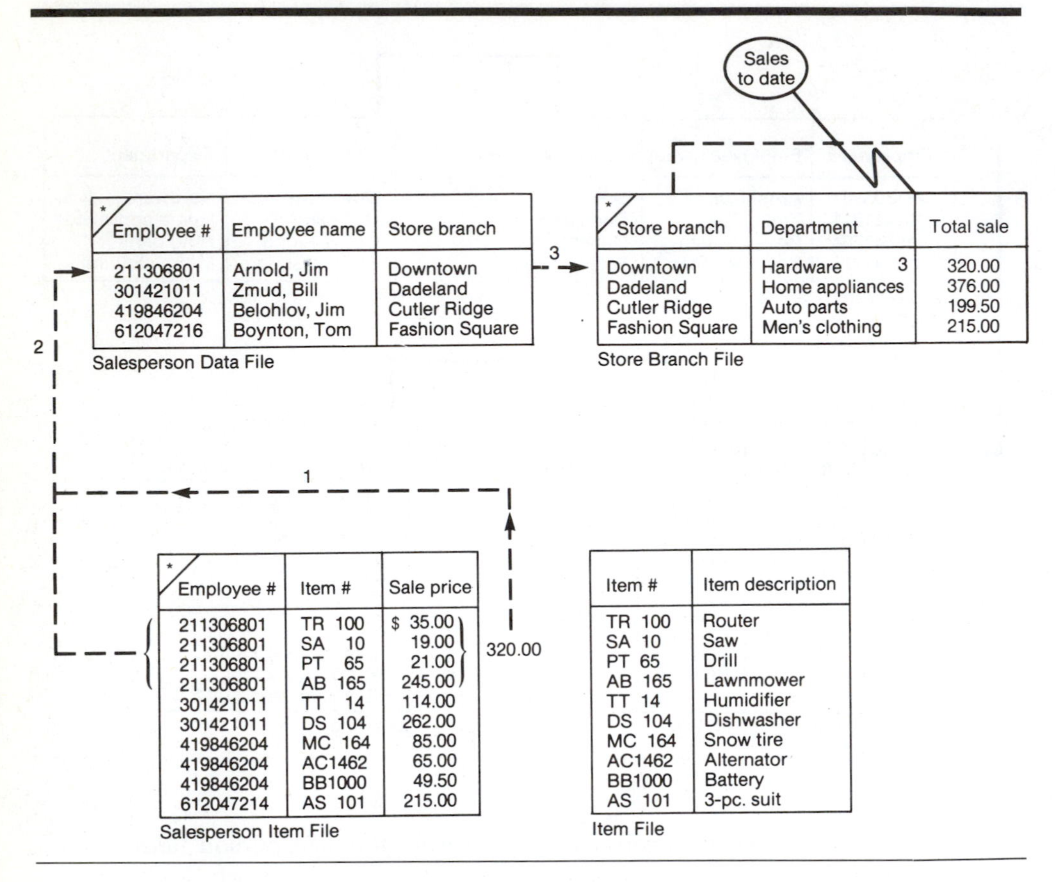

issues that need to be resolved when the data base is in operation. To manage the data base, companies hire a *data base administrator* or DBA to protect and manage the data base on a regular basis.

In addition to resolving user conflicts, the DBA performs maintenance and update tasks—recovery procedures, performance evaluation, data base timing, and new enhancement evaluation. Specifically, the DBA performs three key functions: managing data activities, managing the data base structure, and managing the DBMS.

1. *Managing data activities.* The DBA manages data base activities by providing standards, control procedures, and documentation to ensure each user's independence from other users. Standardization is extremely important in a centralization-oriented environment. Every data base record must have a standard name, format, and unique strategy for access. Standardization, though resisted by users, simplifies reporting and facilitates management control.

In addition to standardization, the DBA is concerned about data access and modification. Deciding who has authorization to modify what data is a job in itself. Locks have to be established to implement this activity. Failures and recovery procedures are added concerns. Failures may be caused by machines, media, communications, or users. The user must be familiar with a recovery procedure for reinputting reports. Training users and maintaining documentation for successful recovery are important responsibilities of the DBA.

2. *Managing data base structure.* This responsibility centers around the design of the schema and special programs for controlling redundancy, maintaining control of change requests, implementing changes in the schema, and maintaining user documentation. In the case of documentation, the DBA must know what changes have been made, how they were made, and when they were made. Data base changes must be backed by a record of test runs and test results.

3. *Managing DBMS.* A third responsibility involves the central processing unit (CPU), compiling statistics on system efficiency, including CPU times and elapsed times of inquiries. CPU time is the amount of time the CPU requires to process a request. Elapsed time is the actual clock time needed to process the activities and return a result (output). Much of this time depends on the nature of the activity, other activities that occur in the interim, and the peak load requirements of the system.

Other elements also affect DBMS management. The DBA investigates user performance complaints and keeps the system's capabilities in tune with user requirements. Modifications may have to be made to the communication network, the operating system, or their interfaces with the DBMS. It is the DBA's responsibility to evaluate changes and determine their impact on the data base environment.

The DBA has a full-time, highly responsible job. In addition to a managerial background, the DBA needs technical knowledge to deal with data base designers. For example, he/she needs to maintain the data dictionary and evaluate new data base features and their implementation. The combination of technical and managerial backgrounds make the job of the DBA unique.

Where does the DBA fit into the organization structure? There is considerable debate about this. Two views are commonly accepted. One proposes that the DBA should not be subordinate to a group that imposes restrictions. The second view is that the DBA should be no more than one level above the prime user that uses the system most frequently. In the long run, the key to the success of the DBA in the organization is the attitude and support of the senior MIS staff and upper management for the DBA function.

Summary

1. The file hierarchy begins with bytes (the smallest addressable units), which make up data items. Data items are records that are grouped to make up a file. Two or more files are a data base.

2. There are four methods of organizing files:
   a. Sequential organization means storing records in contiguous blocks according to a key.
   b. Indexed-sequential organization stores records sequentially but uses an index to locate records. Records are related through chaining using pointers.
   c. Inverted list organization uses an index for each key type. Records are not necessarily in a particular sequence.
   d. Direct-access organization has records placed randomly throughout the file. Records are updated directly and independently of other records.
3. A data base is a collection of interrelated data stored with a minimum of redundancy to serve many applications. Data base design minimizes the artificiality embedded in using separate files. The primary objectives are fast response time to inquiries, more information at low cost, control of redundancy, clarity and ease of use, data and program independence, accuracy and integrity of the system, fast recovery, privacy and security of information, and availability of powerful end user languages.
4. The heart of a data base is the DBMS. It manages and controls the data base file and handles requests from the application program in a data manipulation language (DML). To produce the user's view, the data model represents data structures and describes the view in a data definition language (DDL). DDL simply tells the DBMS how the data must be structured to meet application program requirements. DML, then, manipulates data; DDL describes and identifies data structures; and the DBMS manages and coordinates data according to DML requests and DDL descriptions.
5. A data structure defines relationships among entities. There are three types of relationships: one-to-one, one-to-many, and many-to-many relationships. Data structuring in a DBMS determines whether the system can create a given relationship between entities.
6. Although all DBMSs have a common approach to data management, they differ in the way they structure data. The three types of data structures are hierarchical, network, and relational. A *hierarchical* structure specifies that an entity cannot have more than one parent. A *network* structure allows 1:1, 1:M, or M:M relationships and can best be supported by a network. To simplify the structure, the network is separated into a number of hierarchies with duplicates. A hierarchy, then, becomes a subview of the network structure. A *relational* structure is a flat, two-dimensional table representing data and relationships. It allows the user to update the table's content and provides a powerful inquiry capability.
7. There are four views of data. The first three views are logical: user's view, application program (called *subschema*), and overall logical view (called

schema). The physical view is what the data actually look like in physical storage.

8. The data structure can be refined through a normalization process that groups data in the simplest way possible so that later changes can be made with ease. Normalization is designed to simplify relationships and establish logical links between files without losing information. An inherent problem is data redundancy and the inefficiency it generates.
9. Managing the data base requires a data base administrator (DBA) whose key functions are to manage data activities, the data base structure, and the DBMS. In addition to a managerial background, the DBA needs technical knowledge to deal with data base designers. Important for the success of this important job is the support of the senior MIS staff and upper management for the overall data base function.

Key Words

Attribute
Chaining
Concatenated Key
Data Aggregate (see *Entity*)
Data Base
Data Base Administrator (DBA)
Data Base Management System (DBMS)
Data Definition Language (DDL)
Data Independence
Data Item
Data Manipulation Langauge (DML)
Data Model
Data Structure
Direct-Access Organization
Entity
Field (see *Data Item*)
Hierarchical Structure
Indexed-Sequential Organization
Input/Output Control System (IOCS)
Instance of an Entity
Inverted List Organization
Key
Logical Record
Network Structure
Normalization
Operating System
Physical Record
Pointer
Redundancy
Relation
Relational Structure
Root
Schema
Sequential Organization
Subschema
Tuple

Review Questions

1. Illustrate the hierarchy making up a data base.
2. Explain the difference between:
   a. Logical and physical record.
   b. Data item and field.
   c. File activity and file volatility.
   d. Sequential and indexed-sequential.
   e. Indexed-sequential and inverted list.

3. What methods does the designer consider in file organization? What factors determine the method chosen? Be specific.
4. What is chaining? How does it relate to indexed-sequential file organization? Illustrate.
5. Take a look at Figure 11–6 in the chapter. Produce an inverted list with a brief explanation of its meaning?
6. How does an absolute address differ from a relative address?
7. In your own words, define data base. What is the single most critical aspect of data base? Why?
8. How does the user's view relate to a data model? Explain.
9. Explain how DDL, DML, and DBMS are related.
10. Distinguish between:
    a. Schema and subschema.
    b. Logical and physical view of data.
    c. Entity and attribute.
    d. Relation and tuple.
11. Define data structure. What are the major types? Illustrate.
12. How does hierarchical structuring differ from network structuring?
13. What features does a relational DBMS offer?
14. Give an example, showing entities, attributes, and their values.
15. What is the purpose behind normalization? How does one normalize a file? Illustrate.
16. Write an essay discussing the role of the DBA.

Application Problems

1 Convert the following bubble chart to a set of binary associations.

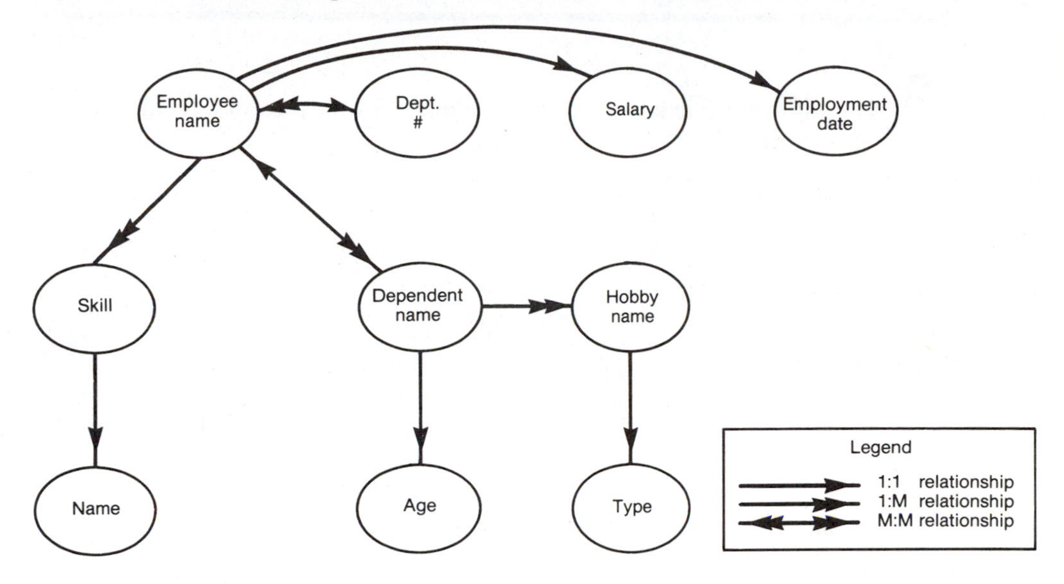

Example:

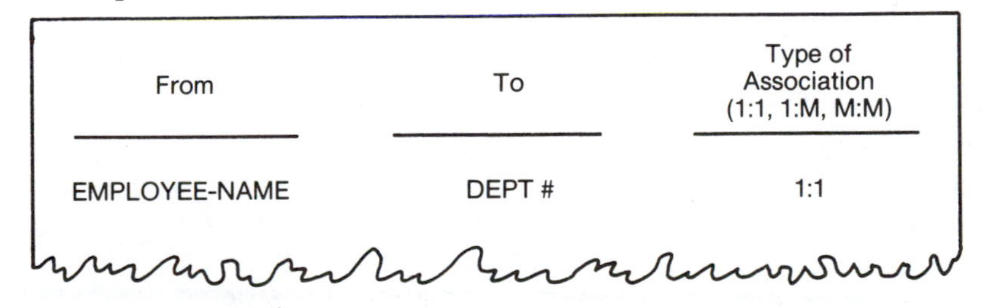

From	To	Type of Association (1:1, 1:M, M:M)
EMPLOYEE-NAME	DEPT #	1:1

2 Review the following schema:

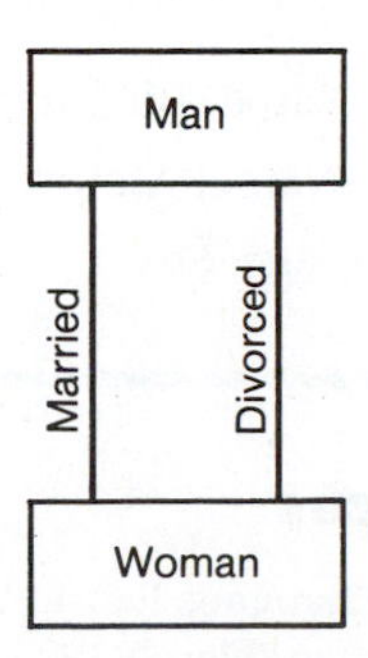

Assignment

a. Explain what the schema means; that is, what types of associations are there between the same two data items?

b. Draw a bubble chart based on the schema.

---

3 The following partial file represents eight employees of a local department store:

Employee Number	Name	Title Code	Years with Firm	Phone Ext.	Salary
7461	Blake, Bill	4	2	225	14,400
7463	Crane, George	2	7	531	16,900
7941	Daleiden, Norb	17	4	236	37,520
8124	Hunt, Mary B.	6	16	294	26,120
8129	Kehoe, William	9	21	333	30,040
9246	Kolb, Debbie	1	3	160	19,500
.	.	.	.	.	.
.	.	.	.	.	.
.	.	.	.	.	.
9940	Ursu, Cindy	2	35	189	18,325
9967	York, Mary	7	5	212	26,890

Assignment

a. How many attributes represent a record? List them.

b. How many values are there of the attribute SALARY?

c. How many tuples represent the partial file?

d. What are the values of the attribute EXTENSION of the following employees: Crane, Kehoe, York?

 Draw a schema, using the following data items:

Person: Employee-Name, Birthdate (day, month, year), Age, Phone

Education: Degree-Name, Major

Child: Child-Name, Age, Sex

Selected References

Adams, Robert G. "The Changing Fabric of Database Technology." *ICP Software Business Review*, February/March 1983, pp. 38–43.

Bradley, James. *File and Data Base Techniques*. New York: Holt, Rinehart & Winston, 1982.

Bronstein, Philip. "DBMS Comes to Small Computers." *Small Systems World*, April 1983, pp. 21–27.

Dumas, Ronald L. "Relational Data Base: What Is It, and What Can It Do?" *Data Management*, January 1981, pp. 15–17.

Gillin, Paul. "Relational Data Base: Pushing for Acceptance." *Computerworld*, December 26, 1983, pp. 66–70.

Holland, Robert H. "The Executive Guide to Data Base Management Systems." *ICP Software Business Review*, February/March 1983, pp. 46–50ff.

Horton, Forest Woody Jr. "Tapping External Data Sources." *Computerworld*, August 15, 1983, pp. 1–3ff.

Houston, Velina. "Database Idea 20 Years Old, but Still Growing." *Management Information Systems Week*, July 13, 1983, pp. 25–26.

Kent, William. "A Simple Guide to Five Normal Forms in Relational Database Theory." *Communications of the ACM* 26, no. 2 (February 1983), pp. 120–25.

Levering, Robert. "Data Management without Programming." *PC World* 1, no. 3, pp. 113–17.

Meng, Douglas and Lenore. "What's in a (Data Item) Name?" *Computerworld*, September 12, 1983, pp. 49–50ff.

Pembroke, Jill. "Public Database: Your Electronic Library." *Small Systems World*, September 1983, 30–34ff.

Ross, Ron. "Data Base Machines: Still a Question Mark." *Computerworld*, November 8, 1982, pp. 29–36.

Siwolop, Sana. "Touching All the Data Bases." *Discover*, March 1983, pp. 68–71.

Walsh, Myles. "Database Management Systems: An Operational Perspective." *Journal of Systems Management*, April 1983, pp. 20–23.

Part Four

System Implementation

Chapter 12

System Testing and Quality Assurance

At a Glance

No system design is ever perfect. Communication problems, programmers' negligence, or time constraints create errors that must be eliminated before the system is ready for user acceptance testing. A system is tested for online response, volume of transactions, stress, recovery from failure, and usability. Then comes system testing, which verifies that the whole set of programs hangs together. Following system testing is acceptance testing, or running the system with live data by the actual user.

System testing requires a test plan that consists of several key activities and steps for program, string, system, and user acceptance testing. The system performance criteria deal with turnaround time, backup, file protection, and the human factor.

For systems to be viable, controls have to be developed to ensure a quality product. Quality assurance cuts across the system life cycle, and is especially involved in implementation. Quality assurance specialists go through system testing and validation before they grant certification. Quality assurance and the DP audit go hand in hand. The role of the auditor is to make sure that adequate controls are built into the system for integrity and reliability.

By the end of this chapter, you should know:

1. Why systems are tested.
2. The activity network for system testing.
3. What steps are taken to test systems.
4. The goals of quality assurance in the system life cycle.
5. The role of the DP auditor in system testing.

INTRODUCTION

No program or system design is perfect; communication between the user and the designer is not always complete or clear, and time is usually short. The result is errors and more errors. The number and nature of errors in a new design depend on several factors:

1. Communications between the user and the designer.
2. The programmer's ability to generate a code that reflects exactly the system specifications.
3. The time frame for the design.

Theoretically, a newly designed system should have all the pieces in working order, but in reality, each piece works independently. Now is the time to put all the pieces into one system and test it to determine whether it meets the user's requirements. This is the last chance to detect and correct errors before the system is installed for user acceptance testing. The purpose of system testing is to consider all the likely variations to which it will be subjected and then push the system to its limits. It is a tedious but necessary step in system development.

This chapter reviews the process of system testing and the steps taken to validate and prepare a system for final implementation. First, we need to be familiar with the following basic terms:

1. *Unit testing* is testing changes made in an existing or a new program.
2. *Sequential or series testing* is checking the logic of one or more programs in the candidate system, where the output of one program will affect the processing done by another program.
3. *System testing* is executing a program to check logic changes made in it and with the intention of finding errors—making the program fail. Effective testing does not guarantee reliability. Reliability is a design consideration.
4. *Positive testing* is making sure that the new programs do in fact process certain transactions according to specifications.
5. *Acceptance testing* is running the system with live data by the actual user.

WHY SYSTEM TESTING?

Testing is vital to the success of the system. System testing makes a logical assumption that if all the parts of the system are correct, the goal will be successfully achieved. Inadequate testing or nontesting leads to errors that may not appear until months later. This creates two problems: (1) the time lag between the cause and the appearance of the problem (the longer the time interval, the more complicated the problem has become), and (2) the effect of system errors on files and records within the system. A small system error can conceivably explode into a much larger problem. Effective testing early in the process translates directly into long-term cost savings from a reduced number of errors.

Another reason for system testing is its utility as a user-oriented vehicle before implementation. The best program is worthless if it does not meet user needs. Unfortunately, the user's demands are often compromised by efforts to facilitate program or design efficiency in terms of processing time or memory utilization. Often the computer technician and the user have communication barriers due to different backgrounds, interests, priorities, and perhaps languages. The system tester (designer, programmer, or user) who has developed some computer mastery can bridge this barrier.

WHAT DO WE TEST FOR?

The first test of a system is to see whether it produces the correct outputs. No other test can be more crucial. Following this step, a variety of other tests are conducted:

1. *Online response.* Online systems must have a response time that will not cause a hardship to the user. One way to test this is to input transactions on as many CRT screens as would normally be used in peak hours and time the response to each online function to establish a true performance level.

2. *Volume.* In this test, we create as many records as would normally be produced to verify that the hardware and software will function correctly. The user is usually asked to provide test data for volume testing.

3. *Stress testing.* The purpose of stress testing is to prove that the candidate system does not malfunction under peak loads. Unlike volume testing, where time is not a factor, we subject the system to a high volume of data over a short time period. This simulates an online environment where a high volume of activities occurs in spurts.

4. *Recovery and security.* A forced system failure is induced to test a backup recovery procedure for file integrity. Inaccurate data are entered to see how the system responds in terms of error detection and protection. Related to file integrity is a test to demonstrate that data and programs are secure from unauthorized access.

5. *Usability documentation and procedure.* The usability test verifies the user-friendly nature of the system. This relates to normal operating and error-handling procedures, for example. One aspect of user-friendliness is accurate and complete documentation. The user is asked to use only the documentation and procedures as a guide to determine whether the system can be run smoothly.

The Nature of Test Data

The proper choice of test data is as important as the test itself. If test data as input are not valid or representative of the data to be provided by the user, then the reliability of the output is suspect.

Test data may be artificial (created solely for test purposes) or live (taken from the user's actual files). Properly created artificial data should provide all combinations of values and formats and make it possible to test all logic and transaction path subroutines. Unlike live data, which are biased toward typical values, artificial data provide extreme values for testing the limits of the candidate system.

For large, complex systems, a computer program is used to generate the necessary test data. Data-generating programs save substantial time for both the programmer and the test itself. A familiarity with system files and parameters, however, is necessary for writing an effective data-generating program.

THE TEST PLAN

The first step in system testing is to prepare a plan that will test all aspects of the system in a way that promotes its credibility among potential users. There is psychology in testing:

1. Programmers usually do a better job in unit testing because they are expected to document and report on the method and extent of their testing.
2. Users are involved, which means communication is improved between users and the designer group.
3. Programmers are involved when they become aware of user problems and expectations. The user also becomes more aware (and appreciative) of the complexity of programming and testing. The upshot of all this is a more realistic and cooperative user for successful testing.

Activity Network for System Testing

A test plan entails the following activities (see Figure 12–1):

1. Prepare test plan.
2. Specify conditions for user acceptance testing.
3. Prepare test data for program testing.
4. Prepare test data for transaction path testing.
5. Plan user training.
6. Compile/assemble programs.
7. Prepare job performance aids.
8. Prepare operational documents.

Prepare Test Plan

A workable test plan must be prepared in accordance with established design specifications. It includes the following items:

FIGURE 12–1 Activity Network for System Testing

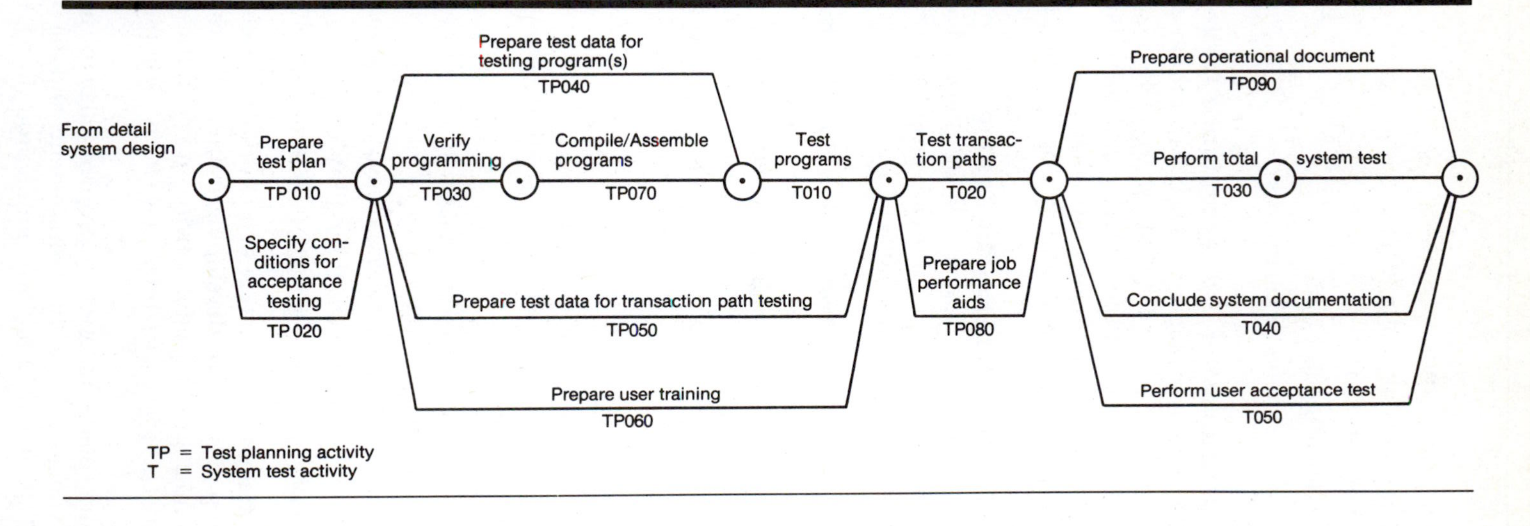

1. Outputs expected from the system.
2. Criteria for evaluating outputs.
3. A volume of test data.
4. Procedure for using test data.
5. Personnel and training requirements.

Specify Conditions for User Acceptance Testing

Planning for user acceptance testing calls for the analyst and the user to agree on the conditions for the test. Many of these conditions may be derived from the test plan. Others are an agreement on the test schedule, the test duration, and the persons designated for the test. The start and termination dates for the test should also be specified in advance.

Prepare Test Data for Program Testing

As each program is coded, test data are prepared and documented to ensure that all aspects of the program are properly tested. After the testing, the data are filed for future reference.

Prepare Test Data for Transaction Path Testing

This activity develops the data required for testing every condition and transaction to be introduced into the system. The path of each transaction from origin to destination is carefully tested for reliable results. The test verifies that the test data are virtually comparable to live data used after conversion.

Plan User Training

User training is designed to prepare the user for testing and converting the system. User involvement and training take place parallel with programming for three reasons:

1. The system group has time available to spend on training while the programs are being written.
2. Initiating a user-training program gives the systems group a clearer image of the user's interest in the new system.
3. A trained user participates more effectively in system testing.

For user training, preparation of a checklist is useful (see Figure 12–2). Included are provisions for developing training materials and other documents to complete the training activity. In effect, the checklist calls for a commitment of personnel, facilities, and efforts for implementing the candidate system.

The training plan is followed by preparation of the user training manual and other text materials. Facility requirements and the necessary hardware are specified and documented. A common procedure is to train supervisors

FIGURE 12-2 User Training Checklist

Company ______________________ Analyst ______________

Project Name ______________________ Date ______________

Activity	Start Date	Completion Date	Staff in Charge	Department in Charge	Comments
1. Notification					
Announcement to officers	10/06	10/20	P. Solen	Personnel Mgr.	
Announcement to employees	10/06	10/20	J. Hill	Auditing	
Coordinated customer activities	10/06	10/29	D. Stang	Cashier	
Coordinate computer service	10/06	10/29	C. Sibley	Sr. Vice Pres.	
2. Procedures					
Interdepartmental	10/14	11/01	A. Blake	Systems	
Interdepartmental	10/14	11/01	J. Hill	Auditing	
3. Forms					
Design	11/01	11/14	A. Blake	Systems	
Printing	11/01	11/20	A. Blake	Systems	
4. Equipment					
Terminals	11/01	12/15			
5. Training and orientation					
Manuals	12/01	21/16	A. Blake	Systems	
Training aids	12/01	12/16	A. Blake	Systems	
Special workshops	12/10	12/14	A. Blake	Systems	
6. Lobby layout	12/10	12/30	D. Stang	President	
7. Supplies	12/10	12/15	M. Steed	Purchasing Agent	
8. Personnel					
Transfers	12/10	12/12	P. Solen	Personnel Mgr.	
New hires	12/15	12/30	P. Solen	Personnel Mgr.	

Approved by: ______________________ ______________________

(Project leader) *(Systems Department)*

and department heads who, in turn, train their staff as they see fit. The reasons are obvious:

1. User supervisors are knowledgeable about the capabilities of their staff and the overall operation.
2. Staff members usually respond more favorably and accept instructions better from supervisors than from outsiders.
3. Familiarity of users with their particular problems (bugs) makes them better candidates for handling user training than the systems analyst. The analyst gets feedback to ensure that proper training is provided.

Compile/Assemble Programs

All programs have to be compiled/assembled for testing. Before this, however, a complete program description should be available. Included is the purpose of the program, its use, the programmer(s) who prepared it, and the amount of computer time it takes to run it. Program and system flowcharts of the project should also be available for reference.

In addition to these activities, *desk checking* the source code uncovers programming errors or inconsistencies. Before actual program testing, a run order schedule and test scheme are finalized. A *run order schedule* specifies the transactions to test and the order in which they should be tested. High-priority transactions that make special demands on the candidate system are tested first. In contrast, a *test scheme* specifies how program software should be debugged. A common approach, called *bottom-up* programming, tests small-scale program modules, which are linked to a higher-level module, and so on until the program is completed. An alternative is the *top-down* approach, where the general program is tested first, followed by the addition of program modules, one level at a time to the lowest level.

Prepare Job Performance Aids

In this activity the materials to be used by personnel to run the system are specified and scheduled. This includes a display of materials such as program codes, a list of input codes attached to the CRT terminal, and a posted instruction schedule to load the disk drive. These aids reduce the training time and employ personnel at lower positions.

Prepare Operational Documents

During the test plan stage, all operational documents are finalized, including copies of the operational formats required by the candidate system. Related to operational documentation is a section on the experience, training, and educational qualifications of personnel for the proper operation of the new system.

System Testing

The purpose of system testing is to identify and correct errors in the candidate system. As important as this phase is, it is one that is frequently compromised. Typically, the project is behind schedule or the user is eager to go directly to conversion.

In system testing, performance and acceptance standards are developed. Substandard performance or service interruptions that result in system failure are checked during the test. The following performance criteria are used for system testing:

1. *Turnaround time* is the elapsed time between the receipt of the input and the availability of the output. In online systems, high-priority processing is handled during peak hours, while low-priority processing is done later in the day or during the night shift. The objective is to decide on and evaluate all the factors that might have a bearing on the turnaround time for handling all applications.

2. *Backup* relates to procedures to be used when the system is down. Backup plans might call for the use of another computer. The software for the candidate system must be tested for compatibility with a backup computer.

In case of a partial system breakdown, provisions must be made for dynamic reconfiguration of the system. For example, in an online environment, when the printer breaks down, a provisional plan might call for automatically "dumping" the output on tape until the service is restored.

3. *File protection* pertains to storing files in a separate area for protection against fire, flood, or natural disaster. Plans should also be established for reconstructing files damaged through a hardware malfunction.

4. *The human factor* applies to the personnel of the candidate system. During system testing, lighting, air conditioning, noise, and other environmental factors are evaluated with people's desks, chairs, CRTs, etc. Hardware should be designed to match human comfort. This is referred to as *ergonomics*. It is becoming an extremely important issue in system development.

Types of System Tests

After a test plan has been developed, system testing begins by testing program modules separately, followed by testing "bundled" modules as a unit. A program module may function perfectly in isolation but fail when interfaced with other modules. The approach is to test each entity with successively larger ones, up to the system test level.

System testing consists of the following steps:

1. Program(s) testing.
2. String testing.
3. System testing.
4. System documentation.
5. User acceptance testing.

Each step is briefly explained here.

Program Testing. A program represents the logical elements of a system. For a program to run satisfactorily, it must compile and test data correctly and tie in properly with other programs. Achieving an error-free program is the responsibility of the programmer. Program testing checks for two types of errors: syntax and logic. A *syntax* error is a program statement that violates one or more rules of the language in which it is written. An improperly defined field dimension or omitted key words are common syntax errors. These errors are shown through error messages generated by the computer. A *logic* error, on the other hand, deals with incorrect data fields, out-of-range items, and invalid combinations. Since diagnostics do not detect logic errors, the programmer must examine the output carefully for them.

When a program is tested, the actual output is compared with the expected output. When there is a discrepancy, the sequence of instructions must be traced to determine the problem. The process is facilitated by breaking the program down into self-contained portions, each of which can be checked at certain key points. The idea is to compare program values against desk-calculated values to isolate the problem.

String Testing. Programs are invariably related to one another and interact in a total system. Each program is tested to see whether it conforms to related programs in the system. Each portion of the system is tested against the entire module with both test and live data before the entire system is ready to be tested.

System Testing. System testing is designed to uncover weaknesses that were not found in earlier tests. This includes forced system failure and validation of the total system as it will be implemented by its user(s) in the operational environment. Generally, it begins with low volumes of transactions based on live data. The volume is increased until the maximum level for each transaction type is reached. The total system is also tested for recovery and fallback after various major failures to ensure that no data are lost during the emergency. All this is done with the old system still in operation. After the candidate system passes the test, the old system is discontinued.

System Documentation. All design and test documentation should be finalized and entered in the library for future reference. The library is the central location for maintenance of the new system. The format, organization, and language of each documentation should be in line with system standards.

User Acceptance Testing. An acceptance test has the objective of selling the user on the validity and reliability of the system. It verifies that the system's procedures operate to system specifications and that the integrity of vital data is maintained. Performance of an acceptance test is actually the user's show. User motivation and knowledge are critical for the successful performance of the system. Then a comprehensive test report is prepared. The report indicates the system's tolerance, performance range, error rate, and accuracy.

QUALITY ASSURANCE

The amount and complexity of software produced today stagger the imagination. Software development strategies have not kept pace, however, and software products fall short of meeting application objectives. Consequently, controls must be developed to ensure a quality product. Basically, quality assurance defines the objectives of the project and reviews the overall activities so that errors are corrected early in the development process. Steps are taken in each phase to ensure that there are no errors in the final software.

Quality Assurance Goals in the Systems Life Cycle

The software life cycle includes various stages of development, and each stage has the goal of quality assurance. The goals and their relevance to the quality assurance of the system are summarized next.

Quality Factors Specifications

The goal of this stage is to define the factors that contribute to the quality of the candidate system. Several factors determine the quality of a system:

1. *Correctness*—the extent to which a program meets system specifications and user objectives.
2. *Reliability*—the degree to which the system performs its intended functions over a time.
3. *Efficiency*—the amount of computer resources required by a program to perform a function.
4. *Usability*—the effort required to learn and operate a system.
5. *Maintainability*—the ease with which program errors are located and corrected.
6. *Testability*—the effort required to test a program to ensure its correct performance.
7. *Portability*—the ease of transporting a program from one hardware configuration to another.
8. *Accuracy*—the required precision in input editing, computations, and output.
9. *Error tolerance*—error detection and correction versus error avoidance.

10. *Expandability*—ease of adding or expanding the existing data base.
11. *Access control and audit*—control of access to the system and the extent to which that access can be audited.
12. *Communicativeness*—how descriptive or useful the inputs and outputs of the system are.

Software Requirements Specifications

The quality assurance goal of this stage is to generate the requirements document that provides the technical specifications for the design and development of the software. This document enhances the system's quality by formalizing communication between the system developer and the user and provides the proper information for accurate documentation.

Software Design Specifications

In this stage, the software design document defines the overall architecture of the software that provides the functions and features described in the software requirements document. It addresses the question; How it will be done? The document describes the logical subsystems and their respective physical modules. It ensures that all conditions are covered.

Software Testing and Implementation

The quality assurance goal of the testing phase is to ensure that completeness and accuracy of the system and minimize the retesting process. In the implementation phase, the goal is to provide a logical order for the creation of the modules and, in turn, the creation of the system.

Maintenance and Support

This phase provides the necessary software adjustment for the system to continue to comply with the original specifications. The quality assurance goal is to develop a procedure for correcting errors and enhancing software. This procedure improves quality assurance by encouraging complete reporting and logging of problems, ensuring that reported problems are promptly forwarded to the appropriate group for resolution, and reducing redundant effort by making known problem reports available to any department that handles complaints.

Levels of Quality Assurance

There are three levels of quality assurance: testing, validation, and certification.

In system *testing*, the common view is to eliminate program errors. This is extremely difficult and time-consuming, since designers cannot prove 100 percent accuracy. Therefore, all that can be done is to put the system through a "fail test" cycle—determine what will make it fail. A successful test, then, is one that finds errors. The test strategies discussed earlier are used in system testing.

System *validation* checks the quality of the software in both simulated and live environments. First the software goes through a phase (often referred to as *alpha testing*) in which errors and failures based on simulated user requirements are verified and studied. The modified software is then subjected to phase two (called *beta testing*) in the actual user's site or a live environment. The system is used regularly with live transactions. After a scheduled time, failures and errors are documented and final correction and enhancements are made before the package is released for use.

The third level of quality assurance is to *certify* that the program or software package is currect and conforms to standards. With a growing trend toward purchasing ready-to-use software, certification has become more important. A package that is certified goes through a team of specialists who test, review, and determine how well it meets the vendor's claims. Certification is actually issued after the package passes the test. Certification, however, does not assure the user that it is the best package to adopt; it only attests that it will perform what the vendor claims.

In summary, the quality of an information system depends on its design, testing, and implementation. One aspect of system quality is its reliability or the assurance that it does not produce costly failures. The strategy of error tolerance (detection and correction) rather than error avoidance is the basis for successful testing and quality assurance.

TRENDS IN TESTING

In the future, we can expect unparalleled growth in the development and use of automated tools and software aids for testing. One such tool is the functional tester, which determines whether the hardware is operating up to a minimal standard. It is a computer program that controls the complete hardware configuration and verifies that it is functional. For example, it can test computer memory by performing read/write tests, and it tests each peripheral device individually.

The functional tester is of great value when minute hardware problems are disguised as software bugs. For example, hardware faults are usually repeatable, wheras software bugs are generally erratic. Problems arise when the delicate interaction between hardware and software cause a hardware problem to appear as an erratic software bug. A functional tester determines immediately that the problem is in the hardware. This saves considerable time during testing.

Another software aid is the debug monitor. It is a computer program that regulates and modifies the applications software that is being tested. It can also control the execution of functional tests and automatically patch or modify the application program being tested.

The use of these tools will increase as systems grow in size and complexity and as the verification and insurance of reliable software become increasingly important.

ROLE OF THE DATA PROCESSING AUDITOR

The planned test of any system ought to include a thorough auditing technique and introduce control elements unique to the system. The data processing (DP) auditor should be involved in most phases of the system life cycle, especially system testing. In the past, auditors have audited systems after they have been installed. Then the cost is often too prohibitive to go back and modify the system to incorporate adequate control. Therefore, audit controls must be built into the system design and tested with the cooperation of both the analyst and the DP auditor. The auditor's role is to judge the controls and make recommendations to the system team in charge of the project. The user department should participate in reviewing the control specifications for the system to ensure that adequate control has been provided.

For testing programs, test data must include transactions that are specifically designed to violate control procedures incorporated in the program as well as valid transactions to test their acceptance by the system. The control setup must be carefully examined. At the time programs are tested, all required files are accumulated and set out in the proper order and format for final testing.

The Audit Trail

An important function of control is to provide for the *audit trail*. In designing tests, the auditor is concerned about the changing nature of the audit trail. In an online environment, the form, content, and accessibility of records are such that the auditor often has difficulty following a single transaction completely through the system. Some of the following changes in the audit trail confront the auditor:

1. Source documents are no longer used by the system after they are transcribed onto a machine-readable medium. They are often filed in areas or ways that makes later access difficult. In direct data entry, traditional source documents are simply eliminated.
2. Files stored on tape or disk can be read only by a computer, which limits the auditing function. A data dump is possible, though, to compare the data against a data map.
3. Processing activities are difficult to observe, since most of them are within the system. It is possible, however, to get a trace when required.

One way to maintain a viable audit trail is to keep a detailed file of the transactions as they occur. The file can then be the input for an audit program that extracts the transactions for selected accounts and prints them so that the auditor can trace the status of an account in detail.

Given the important role of the auditor, he/she is expected to be part of the system development team, which includes the user. As an independent adviser, the role is to judge the controls and make recommendations to the team. Three important steps are considered in the evaluations:

1. Define the control objectives as separate design and test requirements. Input preparation and transmission by the user are important control areas that are viewed with an emphasis on audit trails, error-correction procedures, and adequate documentation during testing. These areas should always be present and well documented.

2. Reexamine budget costs to see whether system testing is within the limits.

3. Review specifications. The auditor should evaluate program acceptance test specifications and assist the system/programmer in developing test standards, various levels of testing, and actual test conditions. He/she should also evaluate the actual system acceptance test to ensure an acceptable level of confidence and reliability.

In summary, it is the auditor's responsibility to build controls into candidate systems to ensure integrity, reliability, and confidence of the users at all levels. The auditor should be called in during design as well as testing so that any suggestions he/she has can be considered before implementation, when changes are less costly. Including the auditor in the system development team makes it easier for him/her to monitor testing procedures and consider the acceptance of new controls to replace those changed by the new design.

Summary

1. Inadequate testing or nontesting leads to errors that may be costly when they appear months later. Effective testing translates into cost savings from reduced errors. It also has utility as a user-oriented vehicle before implementation.
2. A candidate system is subject to a variety of tests: online response, volume, stress, recovery and security, and usability tests. Each test has a unique benefit for a successful installation.
3. Test data may be artificial or live (taken from the user's files). In either case, they should provide all combinations of values or formats to test all logic and transaction path subroutines.
4. The activity network for system testing entails the following.
   a. Prepare test plan.
   b. Specify conditions for user acceptance testing.
   c. Prepare test data for program testing.

d. Prepare test data for transaction path testing.
e. Plan user training.
f. Compile/assemble programs.
g. Prepare job performance aids.
h. Prepare operational documents.

Established design specifications for a test plan cover:

a. Expected output.
b. Criteria for evaluating the output.
c. A volume of test data.
d. Explanation of how test data will be used.
e. Personnel and training requirements.

5. The main purpose of system testing is to identify and correct errors in the candidate system. The performance criteria include:
   a. Turnaround time.
   b. Backup.
   c. File protection.
   d. The human factor with a focus on ergonomics.
6. System testing begins with testing the programs, followed by string and system testing, system documentation verification, and user acceptance testing. The latter test involves the user directly and verifies that the system's procedures operate to standards.
7. Quality assurance has one objective: to make certain that the user receives a quality system tailored to the requirements set in advance. Quality assurance plays a role in each phase of the system's life cycle.
8. The factors that determine system quality include correctness, reliability, efficiency, usability, maintainability, testability, and portability. For each quality factor, the software must meet certain criteria, including completeness, consistency, accuracy, error tolerance, access control, and machine independence.
9. Quality assurance specialists use three levels of quality assurance:
   a. *Testing* a system to eliminate errors.
   b. *Validation* to check the quality of software in both simulated and live environments.
   c. *Certification* that the program or software package is correct and conforms to standards.
10. The planned test of a system should include a thorough auditing technique and introduce control elements unique to the system. The DP auditor should be involved in most phases of the system life cycle, especially testing. The auditor's role is to judge the controls of the system and maintain an audit trail to ensure the integrity, reliability, and confidence of the user at all levels. Being included in the system development team makes it easier for the auditor to monitor testing procedures and consider the acceptance of new controls to replace those changed by the new design.

Key Words

Alpha Testing
Audit Trail
Beta Testing
Desk Checking
Ergonomics
Quality Assurance
Sequential Testing
Stress Testing
String Testing
System Testing
Turnaround Time
Unit Testing
User Acceptance Testing
Validation

Review Questions

1. Why do we test systems? How important is testing? Elaborate.
2. A variety of tests are used to test systems. Discuss three such tests.
3. How is stress testing different from volume testing?
4. What types of test data are used in system testing?
5. Distinguish between the following:
   a. String and system testing.
   b. Quality assurance and DP audit.
   c. Compiling and assembling.
   d. Artificial and live test data.
   e. Error tolerance and accuracy.
6. The first step in system testing is to lay out a test plan. Outline the activities that represent a test plan.
7. What design specifications are considered in preparing a test plan? Explain.
8. There are two ways of debugging program software: bottom-up and top-down. How do they differ?
9. Specify the purpose of system testing. What performance criteria are used for system testing? Discuss.
10. Elaborate on the steps taken in system testing that lead to the user's acceptance of the system.
11. What is a syntax error? How does it differ from a logic error? Give an example.
12. Define quality assurance. From your understanding of this concept, why the fuss over it?
13. List and briefly describe the factors that affect the quality of a system.
14. What levels of quality assurance must a system meet? Explain.
15. Write an essay on the role of the DP auditor in system testing.

Application Problems

1 A supermarket contracted with a local computer service to process daily sales. The arrangement called for the installation of nine terminals at the check-out counters hooked remotely to a computer facility six miles away. The store clerk enters the code number, quantity, and price of each item. The data are transmitted directly to the processing center for inventory update. Other status reports reflecting the pattern of sales, leader items, and other data are also produced on a daily basis.

When the terminals were being installed, the store manager asked the 10 store clerks to be available for training beginning Wednesday of the following week. However, by the time the terminals were checked out and connected to the computer facility, two of the three days reserved for training had gone by. Rather than delaying the final conversion, intensive training was offered on Friday and Saturday morning. The employees went through training without much complaint.

When the system was ready the following Monday, the clerks had difficulty correcting errors. For example, if a grocery item was keyed in as "meat" instead of "fruit," two consecutive keys had to be depressed to allow a correction. The clerks claimed that all they were told to do was depress the "clear" key. Consequently, any other data related to the item in question were wiped out.

Another problem the clerks had was about the sales tax on nonfood items, such as detergents, cigarettes, and magazines. They could not remember exactly how to code or handle these items for sales tax computation. One clerk stated that even when she entered nonfood items properly, the system did not compute the tax.

After two days of operation, a consultant was hired to take a look. He found the entire training pogram to be inadequate. Training was done by someone who knew very little about the software changes that had recently been made in the system. Futhermore, no meaningful test data were used during training to help the clerks understand the relevance of their work on the terminal to the store's operation. The sales tax problem was found to be a software bug. It was easily corrected at the computer's end of the operation.

Assignment

a. What specifically was wrong with the training program? Who was to blame? Why?

b. If you were the store's manager, what would you do?

c. Outline a training program and a test procedure for this type of conversion.

2 ACME BOOKSTORES

Acme Bookstores is a tightly held bookstore chain. Established in 1947 by Joseph Acme, the company has grown from a small bookstore to 17 stores serving 17,000 students of a major university. Mr. Acme, Jr. is president, treasurer, and general manager. He is directly responsible to the board of trustees. The remaining three officers are an executive vice president, a vice president, and an assistant treasurer. There are also five department managers; the corporate comptroller, the director of purchasing, the manager for college textbooks, the manager for law, nursing, and graduate business textbooks, and the director of shipping and receiving (see Exhibit 12–1).

The study focuses on the accounts payable in the accounting department. The accounts payable staff consists of an invoice processing clerk and three accounts payable clerks. Each clerk has specific work; one processes publishers whose names begin with A through L,

EXHIBIT 12–1 Organization Structure

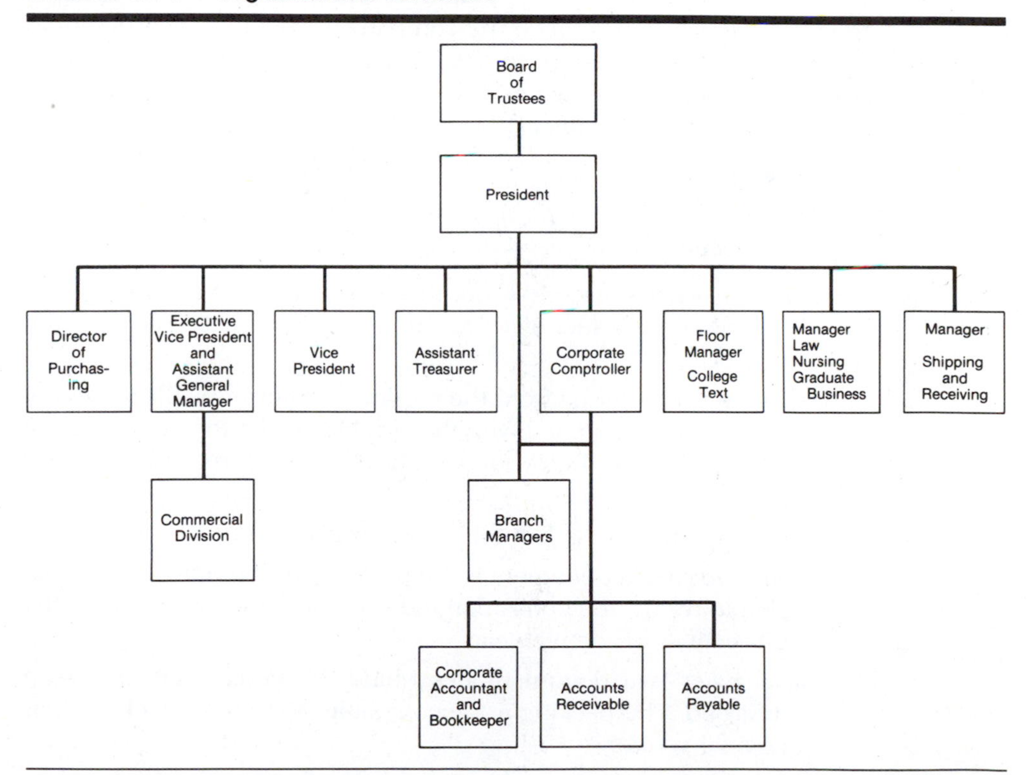

another is responsible for M through Z, and the third handles discount accounts, manufacturers' accounts, and check writing. All clerks are located in the same room and can perform one another's work.

The Problem

Acme Bookstores presently operates a manual accounting system which is preventing the company from expanding or efficiently handling invoices. Because of the volume of transactions that must be handled monthly by the accounts payable clerks, they are unable to locate companies who offer discounts for early payment and, therefore, are losing approximately $10,000 per year in discounts alone. Since they do not have time to verify all the billing statement figures, bad pricing and billing errors are ignored and payments are sent without being reconciled with the firm's records. Because of the time constraints, the accounting records are not being balanced monthly and human errors are not caught.

In addition, the present system does not provide a breakdown by branch of the operating results. This has resulted in a control problem for Acme Bookstores. For example, a branch manager was suspected of embezzling funds, yet by the time the company realized this problem, the audit trail was lost and he could not be prosecuted. Lack of specific branch information has made the company unable to detect potential operating problems relating to that branch. There is also no possible way for the firm to compile monthly financial statements to summarize and examine operating results.

The Present System

The present accounts payable system is run entirely by hand. The following documents are used in the department:

Purchase order is used to order the books. Details of the supplies to be ordered are included on the order form that is sent to the vendor.

The *invoice* is received from the vendor to confirm the placement of an order. It is matched with the purchase order to reassure Acme Bookstores that their order has been properly received.

Receiving report is completed upon arrival of the shipment and used to check whether the supply checks out.

Request to return document is a slip asking for the vendor's permission to return the books that did not sell at the beginning of the university's academic year.

Packing list is a document describing the books that are being returned. The packing list accompanies the books back to their original vendor.

The accounts payable system is illustrated in Exhibit 12–2.

EXHIBIT 12-2 Accounts Payable System—Data Flow Diagram

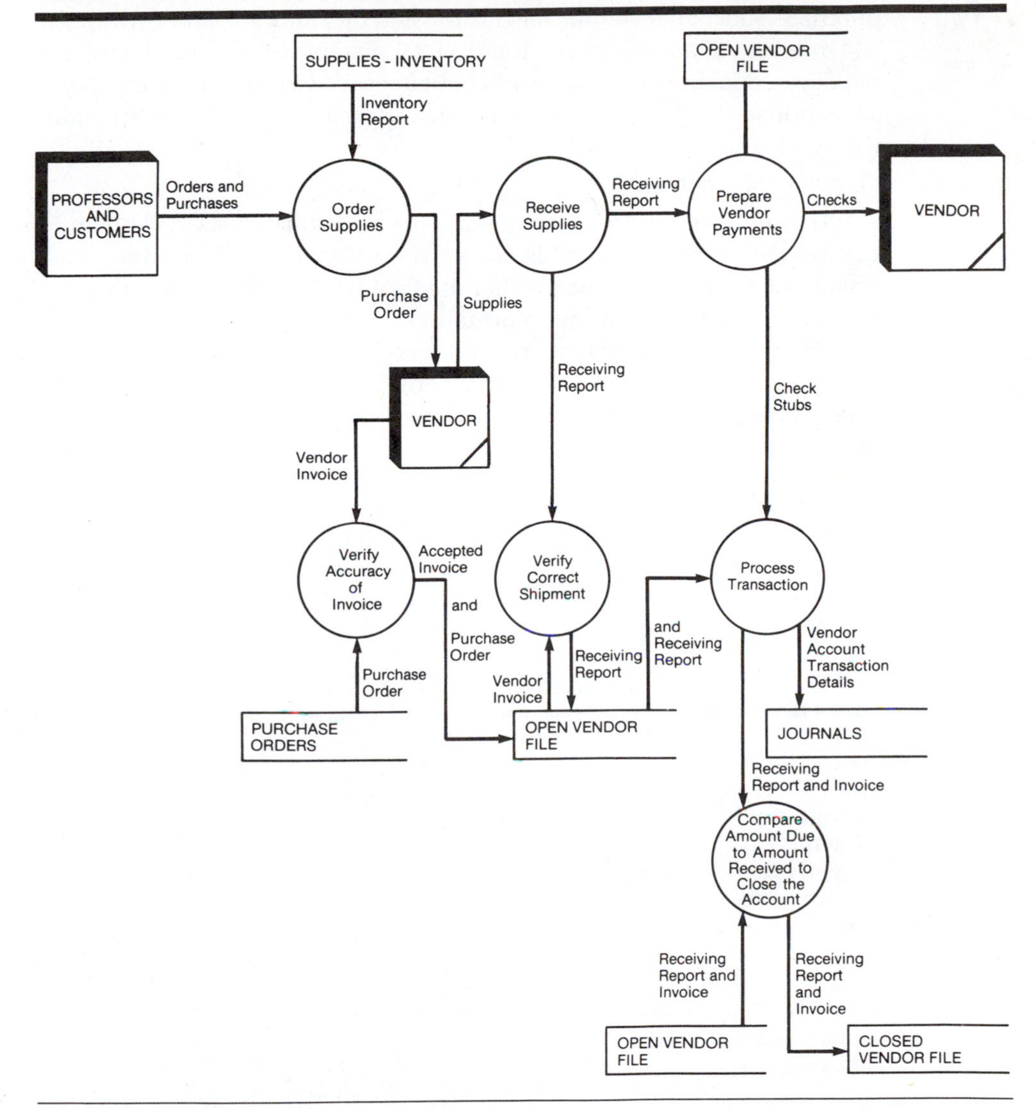

Hardware/Software Selection

To alleviate the problem, the appropriate hardware for the ideal system had to be selected and installed. In the selection process, three decisions were made: the software to do the job, the brand of hardware, and the physical site. Software packages for accounts payable are numerous, depending on the brand of computer. It was found that most of the software was available on IBM systems. Since emphasis has been on the microcomputer, an accounts payable package was located that met the user's requirements, including a check-printing facility, a purchase journal, a vendor file, an invoice file, and a cash flow report.

In terms of the hardware, the IBM/XT system was found to have the memory capacity, flexibility, and cost effectiveness. Other factors, such as response time, communication links, control, and service, also made the system look good.

Cost/Benefit Analysis

The next step in the project was cost/benefit analysis. As shown in Exhibit 12–3, the total tangible costs were $19,092, and there was a cost savings of $18,316. This means that the firm would break even in almost one year. In addition, more information could be handled quickly and accurately. The timeliness ensured making payment deadlines and benefiting from discounts. Security control would be vastly improved. These benefits were expected to improve employee morale and motivation.

The Software Package

The accounts payable software is a completely integrated system that is totally menu-driven and uses clearly formatted screen displays to enter and process data. The system contains a master menu that

EXHIBIT 12–3 Cost/Benefit Analysis

Tangible costs:		
Hardware		
System XT with one drive	$ 2,104	
Extra memory	375	
Monitor	680	
70 megabytes with backup tapes	7,495	
Interface board	149	
Printer	1,000	
Total		$11,803
Software		
DOS 2.1 operating system	65	
Software package (complete)	2,780	
Basic interpreter	195	
Total		3,040
Maintenance		
12 percent of hardware cost = $1,416/year for three years		4,249
Total costs		$19,092
Cost Savings:		
Ability to pick up discounts	$10,000	
Reduction in employee time	3,840	
Check-writing time	2,160	
Processing transactions time	927	
Closing of the books	1,344	
Total Savings		$18,316

allows the user to select from the several options: general ledger, accounts payable, accounts receivable, and inventory control. After an option is selected, the package displays a second menu that contains the options of the particular processing system or utility function selected (see Exhibit 12–4).

The proposed accounts payable system provides the user instant accessibility to his/her cash flow position. It automatically posts debits and credits to vendor invoices. All account distribution information is automatically posted to the general ledger system. The system operates in an online interactive mode. All disbursements, debit/credit entries, and vendor maintenance are accomplished through the console with

EXHIBIT 12–4 Software Package—Main Menu

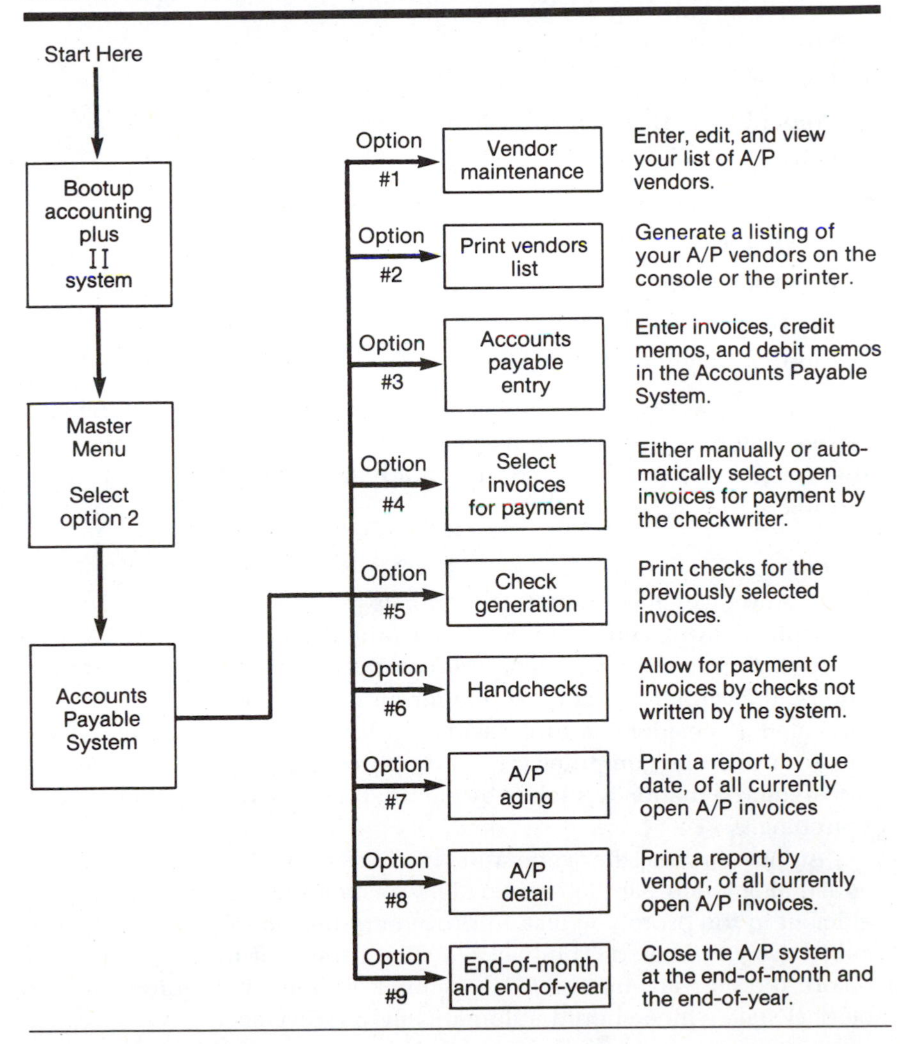

instant updating of data files. A detailed audit trail is maintained while each entry is posted to the accounts payable debit file. Major reports available are vendor listing, cash requirements, check register, hand check register, accounts payable detail, and accounts payable aging. Checks are printed automatically or entered manually, including detail on the check stub. A complete audit trail is provided via the detail journal listing feature of the general ledger module. A data flow diagram of the proposed system is shown in Exhibit 12–5.

Test Procedure

To test the package, a prototype test was conducted on the IBM/XT system in an attempt to reproduce a typical date in the accounts payable department. A representative sample of data collected from the firm included the following:

- Purchase orders.
- Invoices.
- Prepaid request purchase orders.
- Petty cash requests.
- Miscellaneous bills (advertising, rent, etc.).
- Packing lists.
- Receiving reports.
- Credit and debit memos.
- Monthly statements from vendors.

The master vendor list and all account balances were entered prior to the prototype test. A large number of invoices was also entered into the system so that each account would start with different amounts payable.

Results

The average time required to manually process each document was determined from actual observations and then compared with the amount of time required using the computer, which included determining the vendor key and waiting for files to be updated. The results indicated a considerable time savings with the candidate system. Paying an invoice by computer required approximately two thirds of the time required manually, mainly by eliminating the time spent filing and searching.

Automatic computer generation of checks, which can be very time-consuming when handled manually, was about 90 percent more time efficient in the prototype test. Improvement in accuracy could be only subjectively estimated. The package allows the user to edit any errors before the files are updated. No manual addition is required by the clerical staff, which should automatically improve accuracy.

EXHIBIT 12-5 Data Flow Diagram—Candidate System

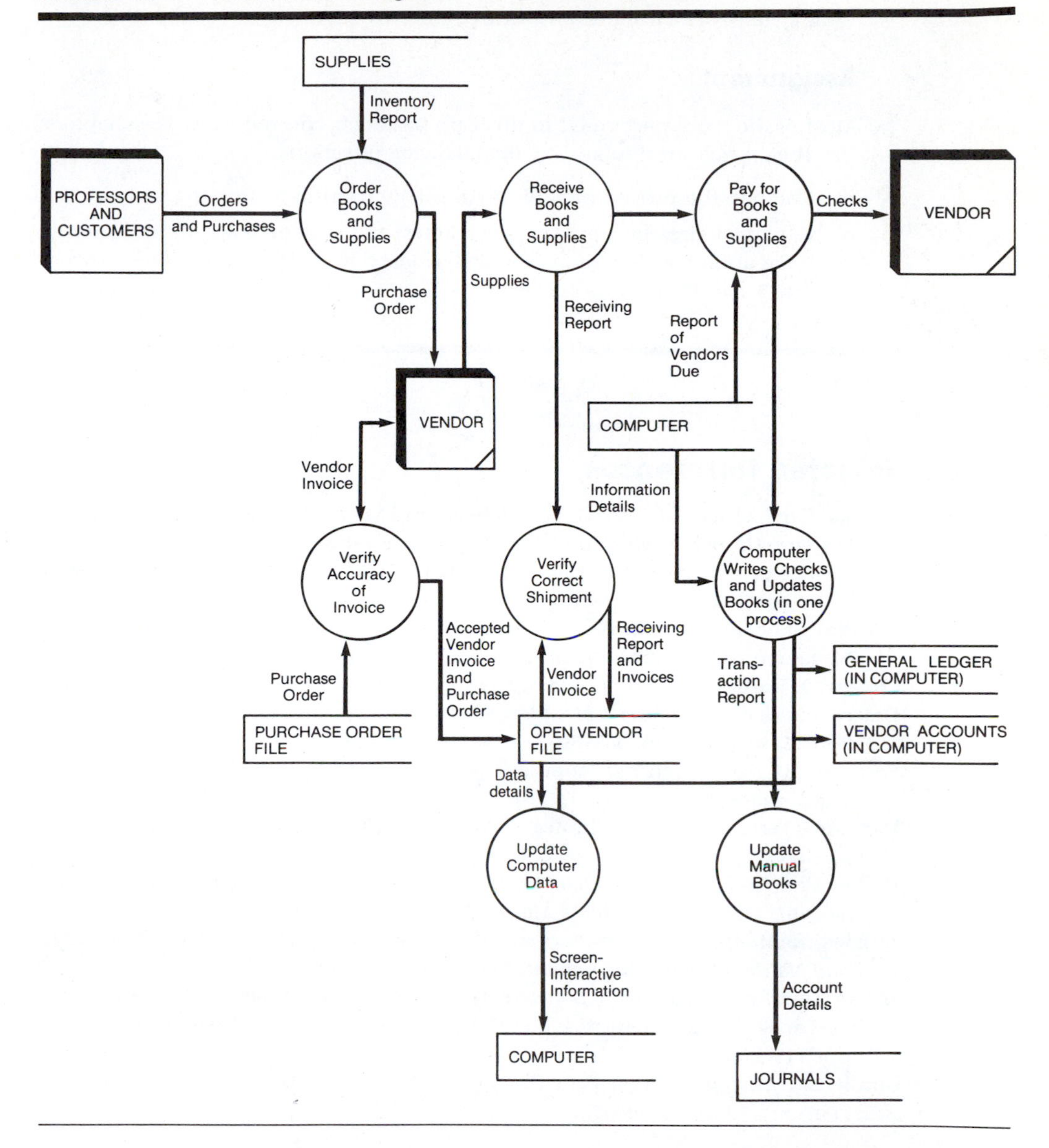

There were some drawbacks to the system, however. The amount of time required to read and write from the disk drives represented idle human time. This could be slightly annoying for the user. In addition, some of the character fields were not long enough to include all the characters that the user wanted to enter into the system. For example, a few invoice numbers exceeded the allowable maximum of eight characters. This required the user to stop and think about what characters should be omitted. Such problems could be frustrating because they take a disproportionate amount of time to solve.

Assignment

Most of the work performed in this case has been covered in earlier chapters. Use this background to answer the following questions:

a. Evaluate the procedure and methodology used in testing the system.

b. What test strategy would you use to test the system? Explain in detail.

c. What audit considerations should be looked into? How would one incorporate an audit trail? Be specific.

Selected References

Brooks, Cyril; Phillip Grouse; D. Ross Jeffery; and Michael J. Lawrence. *Information Systems Design*. Englewood Cliffs, N.J.: Prentice-Hall, 1982, pp. 366–71.

Brown, Foster. "Audit Control and Systems Design." *Journal of Systems Management*, April 1975, pp. 24–31.

Cerullo, Michael J. "Controls for On-Line Real-Time Computer Systems." (1) *CA Magazine*, March 1980, pp. 58–61; (2) *CA Magazine*, May 1980, pp. 54–58.

Couger, J. Daniel. *The Art of Software Testing*. New York: John Wiley & Sons, 1979.

Dolan, Kathleen. *Business Computer Systems Design*. Santa Cruz, Calif.: Mitchel Publishing, 1984, pp. 302–305.

Foss, W. B. "A Structured Approach to Computer System Testing." *Canadian Datasystems*, September 1977, pp. 28–32.

Harrison, Warren. "Testing Strategy." *Journal of Systems Management*, May 1981, pp. 34–37.

Holley, Charles L., and Frederick Millar. Auditing the On-Line, Real-Time Computer System." *Journal of Systems Management*, January 1983, pp. 14–19.

Jeffries, Kenneth R. "Auditing Advanced EDP Systems—The Problems Still Exist." *The National Public Accountant*, July 1981, pp. 16–21.

Myers, Glenford J. *The Art of Software Testing*. New York: Wiley Interscience, 1979.

Ogdin, Carole A. "Software Aids for Debugging." *Mini-Micro Systems*, July 1980, pp. 115–20.

Smolenski, Robert J. "Test Plan Development." *Journal of Systems Management*, February 1981, pp. 32–37.

Chapter 13

Implementation and Software Maintenance

At a Glance

A crucial phase in the systems life cycle is the successful implementation of the new system design. Implementation simply means converting a new system design into operation. This involves creating computer-compatible files, training the operating staff, and installing hardware, terminals, and telecommunications network (where necessary) before the system is up and running. A critical factor in conversion is not disrupting the functioning of the organization.

In system implementation, user training is crucial for minimizing resistance to change and giving the new system a chance to prove its worth. Training aids, such as user-friendly manuals, a data dictionary, job performance aids that communicate information about the new system, and "help" screens, provide the user with a good start on the new system.

Following conversion, it is desirable to review the performance of the system and to evaluate it against established criteria. Software maintenance follows conversion to the extent that changes are necessary to maintain satisfactory operation relative to changes in the user's environment. Maintenance often includes minor enhancements or corrections to problems that surface late in the system's operation.

By the end of this chapter, you should know:

1. What activities are considered in system conversion.
2. The do's and don'ts of user training.
3. How to handle user resistance to change.
4. The makeup of the post-implementation review.
5. The role, elements, and importance of software maintenance.

INTRODUCTION

An important aspect of a systems analyst's job is to make sure that the new design is implemented to established standards. The term *implementation* has different meanings, ranging from the conversion of a basic application to a complete replacement of a computer system. The procedure, however, is virtually the same. *Implementation* is used here to mean the *process of converting a new or a revised system design into an operational one.* Conversion is one aspect of implementation. The other aspects are the post-implementation review and software maintenance. These topics are covered later in the chapter.

There are three types of implementation:

1. *Implementation of a computer system to replace a manual system.* The problems encountered are converting files, training users, creating accurate files, and verifying printouts for integrity.
2. *Implementation of a new computer system to replace an existing one.* This is usually a difficult conversion. If not properly planned, there can be many problems. Some large computer systems have taken as long as a year to convert.
3. *Implementation of a modified application to replace an existing one, using the same computer.* This type of conversion is relatively easy to handle, provided there are no major changes in the files.

This chapter discusses the activities involved in conversion and the changes made in the system following implementation. Software maintenance is becoming an important aspect of building and maintaining systems.

CONVERSION

Conversion means changing from one system to another. The objective is to put the tested system into operation while holding costs, risks, and personnel irritation to a minimum. It involves (1) creating computer-compatible files, (2) training the operating staff, and (3) installing terminals and hardware. A critical aspect of conversion is not disrupting the functioning of the organization.

Conversion should be exciting because it is the last step before the candidate system begins to show results. Unfortunately, the results of conversions have been chaotic and traumatic for many firms. Unforeseen difficulties crop up as the system breaks down, data files are damaged, and tempers grow short. The training package is frequently not complete and people are trying to figure out what to do. Much of this stems from poor planning or no planning at all. Let us examine the steps that preceded conversion:

1. The user recognizes a need.

2. The user solicits help from the analyst, who does an $8,000 study that leads to a proposal.
3. The proposal has general specifications of software and hardware. A vendor is selected and a firm conversion date is set.
4. The vendor or the project team begins to fit its solution to the user's specification.
5. Delays occur; bargaining for trade-offs proceeds, but eventually the system is ready to install—often with little user involvement.
6. The user has not been prepared for duplicate labor costs when both present and candidate systems run during conversion. The user attempts to do much of the "leg work" with his/her staff on an overtime basis.
7. With a combination of poor training and tight deadlines, the user staff grows short-tempered at being in an untenable position.
8. Now the user is angry at the project team or the vendor, and the vendor's staff is getting impatient with the user staff. Everyone is blaming everyone else for this horrible experience.

For many first-time users, this theme is familiar. What went wrong? The fundamental concerns lie with the user, who should know in advance what activities to control through a computer. Users who do not do their homework play right into the hands of vendors who have done theirs. So it is understandable that conversions are often a fiasco.

Activity Network for Conversion

Several procedures and documents are unique to the conversion phase. As shown in Figure 13–1, they include the following:

1. Conversion begins with a review of the project plan, the system test documentation, and the implementation plan. The parties involved are the user, the project team, programmers, and operators.
2. The conversion portion of the implementation plan is finalized and approved.
3. Files are converted.
4. Parallel processing between the existing and the new systems is initiated.
5. Results of computer runs and operations for the new system are logged on a special form.
6. Assuming no problems, parallel processing is discontinued. Implementation results are documented for reference.
7. Conversion is completed. Plans for the post-implementation review are prepared. Following the review, the new system is officially operational.

FIGURE 13–1 Procedures and Documents for Conversion

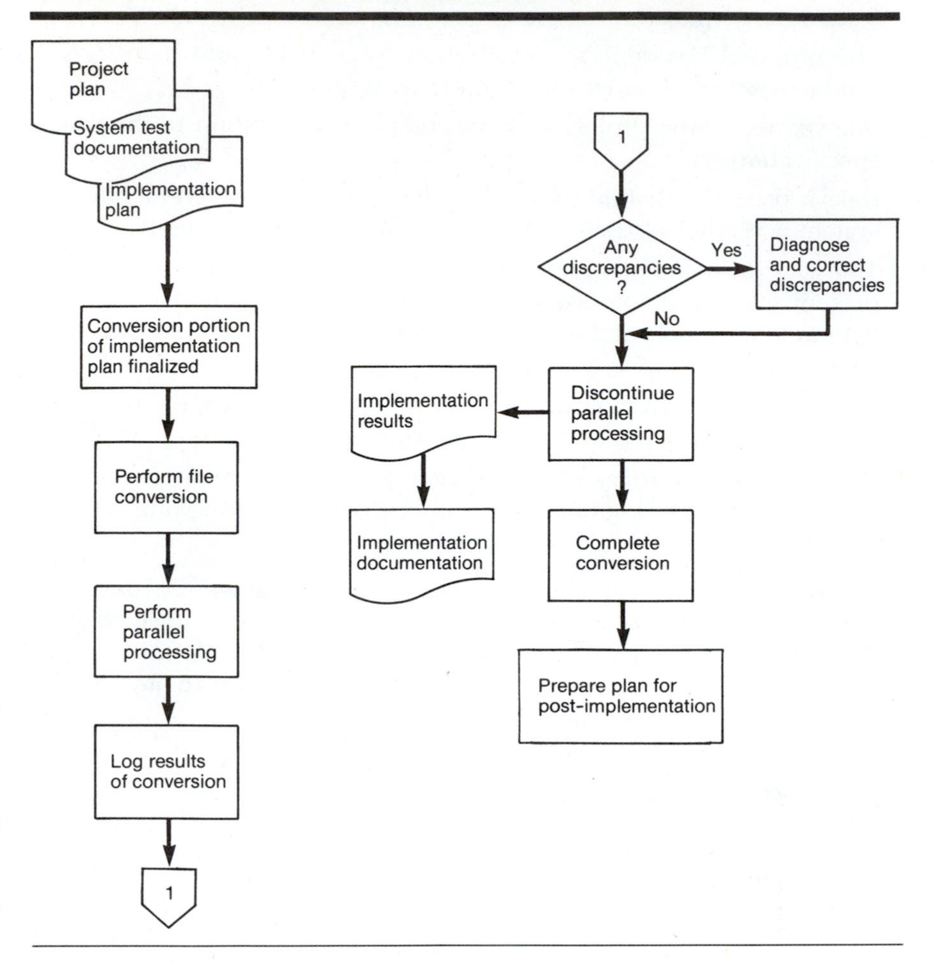

Accompanying the procedures are activities that are unique to conversion (see Figure 13–2). Before we discuss each activity, we need to note that during the transition, the user should be discouraged from making changes or enhancements. To illustrate, a Miami-based commercial bank contracted to convert to a real-time teller system with new terminals linked to a computer service. Forty-eight days were spent testing the software, terminals, controllers, and other features. Two weeks before the final conversion, it was found that when the system was down, the terminals lost all the transaction balances accumulated in memory. To restart the system, the teller had to rekey all items to date before handling new transactions. The solution was to install a backup system, which meant more money and weeks of delays. The request was vetoed in preference for going ahead with the conversion. Backup considerations were shelved until later inclusion as enhancements.

FIGURE 13-2 Activity Network for Conversion

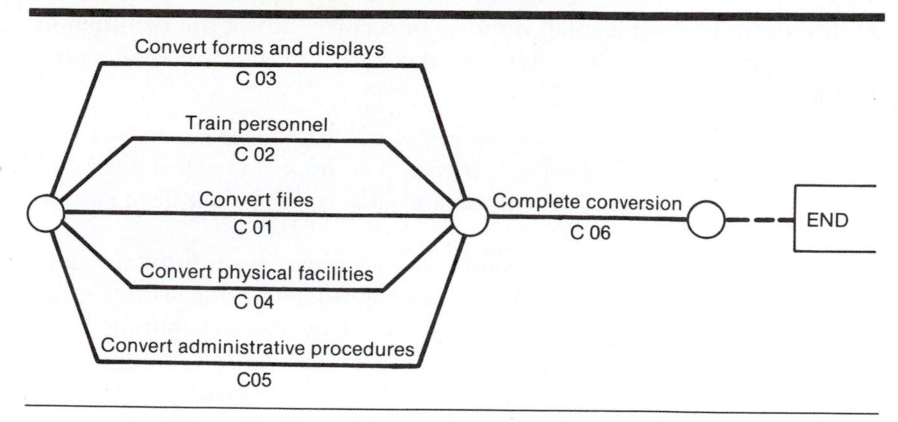

File Conversion

File conversion involves capturing data and creating a computer file from existing files. Problems are staff shortages for loading data and the specialized training necessary to prepare records in accordance with the new system specifications. In most cases, the vendor's staff or an outside agency performs this function for a flat fee.

Copying the "old" files intact for the new system is the prime concern during conversion. The programs that copy the files should produce identical files to test programs on both systems. At the outset, a decision is made to determine which files need copying. Personnel files must be kept, of course, but an accounts receivable file with many activities might not need copying. Instead, new customer accounts might be put on the new system, while running out the old accounts on the old system.

Once it is determined that a particular file should be transferred, the next step is to specify the data to be converted: current files, year-end files, and so on. The files to be copied must be identified by name, the programmer who will do the copying, and the methods by which the accuracy of the copying will be verified. A file-comparison program is best used for this purpose.

Creating Test Files. The best method for gaining control of the conversion is to use well-planned test files for testing all new programs. Before production files are used to test live data, test files must be created on the old system, copied over to the new system, and used for the initial test of each program. The test file should offer the following:

1. Predictable results.
2. Previously determined output results to check with a sampling of different types of records.
3. Printed results in seconds.

4. Simplified error-finding routines.
5. Ability to build from a small number of records out of the production files and then progressively alter the records until they are challenging to the programs.

Selecting the right records for testing is not easy. The user's key staff should get involved in the selection process. The trick is to select a reasonable number of uncomplicated records and also records that have caused problems in the past.

What about security? No user wants employee salaries printed for all to see. Information such as salary, ethnic background, and trade secrets must be altered so that test file results can be read by the programmer, the project leader, and other involved in the conversion without compromise.

A good test file should show how the new system handles the most difficult tasks. A health insurance test file, for instance, would include an employee with every possible type of insurance: worker's compensation, disability, dental, cancer, life, and accident insurance. If the new system programs process this file correctly, it is likely to handle the ordinary employee's insurance.

Data Entry and Audit Control. Many systems are prone to errors because of insufficient attention given to data entry control or protective features such as audit control trails. These items must be part of the overall plan for conversion. Before a data entry clerk begins to enter changes to items in the files, the count of the number of documents to be keyed in should be entered. The data entry program should be designed so that these summary totals are entered first, with the computer keeping a running count of the items entered.

A good audit control trail is the key to detecting fraud and mistakes that might indicate problem entries. At a minimum, the user should have a copy of the additions or deletions to any file. There should be a daily computer-printed entry list with entries and totals before they are entered and a second list showing actual changes after they have been made in the file(s). To detect fraud, many system designs do not allow any records on the file to be deleted easily. Modifications are made by adjustments involving more than one or two operating personnel or similar routines. Preventing records from being deleted makes it more difficult to cover up problems or conceal fraud.

User Training

An analysis of user training focuses on two factors: user capabilities and the nature of the system being installed. Users range from the naive to the highly sophisticated. Developmental research provides interesting insights into how naive computer users think about their first exposure to a new system. They approach it as *concrete* learners, learning how to use the system without trying to understand which abstract principles determine which function. The distinction between concrete and formal (student-type) learning says much about what one can expect from trainees in general.

The requirements of the system also range from very simple tasks like using a pocket calculator to complex tasks like learning to program a data base system. Tasks that require the user to follow a well-defined, concrete, step-by-step procedure require limited problem solving. This means that the training level and duration are basic and brief. For example, training a person how to operate an automated teller machine to obtain cash is a concrete task. Virtually all the "thinking" is performed by the computer. In contrast, requiring a trainee to analyze a given situation and translate it into a procedure for computer manipulation requires formal training for a relatively long time. In either case, training has to be geared to the specific user based on his/her capabilities and the system's complexity.

Elements of User Training. For more than a decade, various user guides and methods have been used, but no single method has been totally successful. Let us look at a real training experience and explain what happened. This case took place during the implementation of our safe deposit billing system described in earlier chapters. Four clerks had to be trained on more than 30 inquiry and update functions using an IBM personal computer. The most important element of the training was a recently revised user manual produced at a cost of 70 analyst-hours. Each online function was described in a menu-driven format. Each clerk was given verbal explanations of each function and its use was demonstrated on the PC. Clerks were asked to read the manual carefully for an in-depth understanding of the software package. Questions were answered during the first meeting and for two weeks afterward. Phone calls and lengthy discussions also followed.

This may seem to be adequate for a training session; the package had only so much to offer. But here is what happened:

Phase 1: The initial training period. Only one of the four clerks read the manual. Instead, the clerks took notes on legal paper, listed the key functions, and described how they are used on the PC. The list was referred to continually. The clerk who read the manual became the *resident expert* by bailing every other clerk out of a jam. During the training period, most of the questions asked had answers in the manual. The most frequently consulted reference was that sheet of paper with the key functions and the name and phone number of the analyst.

Phase 2: Two months later. None of the three clerks had read the manual. A six-page update of the manual from the software house had not even been filed. In fact, two of the clerks had misplaced their copies. Fewer questions were not posed to the analyst out of embarrassment. The questions were answered somehow within the group. The list of functions from the initial training period was beginning to show age from regular use. No one bothered to type it or make backup copies in case it is lost. Worse yet, clerks who did not understand a function made no effort to look it up in the manual as long as it did not affect their job.

Phase 3: Six months later. Training was by the resident expert—word of mouth and a quick two-hour "show and tell" approach to generating reports for billing purposes. The analyst did not get one call. General functions were deleted or changed so that the manual was no longer up to date.

Three important lessons that pertain to user training can be concluded from this case situation:

1. Users are reluctant to read manuals, but they will learn from demonstrations and through visual aids. Users also tend to be natural teachers. For many users, training is mostly informal. There is no question that some writing is required for user training; however, the challenge is to provide documentation that users will read and refer to over the life of the system. An important document that can be provided for training is a one-page summary of important functions about the system and/or software. Such a "cheat sheet" instructs the user on how to start the system and about the various functions and meanings of various codes. It is actually a fact sheet that contains few words and short sentences (see Figure 13–3).

2. Another user training element is a *training demonstration*. Live demonstrations with personal contact are extremely effective for training users. In a demonstration, a new concept that is shown in many ways (see it, hear it, say it—believe it) is quickly (and permanently) learned. More information is conveyed and discussed verbally than through reading or writing during the same time. Finally, during a training demonstration, the user receives encouragement and attention, which prompts him/her to perform. Hearing others ask questions and make mistakes helps alleviate anxieties and improve self-confidence.

Providing a demonstration is not easy for many analysts. A successful effort requires an expert knowledge of the system and advanced planning and organization of reference materials, procedures, and technology. If you

FIGURE 13–3 Fact Sheet—An Example

Code	Unacceptable	Acceptable
DBMS	This code stands for data base management system, which is the software in a computer	Data base management system
ORDR	The purpose of this function is to provide order inquiry	Order inquiry
BOXDRL	This code represents the box drilled in safe deposit	Box drilled

must say, "That's a good question. Let me follow up on that and I'll let you know tomorrow," you are not ready to handle a demonstration. A rule of thumb is to limit the scope of any single demonstration. It should be conducted in an environment conducive to learning and self-discovery. Keeping the session(s) short reduces fatigue and controls confusion. The outcome of all this is to have the user walk away feeling satisfied that he/she has some knowledge about what the system is capable of doing.

3. The third element of user training is the *resident expert*. In our example, one clerk read the manual carefully, spent time on her own to practice, and ended up being the resident expert—a natural teacher. Such a person, whether a respected supervisor or a peer, can relate much better to the user group, speak the language, and use examples based on common experience to teach (and sell) the new system. Training in this case is more rapid, which means a quicker use of the system on a regular basis.

Training Aids. There are several user training aids available:

1. *The user manual.* Traditionally, the user manual is prepared reflexively because it is an item that must accompany every system. Yet, there are times when a user manual is discretionary. The important point is that the manual should be prepared only if it will serve the user. A manual is necessary when the user is geographically removed from the project team or when he/she cannot attend all the training sessions. Of course, if the user requests a manual, then one must be prepared.

Probably the best reason for preparing a user manual is on the recommendation of the resident expert. If it will be used, the manual should be written, despite the cost. It should discuss, among other things the functions available to the user and what each can do, how they are executed, and how diagnostic messages should be handled. Above all, the manual should be well organized and indexed for quick reference. Graphics, pictures, and line drawings enhance the teaching value of the material.

2. *Help screens.* This feature is now available in virtually every software package, especially when it is used with a menu. Essentially, the user selects a "help" option from a menu. The system accesses the necessary description or information for user reference. Then additional "help" information may be accessed, or the user returns to the menu for further action. HELP offers the advantage of virtually unlimited space and, since it is separate from the program, it does not interfere with system operation.

3. *Data dictionary.* As explained in Chapter 9, a data dictionary is a separate place for describing data elements. It is more like an electronic "one-page" sheet available to the user to assure that functions are interpreted and executed properly.

4. *Job aids.* A job aid communicates essential information about certain jobs. It takes a number of forms, for example:

a. Color printing in forms to accentuate headings.

b. Color to identify pieces of hardware, cables, etc.

c. Wall charts to illustrate schematics of processing runs.

d. Flowcharts to guide the user in detecting and handling errors, restarting the system, etc.

It can be seen, then, that training aids help communicate vital information about the new system. To make these aids useful, it is important to consider to whom, when, and why the user will be communicating this information.

Forms and Displays Conversion

During this activity, old forms and displays are withdrawn and new ones are instituted. Various controls are implemented to ensure the system's reliability, integrity, and security. The activities implemented here were initiated early in the system design phase.

Conversion of Physical Facilities

In conjunction with these activities, the physical facilities are transformed to meet the specifications set in the candidate system design. In a recent online conversion in a medium-size bank, the installation of new teller terminals required a complete redesign of the teller cages—relocating the telephones, replacing ceiling lights with soft lighting, and building drawer space for additional cash storage. The cost of the redesign was 10 percent of the new system's cost.

Other factors included in physical facilities conversion are the communications network, safety and security provisions, and notification to customers of the change (see Figure 13–4). For the most part, these activities were planned in the system design. The conversion phase checks each activity prior to final operation.

Conversion of Administrative Procedures

A final important activity in the conversion phase is setting up administrative procedures for controlling the new system. This includes scheduling, determining job priorities on the system, and implementing personnel policies for managing the system. The user is trained to handle various emergencies and procedures. Most important, supervisors are trained on how the information is gathered, produced, and presented to management.

COMBATING RESISTANCE TO CHANGE

Hazel Jones was 60 years old and had worked for Power House Athletics (a shoe store) for 17 years. Her reputation among the employees is that there is the right way, the company's way, and Hazel's way. Her way is always followed. Her boss, Herman, is a young college graduate. He is always nervous around Hazel.

Hazel's job is to post customer and vendor accounts, keep track of the

FIGURE 13-4 Notice of Conversion to Customers—An Example

To Our Valued Customers

WE NEED YOUR HELP!

On August 1, 1984, we will begin working with a modern "on-line" teller system designed to provide you with the best available service. Please bear with us during our conversion period.

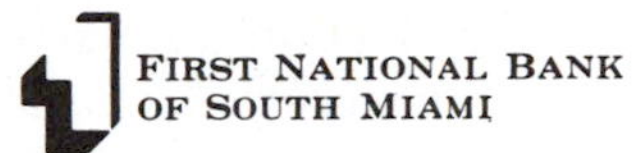

inventory, and balance the books before she leaves for home. One spring day, the main door opened. A husky fellow unloaded a personal computer from the van and set it on Hazel's desk. Hazel first refused to allow the boxes in the office. She even refused to sign the delivery notice. Later, when the computer store representative came to set up the system, she refused to allow him to use the bathroom. He had to go to the Dunkin Donut facilities next door. Her boss arrived at the scene. All we know is that he left to go home sick within 20 minutes.

This scenario might be somewhat exaggerated, but it exemplifies what might happen when a new system arrives and the user staff is ill prepared to meet the challenge. Frequently, behavioral factors are overlooked in system implementation. Change is accompanied by anxiety and other stresses, which induce resistance. Any type of change imposes stress on the people affected by it. According to Holmes and Rahe, after death of a spouse, which they placed highest at 100, changes in job responsibilities as are common in system changes, are located on a scale of 30.[1] This implies that system changes can produce stress which increases unless checked through planning.

Anxiety is produced when a person does not know the outcome of an induced change:

1. Change in job content.

---

[1] T. H. Holmes, and R. H. Rahe, "The Social Readjustment Rating Scale," *Journal of Psychosomatic Research*, 1967, pp. 213–18.

2. How to think along new lines.
3. Change in personal relations patterns.
4. Possible loss of self-esteem relative to the new job.
5. Possible loss of one's job.
6. Loss of control of job content.

Thus, when people experience stress, part of their reaction is resistance to change. Here are examples of the ways they react:

1. Hostility toward peers or others, communication barriers, or as in our example, refusing to sign the delivery notice.
2. Withdrawal from the cause of stress—the supervisor going home sick.
3. Rejection of the stressful situation—refusing to accept the computer.

These are general effects, but there are also specific problems associated with resistance to a new installation. For example, the introduction of the new system has an adverse effect on employee productivity and job satisfaction. It may create an environment of hostility that lasts a long time after implementation. Employees may even sabotage the system to seek relief because they are unable to understand its operation. All these problems stem from inadequate communication between the user staff and the project team, attitudinal factors, or perceptions that the new system inadequately meets the user's requirements.

Several strategies reduce resistance to system change:

1. Identify and discuss the deficiencies of the present system.
2. Do a good, honest job at showing how a changeover will improve the quality of life at work.
3. Establish open communication channels between the user staff and the technical staff to answer questions and follow up on difficulties after implementation.
4. Invite and use employee participation in all phases of the conversion process.

With all the technical sophistication that accompanies conversion, interpersonal relations with the user staff are extremely important in implementation. Those charged with changing the system should be sensitive to these relations and, in so doing, try to minimize the impact of the new system on the user.

POST-IMPLEMENTATION REVIEW

Operational systems are quickly taken for granted. Every system requires periodic evaluation after implementation. A post-implementation review measures the system's performance against predefined requirements. Unlike system testing, which determines where the system fails so that the necessary adjustments can be made, a post-implementation review determines how well the system continues to meet performance specifications. It is after the fact—after design and conversion are complete. It also provides information to determine whether major redesign is necessary.

A post-implementation review is an evaluation of a system in terms of the extent to which the system accomplishes stated objectives and actual project costs exceed initial estimates. It is usually a review of major problems that need converting and those that surfaced during the implementation phase. The primary responsibility for initiating the review lies with the user organization, which assigns special staff for this purpose. Figure 13-5 shows the activity network for the post-implementation review. Each activity is explained briefly.

Request for Review

The initiating study begins with the review team, which gathers and reviews requests for evaluation. It also files discrepancy notices after the system has been accepted. Unexpected change in the system that affects the user or system performance is a primary factor that prompts system review. Once a request is filed, the user is asked how well the system is functioning to specifications or how well the measured benefits have been realized. Suggestions regarding changes and improvements are also sought. This phase sets the stage for a formal post-implementation review.

A Review Plan

The review team prepares a formal review plan around the objective(s) of the review, the type of evaluation to be carried out, and the time schedule required. An overall plan covers the following areas (see Figure 13-5):

FIGURE 13-5 Activity Network for Post-implementation Review

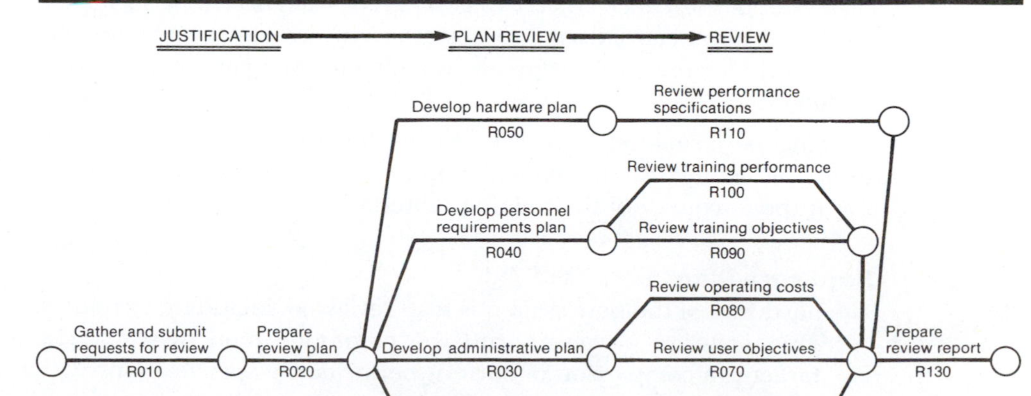

1. *Administrative plan*—Review area objectives, operating costs, actual operating performance, and benefits.
2. *Personnel requirements plan*—Review performance objectives and training performance to date.
3. *Hardware plan*—Review performance specifications.
4. *Documentation review plan*—Review the system development effort.

Once drafted, the review should be verified and approved by the requester or the end user.

Administrative Plan

The review group probes the effect of the operational system on the administrative procedures of the user. The following activities are reviewed:

1. *User objectives.* This is an extremely critical area since it is possible that over time either the system fails to meet the user's initial objectives or the user objectives change as a reflection of changes in the organizational objectives. We need to think in terms of problems and of further opportunities. The results of the evaluation are documented for future reference.
2. *Operating costs and benefits.* Under the administrative plan, the cost structure of the system is closely reviewed. This includes a review of all costs and savings, a review and update of the noncost benefits of the system, and a current budget designed to manipulate the costs and savings of the system.

Personnel Requirement Plan

This plan evaluates all activities involving system personnel and staff as they deal directly with the system. The emphasis is on productivity, morale, and job satisfaction. After the plan is developed, the review group evaluates the following:

1. *Personnel performance objectives compared with current performance levels.* Turnover, tardiness, and absenteeism are also evaluated. The results are documented and made available to the maintenance group for follow-up.
2. *Training performance.* Through testing, interviews, and other data-gathering techniques, the review group attempts to answer questions about the adequacy of the training materials.

Hardware Plan

The hardware of the new system is also reviewed, including terminals, CRT screens, software programs, and the communication network. The primary target is a comparison of current performance specifications with design specifications. The outcome of the evaluation indicates any differences between expectations and realized results. It also points to any necessary modifications to be made.

FIGURE 13-6 **Maintenance Represents up to 70 Percent of the Total Life Cycle Cost of Software**

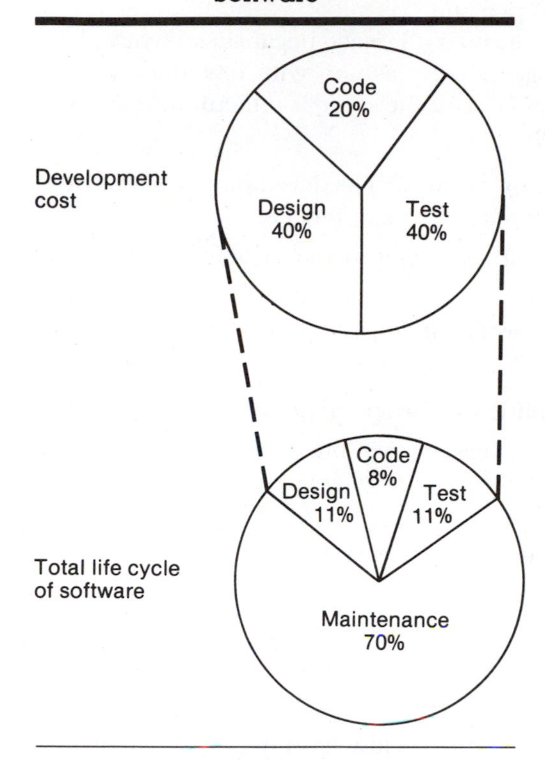

Source: Adapted from Wm. W. Agresti, "Managing Program Maintenance," *Journal of Systems Management,* February 1982, p. 34.

Documentation Review Plan

The reason for developing a documentation review plan is to evaluate the accuracy and completeness of the documentation compiled to date and its conformity with pre-established documentation standards. Irregularities prompt action where changes in documentation would improve the format and content.

SOFTWARE MAINTENANCE

Maintenance is the enigma of system development. It holds the software industry captive, tying up programming resources. Analysts and programmers spend far more time maintaining programs than they do writing them. Maintenance accounts for 50–80 percent of total system development (see Figure 13-6). The federal government alone spends more than $1.3 billion a year on software maintenance.[2] Whereas the cost of hardware has steadily

[2] General Accounting Office, "Federal Agencies Maintenance of Computer Programs: Expensive and Undermanaged," AFMD-81-25, February 26, 1981.

declined, the cost of producing programs has skyrocketed. For example, 10 years ago the development cost of one Department of Defense (DOD) project averaged $75 per line of programming instruction, whereas maintenance costs ran as high as $4,000 per line.

This problem occurs across industry largely because software is a handmade product designed in an ad hoc fashion with few standards; it comes out late, is poorly documented, and therefore is difficult to maintain. There are other problems as well:

1. Maintenance is not as rewarding as exciting as developing systems. It is perceived as requiring neither skill nor experience.
2. Users are not fully cognizant of the maintenance problem or its high cost.
3. Few tools and techniques are available for maintenance.
4. A good test plan is lacking.
5. Standards, procedures, and guidelines are poorly defined and enforced.
6. Maintenance is viewed as a necessary evil, often delegated to junior programmers. There is practically no maintenance manager job classification in the MIS field.
7. Programs are often maintained without care for structure and documentation.
8. There are minimal standards for maintenance.
9. Programmers expect that they will not be in their current commitment by the time their programs go into the maintenance cycle.

Most programmers view maintenance as low-level drudgery. After they develop an application, they spend years locked into maintaining it. Eventually, boredom sets in, with subsequent turnover and loss of expertise necessary to maintain the system. It is obvious that the more carefully a system is thought out and developed, with attention paid to external influence over a reasonable lifetime, the less maintenance will be required.

Maintenance or Enhancement?

Maintenance can be classified as corrective, adaptive, or perfective. *Corrective maintenance* means repairing processing or performance failures or making changes because of previously uncorrected problems or false assumptions. *Adaptive maintenance* means changing the program function. *Perfective maintenance* means enhancing the performance or modifying the program(s) to respond to the user's additional or changing needs.[3] Of these types, more time and money are spent on perfective than on corrective and adaptive maintenance together.

[3] See Wm. W. Agresti, "Managing Program Maintenance," *Journal of Systems Management*, February 1982, pp. 34–37.

Maintenance covers a wide range of activities, including correcting coding and design errors, updating documentation and test data, and upgrading user support. Many activities classified as maintenance are actually enhancements. *Maintenance* means restoring something to its original condition. Unlike hardware, however, software does not wear out; it is corrected. In contrast, *enhancement* means adding, modifying, or redeveloping the code to support changes in the specifications. It is necessary to keep up with changing user needs and the operational environment.

Although software does not wear out like a piece of hardware, it "ages" and eventually fails to perform because of cumulative maintenance. Over time, the integrity of the program, test data, and documentation degenerates as a result of modifications. Eventually, it takes more effort to maintain the application than to rewrite it.

A major problem with software maintenance is its labor-intensive nature (and therefore the likelihood of errors). Consider a change in the code. Altering the code, no matter how slight, must be manually introduced into each program because there is no easy way of making sure that the changes will interface with all the programs. Reusing the old codes depends heavily on the programmer's ability to judge what code can and cannot be reused. It is an error-prone process that is still perceived by many as more cost effective than writing programs from scratch.

Primary Activities of a Maintenance Procedure

Maintenance activities begin where conversion leaves off. Maintenance is handled by the same planning and control used in a formal system project. The flowchart in Figure 13–7 shows the basic activities. Documentation is as much a part of maintenance as it is of system development. Briefly, the maintenance staff receives a request for service from an authorized user, followed by a definition of the required modifications. The source program and written procedures for the system are acquired from the programming library. Program changes are then tested and submitted to the user for approval. Once approved, the modified documentation is filed with the library and a project completion notice is sent to the user, signaling the termination of the project.

Reducing Maintenance Costs

From the discussion so far, it is clear that some plan is needed to attack the ever-growing problem of software maintenance. Several MIS organizations have done this through a maintenance reduction plan that consists of three phases:

1. *Maintenance management audit,* which through interviews and questionnaires evaluates the quality of the maintenance effort. Some of the questions asked are:

FIGURE 13–7 Primary Activities of a Maintenance Procedure

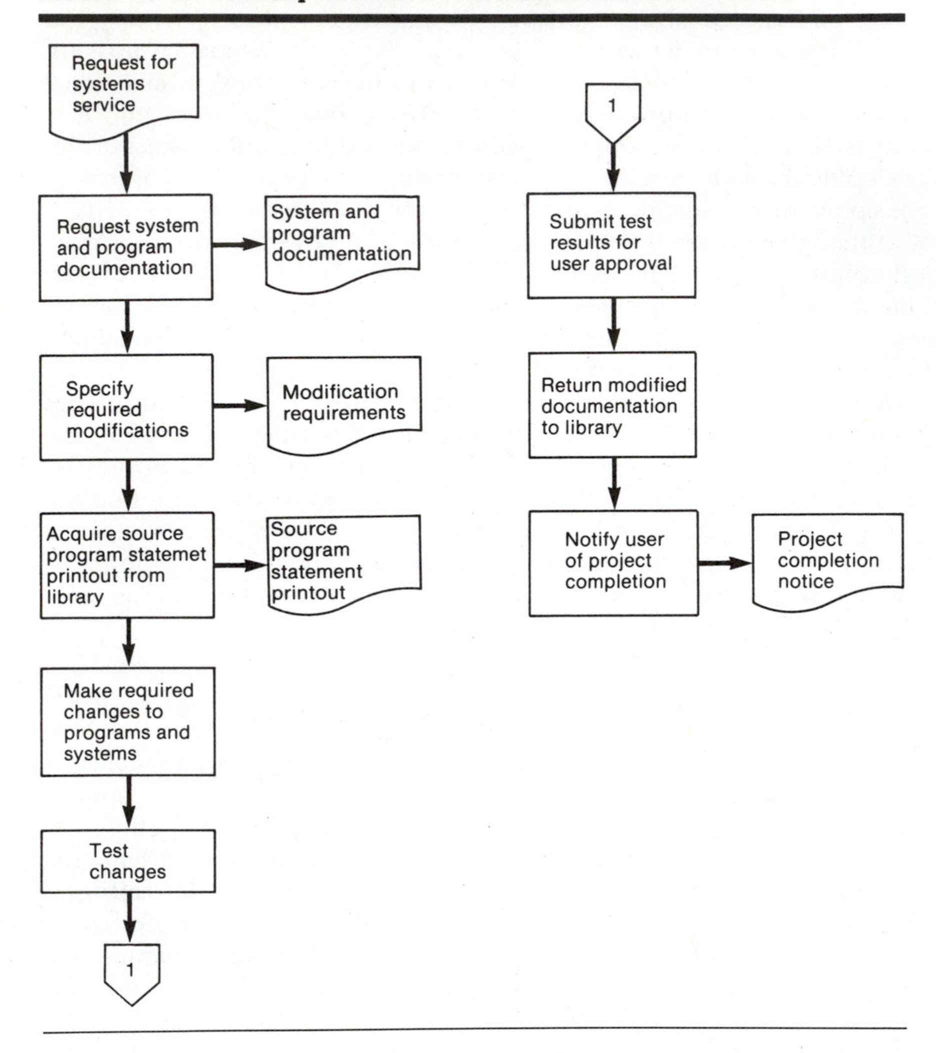

a. Are maintenance requests logged in a maintenance request log?

b. What percent of total hours worked are spent on error corrections, additions/changes/deletions, and improvements?

c. Does your organization currently have a well-defined maintenance reduction program?

The data gathered are used to develop a diagnostic study to provide management with an assessment of the software maintenance function.

2. *Software system audit,* which entails:

   a. An overall view of the system documentation and an assessment of the quality of data files and data bases and system maintainability, reliability, and efficiency.

   b. Functional information gathered on all the programs in the system to determine how well they do the job. Each program is assigned a preliminary ranking value.
   c. A detailed program audit, which considers the ranking value, mean time between failure (MTBF), and size of the maintenance backlog. MTBF determines system availability to users.
3. *Software modification*, which consists of three steps:
   a. Program rewrites, which include logic simplification, documentation updates, and error correction.
   b. System level update, which completes system level documentation, brings up to date data flow diagrams or system flowcharts, and cross-references programs.
   c. Reaudit of low-ranking programs to make sure that the errors have been corrected.

The outcome of such a maintenance reduction plan is more reliable software, a reduced maintenance backlog, improved response times in correcting errors, improved user satisfaction, and higher morale among the maintenance staff.

In summary, the work of maintenance is certainly not for neophyte programmers. To put maintenance in its proper perspective requires considerable skill and experience and is an important and ongoing aspect of system development. The ability of the maintenance programmer to make sound judgments depends on his/her technical expertise and ability to identify user needs. This means understanding the operational environment and what the user is trying to accomplish with the software.

An additional factor in the success of the maintenance programmer is the work environment. Maintenance programmers have generally been paid less and received less recognition than other programmers. Little attention has been paid to their training and career plans within the MIS function. Similarly, the need for maintenance programmers to apply standards and methodologies has been overlooked.

Maintenance demands more orientation and training than any other programming activities, especially for entry-level programmers. The environment must recognize the needs of the maintenance programmer for tools, methods, and training.

Summary

1. Implementation is the process of converting a new system design into operation. Conversion entails several steps:
   a. Review the project plan, test documentation, and implementation plan.
   b. Convert the files.
   c. Conduct parallel processing.
   d. Log the computer run for reference.
   e. Discontinue the old system.
   f. Plan for post-implementation review.

2. The prime concern during conversion is copying the old files to the new system. Once a particular file is selected, the next step is to specify the data to be converted. A file comparison program is best used for verifying the accuracy of the copying process.
3. Well-planned test files are important for successful conversion. A test file should contain predictable results, simplified error-finding routines, and printed results in seconds. It should also show how the new system handles the most difficult tasks.
4. A good audit trail is the key to detecting errors and fraud in a new system. To detect fraud, many system designs do not allow the easy delection of any records on the file. This feature makes it difficult to cover up problems or conceal fraud.
5. Users approach their first exposure to a new system as superficial learners or formal (student-type) learners. A superficial learner knows what functions the system performs without understanding the principles behind each function. The formal learner knows the principles and follows the functions as well.
6. During user training, the resident expert emerges. He/she is the one who probably has the most interest in the system, works the hardest to learn it, and consequently benefits from the unique role of being an "expert" in the group.
7. Experience in user training suggests that:
   a. Users are reluctant to read manuals but learn well from demonstrations and visual aids.
   b. Users tend to be natural teachers.
   c. The resident expert is a natural trainer, since he/she speaks the user group's language and uses samples based on common experience to teach the new system.
8. The primary teaching aids in user training are the user manual, "help" screens, data dictionary, and job performance aids. The tools and procedures are designed to minimize the user's resistance to change and motivate the user staff to adopt a constructive attitude for full utilization of the new system.
9. The post-implementation review has the objective of evaluating the system in terms of how well performance meets stated objectives. The study begins with the review team, which gathers requests for evaluation. The team prepares a review plan around the type of evaluation to be done and the time frame for its completion. The plan considers administrative, personnel, and system performance and the changes that are likely to take place through maintenance.
10. Maintenance is actually the implementation of the post-implementation review plan. As important as it is, many programmers and analysts are reluctant to perform or identify themselves with the maintenance effort. There are psychological, personality, and professional reasons for

this. In any case, a first-class effort must be made to ensure that software changes are made properly and in time to keep the system in tune with user specifications.

11. Maintenance is expensive. One way to reduce maintenance costs is through maintenance management and software modification audits. Software modification consists of program rewrites, system level updates, and reaudits of low-ranking programs to verify and correct the soft spots. The outcome should be more reliable software, a reduced maintenance backlog, and higher satisfaction and morale among the maintenance staff.

Key Words

Conversion
Enhancement
Implementation
Maintenance
Mean Time between Failure (MTBF)
Post-implementation Review
Resident Expert

Review Questions

1. What is implementation? How does it differ from conversion? Elaborate.
2. The chapter suggests that conversion has been chaotic and traumatic for many firms. Do you agree? Contact a local firm with a computer facility and discuss the extent to which the statement is true.
3. Explain the major activities in conversion. Which activity is the most important? Why?
4. Distinguish between the following:
   a. Parallel processing and system testing.
   b. Maintenance and enhancement.
   c. Software modification and software system audit.
5. What is involved in converting files? Be specific.
6. "A good test file should show how the new system handles the most difficult tasks." Do you agree? Discuss.
7. What is the role of the audit control trail in conversion? Who performs it? Explain.
8. Suppose you were asked to prepare a plan for training the user staff on a newly acquired microcomputer system.
   a. What factors do you consider in preparing the plan?
   b. How would you design the plan?
   c. What objective(s) are considered as a basis for the plan?
9. If new system design is likely to meet user specifications, why do users resist change? How would one reduce resistance to change? Explain in detail.
10. What personality and professional attributes qualify or motivate a per-

son to become a resident expert? Discuss this question with a local manager of an MIS facility.

11. Several training aids are used for training users on a new system. Summarize the key aids with examples?
12. Briefly explain the procedure and makeup of the post-implementation review. Can one perform maintenance on a system without a post-implementation review? Why?
13. Review the primary activities of a maintenance procedure.

Application Problems

1 A rug manufacturer and importer wanted to have his own computer and run applications away from the parent company's mainframe. A computer consultant (with no knowledge of the rug business) did a feasibility study. He recommended a system with which he had design experience.

The firm leased the system and signed a contract with the consultant to do the installation and training. The consultants brought in two programmers and a data base specialist to convert the files and train personnel. For weeks, everyone was busy with what was bound to be a successful system.

As a first step, the company decided to compare the reports generated by the new system with those available from the main computer. The procedure was to run invoices first, followed by accounts receivable and payable. The first inventory reports seemed way out of line with reality. Both the format of the reports and the data were off. The few invoices sent out brought hostile complaints from customers who were overcharged for their orders. Further attempts to correct the errors only generated more inaccuracies. The company decided to go back to the old system and cancel the whole project.

The matter ended up in court with the consultant demanding the balance due him on the project. The company filed a countersuit claiming irreparable damage to the firm. An investigation discovered that despite management's lack of experience with computers, they decided to convert three major applications at the same time. The employees, not having been forewarned of the conversion, panicked. Prior to the computer, they had undocumented methods of invoicing, keeping track of inventory, and billing—procedures that the consultant never knew of or inquired about. To make matters worse, he did not even know that the parent company's warehouse system had a terminal that used the mainframe to update inventory.

System testing was also a disaster. Only real data were used. The resulting output was so unwieldy that no one could audit or verify its accuracy until it was too late. With no interface between the system being tested and the mainframe, there was no way the files could be copied. The consultant decided to go ahead with incoming data only and to worry later about copying the files on the mainframe.

Documentation and audit procedures were virtually nonexistent. No one seemed to know who changed what. There was no way of telling whether errors were caused by the software or by incorrectly entered data.

The contract was well written. It simply committed the consultant to install a computer system and the company to pay the consultant $75 per hour plus out-of-pocket expenses. The consultant never really knew what the company wanted, and the company had no one to work with the consultant. The employees stayed out of the ways, since they had not been consulted and were not knowledgeable about computers. The programmers, in their opinions, were simply obnoxious. Another consultant who came in to evaluate the mess thought the whole installation was primitive and lacked state-of-the-art software.

The company continued operating after its employees volunteered hours of their time to maintain accurate records by hand. The outcome was an out-of-court settlement. The company dreaded costly litigation. The consultant decided to cut his losses and protect future business.

Assignment

a. What went wrong in this case? Be specific.

b. Elaborate on the importance of a computer contract. What elements would you have emphasized in the contract? Why?

c. Does a contract save an installation from failure? In what way?

d. What testing procedure should have been followed? Explain.

e. What aspects of implementation were absent or deficient?

f. Where did the company fail in the implementation phase? Discuss.

g. List the major recommendations you would make for a similar installation.

---

2 A large mail-order house decided to implement a real-time order-processing system. An outside consultant was hired to do the work. One package that he considered was a software package capable of processing orders as well as monitoring sales by salesperson, product, brand, and the like. The success of this system, implemented in more than 40 corporations, had varied widely. In some organizations, it was disastrous—long, drawn-out training periods, behind schedule, and expensive. In others, everything went smoothly.

Since the candidate system will be used by the operations and sales staff, it is essential that implementation involves both divisions. However, there is a culture gap. Sales personnel know very little about computers and consider themselves independent of operations. Operations personnel, on the other hand, have used first-generation computerized order processing. They consider themselves psychologically isolated from sales, since they are located in the back room and take orders by mail.

This gap had to be bridged. The sales staff was not eager to participate. The question was: What should management do to implement the system? Earlier attempts within the organization at changing employees' attitudes have been largely ineffective.

Assignment

Develop a workshop to do the following:

a. Close the culture gap.

b. Help employees understand the implementation plans and the need to initiate action on their own.

---

3 The stock orders of a large appliance store are kept in a card index. We need to convert them into a magnetic tape file.

Assignment

a. What steps are required in the conversion?

b. State the controls that would be included at each stage.

---

4 A new online teller system design for a medium-size bank was approved by the president, signaling the beginning of implementation. The project leader devised a master plan to specify who is to perform each task and in what order. New deposit and withdrawal slips were ordered and delivered three weeks before implementation. In the interim, copies of the user manual were prepared for the lobby and drive-in tellers.

Soon after the terminals were installed, the tellers began to learn how to enter various transactions. After training sessions were over, they had a chance to ask questions and inquire about the new system.

Once completed, the telephone company and the computer service representative hooked up the terminal online with the master system.

The following Monday (a week before actual conversion), the analyst asked the head teller whether the tellers would come in on Saturday to catch up on their work and run test data to reinforce recent training. The head teller agreed to the overtime, but on Saturday, only 12 of 17 tellers showed up. During that time, the entire system was checked out and functioned as expected.

The bank opened the following Monday. The online system operated normally. Customers were greeted at the door by the president. Coffee and cake were served in the lobby. As the end of the day, the analyst sent a report to the board of directors, informing them that the system was now in operation and all user requirements had been met.

Three weeks later, the analyst was called to the board meeting. The chairman criticized the analyst for exceeding the budgeted amount approved by the board. Furthermore, the authorization the analyst gave the terminal vendor to bring in two CRT screens to expedite information retrieval exceeded his authority to implement the system. The bank's auditor also estimated it would take 3.8 years rather than the initial estimate of 2.1 years to break even on the total cost of the installation. Not knowing what to say, the analyst left the board room with a feeling of total failure.

Assignment

a. What are the major problems in the case? Who is to blame? Why?

b. Was the board chairman justified in his criticism of the analyst? Explain.

c. Where did the analyst fail in handling implementation? Be specific.

---

5 The manager of a jewelry store in Coral Gables, Florida, is considering a computer installation for sales analysis and inventory control. The system will give a profile of sales activities by sales item and salesperson and also analyze sales fluctuations. It will provide an online update of the inventory (watches, bracelets, diamond rings, etc.). This area is extremely important to make sure that as items are sold, they are deleted from the inventory file. At the same time, the system will generate orders to replenish stock items and show the dates of arrival.

The jewelry store has a good reputation among tourists during the winter season from November 1 to February 1. Almost two thirds of the business is done during that time. Obviously, if various items are not available, there is the risk of losing sales and customers. To date, inventory and reordering have been done by two employees on a full-time basis—a job that is both tedious and fraught with inaccuracies.

It is now June 1, well into the off-season for tourism. Assume that hardware and software are available within 60 days. It will take one week to test and install the system and one week to train the stockroom clerks. Loading the entire inventory is estimated to take about 1.5 weeks.

Assignment

a. Use a bar chart or a Gantt chart to show the sequence of activities (by week).

b. When should the firm sign a contract to implement the system? Why? Discuss.

ACME BOOKSTORES *(continued)*

The case involving Acme Bookstores was described at the end of Chapter 12. This part focuses on implementation procedures and training.

Conversion Procedure

Except for the installation of hardware, the main problem faced in the implementation of the accounts payable system was the user's staff. Changes in personnel, orientation to the new system, training, and file conversions were problems encountered soon after testing. The installation of the software package was straightforward. At the accounts payable department, employees were assured that no one would be replaced by the system. The system eventually would allow the department to grow; thus, new employees would be needed as the store expanded.

The orientation of present employees to the new system was important. Some employees had a positive attitude and looked forward to the benefits of the system; others had a negative attitude, fearing loss of job or intimidation by the computer.

After selling the system, the next step was personnel training. The computer store worked with the user's staff and trained them in computer operation and use of the package. The *Guide to Setting Up* book and the Reference Manual were quite informative and user-friendly.

The next task was file conversion. This step encompassed transfer of manual records (invoices, packing lists, and purchase orders) onto diskettes. The job required extra help on a temporary basis and took three weeks to complete.

The final component was parallel operation, using both systems simultaneously. The time and effort in this step were enormous, because everything had to be processed twice, once by the computer and once by hand. The parallel operation continued until a full cycle of the business was completed—six months.

Assignment

a. Evaluate the implementation procedure.

b. How would you have handled employees resistance to change? Explain.

c. Was parallel processing necessary in this case? Why?

Selected References

Agresti, William W. "Managing Program Maintenance." *Journal of Systems Management*, February 1982, pp. 34–37.

Bronsema, Gloria S., and Peter G. W. Keen. "Education before Change in Systems Implementation." *Computerworld*, October 3, 1983, p. 27ff.

Burrows, James H. "Actual Conversion Experience." *Data Base*, Summer–Fall 1981, pp. 20–23.

Cerullo, Michael. "Determining Post-Implementation Audit Success." *Journal of Systems Management*, March 1979, pp. 27–31.

Glass, R. L., and R. A. Noiseux. *Software Maintenance Guidebook*, Englewood Cliffs, N.J.: Prentice-Hall, 1981.

Guimares, Tor. "Understanding Implementation Failure." *Journal of Systems Management*, March 1981, pp. 12–19.

Kapur, Gopal. "Software Maintenance." *Computerworld*, September 3, 1983, pp. 13–22.

Koogler, Paul; Frank Collins; and Donald K. Clancy. "The New System Arrives." *Journal of Systems Management*, November 1981, pp. 32–37.

Lee, Len. "Preimplementation Activities." *Journal of Systems Management*, May 1981, pp. 24–27.

Martin, George. "Managing Systems Maintenance." *Journal of Systems Management*, July 1978, pp. 30–33.

Martin, James, and Carma McClure. *Software Maintenance: The Problem and Its Sources*. Englewood Cliffs, N.J.: Prentice-Hall, 1983.

Molnar, Joseph. "Preventive Maintenance for Small Business Systems." *Mini-Micro Systems*, July 1980, p. 110ff.

Rich, Robert J. "Pros and Cons of Third Party Maintenance." *Small Systems World*, January 1981, pp. 32–35.

Rhodes, Wayne L. "Preventive Maintenance: Funding a Cheaper Way." *Infosystems*, May 1981, p. 86ff.

Scharer, Laura L. "User Training: Less Is More." *Datamation*, July 1983, pp. 175–76ff.

Chapter 14

Hardware/Software Selection and the Computer Contract

At a Glance

Building systems culminates in the selection of hardware and compatible software as a unit. In today's ever-growing technology, the analyst must be familiar with computer products, software packages, services, and suppliers.

Selecting a system begins with requirements analysis, which draws up the request for proposal submitted to selected vendors for bids. Consultants may be hired to assist in the selection process. With software, selection focuses on reliability, functionality, user-friendliness, and other criteria. Evaluating a software package or a computer system involves acceptance tests before adoption. Once a system is approved, a decision has to be made whether to purchase, rent, or lease it. Finally, a contract is negotiated to ensure against default or nonperformance. A list of responsibilities and damages in the contract obligates the vendor to deliver and operate according to plans.

By the end of this chapter, you should know:

1. The various types of suppliers and software.
2. The phases in hardware/software selection.
3. The financial considerations in selection.
4. The makeup and basics of computer contract negotiations.

INTRODUCTION

A major element in building systems is selecting compatible hardware and software. The systems analyst has to determine what software package is best for the candidate system and, where software is not an issue, the kind of hardware and peripherals needed for the final conversion. To do the job well, the analyst must be familiar with the computer industry in general, what various computers can and cannot do, whether to purchase or lease a system, the vendors and their outlets, and the selection procedure.

Hardware/software selection begins with requirements analysis, followed by a request for proposal and vendor evaluation. The final system selection initiates contract negotiations. It includes purchase price, maintenance agreements, and the amount of updating or enhancements to be available by the vendor over the life of the system. Contract negotiations, seemingly too legal for an analyst, require finesse and strategies designed to get the best deal for the user and protect the user's interests in the acquired system. This chapter focuses on these elements and provides background on the makeup and ramifications of software and hardware selection.

THE COMPUTER INDUSTRY

More than 2,000 U. S. vendors provide a wide range of computer products and services. Figure 14–1 lists some of the major firms in the industry. They can be classified into three groups: (1) hardware suppliers, (2) software suppliers, and (3) service suppliers.

Hardware Suppliers

This group includes mainframe manufacturers, peripheral vendors, supplies vendors, computer leasing firms, and used systems dealers. IBM is the major supplier of mainframe computers. In microcomputers, IBM, Apple, and Tandy Corporations top the list.

Peripheral manufacturers supply tape drives, disk and diskette drives, printers, and other components. Vendors of supplies provide *consumable* supplies such as diskettes and printer forms and *nonconsumable* supplies such as disk packs, tape reels, tape library shelves, and fireproof vaults. Hundreds of independent vendors are in this field. Used computer dealers purchase secondhand equipment from computer users, rebuild them, and sell them at attractive prices. Computer leasing firms generally finance hardware and software acquisition. Leasing companies may also underwrite or insure the development of a computer system. The acquisition of used computers is discussed later in the chapter.

Software Suppliers

In today's market, 17,000 firms offer more than 5,000 systems and applications. In the microcomputer area, over 30,000 software packages are avail-

FIGURE 14–1 Selected Computer Firms

Computer Systems

Amdahl Corp.
Burrough Corp.
Computer Automation
Control Data Corp.
Cray Research Inc.
Data General Corp.
Datapoint Corp.
Digital Equipment
Four-Phase Systems
Foxboro
General Automation
GRI Computer Corp.
Hewlett-Packard Co.
Honeywell Inc.
IBM
Microdata Corp.
Mini-Computer Systems
NCR
Perkin-Elmer
Prime Computer Inc.
Sperry Univac
Systems Engineering Labs
Tandem Computers Inc.
Wang Labs

Leasing Companies

Booth Courier Corp.
Comdisco Inc.
Commerce Group Corp.
Computer Investers Group
Continental Info. Systems
Datronic Rental
DCL Inc.
DPF Inc.
Itel
Leasco Corp.
Leaspac Corp.
Pioneer Tex Corp.
U.S. Leasing

Software and EDP Services

Advanced Comp Tech.
Anacomp Inc.
Applied Data Research
Automatic Data Processing
Compu-Serv Network
Computer Horizons
Computer Network
Computer Sciences
Computer Task Group
Computer Usage
Comput Auto Rep. SVC
Comshare
Cullinane Corp.
Data Dimensions Inc.
Datatab
Electronic Data Systems
Keane Associates
Keydata Corp.
Logicon
National Data Corp.
On Line Systems Inc.
Planning Research
Programming & Systems
Rapidata Inc.
Reynolds & Reynolds
Scientific Computers
Tymeshare Inc.
Wyly Corp.

Peripherals and Subsystems

Advanced Memory Systems
Ampex Corp.
Anderson Jacobson
Ampex Corp.
Anderson Jacobson
Applied Dig Data Sys.
Bunker-Ramo
Calcomp
Cambridge Memories
Centronics Data Comp
Cognitronics
Computer Commun.
Computer Consoles
Computer Equipment
Computer Transceiver
Computervision Corp.
Comten
Data Access Systems
Data 100
Data Products Corp.
Dataram Corp.
Datum Inc.
Delta Data Systems
Documation Inc.
Electronic M&M
General Computer Systems
General Datacomm Ind.
Harris Corp.
Hazeltine Corp.
Inforex Inc.
Information Intl. Inc.
Infoton
Intel Corp.
Lundy Electronics
Memorex
Mohawk Data Sciences
MSI Data Corp.
Paradyne Corp.
Potter Instrument
Quantor Corp.
Recognition Equipment
Scan Data
Storage Technology
Tally Corp.
Tech Inc.
Tektronix Inc.
Telex
Wiltex Inc.

Supplies and Accessories

American Business Products
Baltimore Bus Forms
Cybermatics Inc.
Duplex Products Inc.
Ennis Business Forms
3M Company
Moore Corp. Ltd.
Nashua Corp.
Standard Register
Tab Products Co.
UARCO
Wabash Magnetics
Wallace Business Forms

able. Computer users can acquire programs from either the vendor or the software house for virtually every application imaginable. Prices vary from a basic payroll program stored on cassette for $10 to mainframe-based inventory control for $35,000. Prices and levels of complexity of software depend on the computer and the state of the competition.

Service Suppliers

Outside computer services are commonly used by small firms or first-time users. Also called *servicers*, they include the following:

1. *Computer manufacturers* supply services such as system design, programming, education and training, and hardware maintenance.
2. *Service bureaus* run "bread and butter" applications for small firms. Larger firms contract for specialized applications or for running jobs during peak volume periods. The primary services are programming, file and system conversion, system design, and user training.
3. *Facilities management* (FM) furnishes specialists to manage a user-installed computer on the user's premises. In some cases, service is limited to developing application programs. The user runs the system but calls on the service organization for developmental work and maintenance.

The FM concept has several benefits. The user pays only for the service rendered. Turnover problems for the user are eliminated when the servicer manages the center. The main drawbacks, however, are loss of control over the operation, vulnerability of information, and the high fees charged for the service.

THE SOFTWARE INDUSTRY

Software has become an industry in itself. Before we discuss it further, we define the types and characteristics of software for system development.

Types of Software

Software is classified according to whether it performs internal computer functions or allows use of the computer for problem solving. The former is called *systems software*—programs designed to control system operations (e.g., operating systems and data base management systems) and system implementation (e.g., assemblers and compilers).[1] The latter classification is called *applications programs*, which perform user-oriented functions. Within this classification we have two groups:

---

[1] Daniel Couger, "Development to Facilitate Managerial Use of the Computer," *Computing Newsletter*, no. 8 (April 1983), pp 2–4.

1. *Cross-industry* applications software, such as accounts receivable and payroll calculations.
2. *Industry-specific* software, such as a safe deposit tracking system, hospital billing system, or airline reservation system.[2]

In general, software can have one or more of the following attributes:

1. *Concurrence of operation.* Software allows simultaneous activities to take place in a computer run. For example, keying in data on the terminal takes place while the system is reading in data from disk and the central processor is operating on a separate activity.
2. *Resource and information sharing.* Different programs share the same hardware resource. Different users may use various programs, or different programs use a centralized data base. This implies multiple interfaces of a system with the outside world.
3. *Modularity.* Software consists of segments connected in various ways, partly due to the separate functions they perform.
4. *Multiplexed operation.* Some software systems rotate the use of a resource, such as an input/output device, among different users, while each user has the impression of being the sole user of the system.

The software industry, represented by time-sharing, facilities management, turnkey systems, packaged software products, and contract programming, is in the midst of an incredible boom. IBM's "unbundling" of its software from hardware in January 1970 sparked an exponential growth in the software industry. The astounding growth of the software industry is outpacing virtually all other areas. The stock prices of software companies continue to rise, and new software products find immediate success. In the microcomputer environment, William Gates' Microsoft Corporation started with two employees in 1981. Since it developed the disk operating system (DOS) for the IBM PC late that year, it has grown to a multimillion-dollar corporation serving the software needs of other computer makers as well.

Industrywide 1984 revenues were $8.6 billion and are estimated to grow by at least 20 percent through the 1980s, especially in the software products area. There are three reasons for this growth:

1. *Programmer shortage.* The demand/supply factor for programming is such that computer service firms are filling the gap with "canned" solutions ranging from utilities and programming to data base management systems and applications.[3]
2. *Hardware/software cost reversal.* Although the price of software relative to hardware is rising, the actual price is decreasing, considering the power and performance of good software. With software geared for the mass market, application developers have scaled down hardware re-

---

[2] Alan M. Hoffberg, "Getting Hardnosed about Software," *Infosystems*, July 1981, p. 37ff.

[3] Ibid., p. 27.

quirements more toward the personal computer, while providing equal capabilities and lower prices than the software a decade ago.

3. *Economies of scale.* Assembling and distributing specialized computer solutions for a vast customer base and potentially cheaper than users acquiring their own skills and facilities to do the same.

Knowledge of the computer and of software is helpful as a basis for selecting hardware/software. Today's maturing market means a wide choice for the user. Searching for the best product requires specialized knowledge and a serious approach. This is when an experienced analyst or outside consultant can contribute to a successful installation.

A PROCEDURE FOR HARDWARE/SOFTWARE SELECTION

Gone are the days when a user calls IBM to order a 360 system. The system then came with hardware, software, and support. Today, selecting a system is a serious and time-consuming business. Unfortunately, many systems are still selected based on vendor reputation only or other subjective factors. The time spent on the selection process is a function of the applications and whether the system is a basic microcomputer or a mainframe. In either case, planning system selection and acquiring experienced help where necessary pay off in the long run.

There are several factors to consider prior to system selection:

1. Define system capabilities that make sense for business. Computers have proven valuable to business in the following areas:
    a. *Cost reduction* includes reduction of inventory, savings on space, and improved ability to predict business trends.
    b. *Cost avoidance* includes early detection of problems and ability to expand operations without adding clerical help.
    c. *Improved service* emphasizes quick availability of information to customers, improved accuracy, and fast turnaround.
    d. *Improved profit* reflects the "bottom line" of the business and its ability to keep receivables within reason.
2. Specify the magnitude of the problem; that is, clarify whether selection entails a few peripherals or a major decision concerning the mainframe.
3. Assess the competence of the in-house staff. This involves determining the expertise needed in areas such as telecommunications and data base design. Acquiring a computer often results in securing temporary help for conversion. Planning for this step is extremely important.
4. Consider hardware and software as a package. This approach ensures compatibility. In fact, software should be considered first, because often the user secures the hardware and then wonders what software is available for it. Remember that software solves problems and hardware drives the software to facilitate solutions.

5. Develop a schedule (a time frame) for the selection process. Maintaining a schedule helps keep the project under control.
6. Provide user indoctrination. This is crucial, especially for first-time users. Selling the system to the user staff, providing adequate training, and preparing an environment conducive to implementation are prerequisites for system acquisition.

Major Phases in Selection

The selection process should be viewed as a project, and a project team should be organized with management support. In larger projects the team includes one or more user representatives, an analyst, and EDP auditor, and a consultant. Several steps make up the selection process; some overlap due to the dynamic nature of selection:

1. Requirements analysis.
2. System specifications.
3. Request for proposal (RFP).
4. Evaluation and validation.
5. Vendor selection.
6. Post-installation review.

Requirements Analysis

The first step in selection is understanding the user's requirements within the framework of the organization's objectives and the environment in which the system is being installed. Consideration is given to the user's resources as well as to finances.

In selecting software, the user must decide whether to develop it in-house, hire a service company or a contract programmer to create it, or simply acquire it from a software house. The choice is logically made after the user has clearly defined the requirements expected of the software. Therefore, requirements analysis sets the tone for software selection.

System Specifications

Failure to specify system requirements before the final selection almost always results in a faulty acquisition. The specifications should delineate the user's requirements and allow room for bids from various vendors. They must reflect the actual applications to be handled by the system and include system objectives, flowcharts, input-output requirements, file structure, and cost. The specifications must also describe each aspect of the system clearly, consistently, and completely.

Request for Proposal (RFP)

After the requirements analysis and system specifications have been determined, a request for proposal (RFP) is drafted and sent to selected vendors for bidding. Bids submitted are based on discussions with vendors. At a minimum, the RFP should include the following:

1. Complete statement of the system specifications, programming language, price range, terms, and time frame.
2. Request for vendor's responsibilities for conversion, training, and maintenance.
3. Warranties and terms of license or contractual limitations.
4. Request for financial statement of vendor.
5. Size of staff available for system support.

Evaluation and Validation

The evaluation phase ranks vendor proposals and determines the one best suited to the user's needs. It looks into items such as price, availability, and technical support. System validation ensures that the vendor can, in fact, match his/her claims, especially system performance. True validation is verified by having each system demonstrated.

Role of the Consultant. For a small firm, an analysis of competitive bids can be confusing. For this reason, the user may wish to contract an outside consultant to do the job. Consultants provide expertise and an objective opinion. A recent survey found, however, that 50 percent of respondent users had unfavorable experiences with the consultants they hired, and 25 percent said they would never hire another consultant. With such findings, a decision to use consultants should be based on careful selection and planning. A rule of thumb is that the larger the acquisition, the more serious should be the consideration of using professional help.

Although the payoffs from using consulting services can be dramatic, the costs are also high. For many small companies that are exploring system acquisition, consulting services may be totally out of reach. During 1984, the average rates of consultants were $600–$1,800 a day, not including travel and out-of-pocket expenses.

The past decade has seen the growth of internal management consultant teams in large organizations, as opposed to external consulting teams. Figure 14–2 outlines the cases where an external or internal consultant is appropriate.

Vendor Selection

This step determines the "winner"—the vendor with the best combination of reputation, reliability, service record, training, delivery time, lease/finance terms, and conversion schedule. Initially, a decision is made on which vendor to contact. The sources available to check on vendors include the following:

1. Users.
2. Software houses.
3. Trade associations.
4. Universities.

FIGURE 14-2 Pros and Cons of Using Consultants

External Consultant	Internal Consultant
Full-time internal consultant is not needed or is beyond the budget of the organization.	An outside consultant is too costly; internal consultants can be much cheaper.
Extra help on a project is needed for a short time; an internal person cannot afford the time.	A fast decision necessitates using an internal consultant.
The internal staff does not possess the expertise or broad knowledge needed for a specific situation.	An external consultant often does not understand the nature of the internal problem.
The political nature of the problem requires an objective, neutral opinion.	An internal consultant already exists who has an objective and technical understanding.
An outside opinion is desired in addition to that of the internal consultant.	An inside opinion is desired in addition to that of the external consultant.

5. Publications/journals.
6. Vendor software lists.
7. Vendor referral directories.
8. Published directories.
9. Consultants.
10. Industry contacts.

For comprehensive applications, the user routinely submits an RFP that specifies the performance requirements and information needed to make an evaluation. Copies of the vendor's annual financial statement are also requested. Once received, each vendor's report is matched against the selection criteria. Those that come the closest are invited to give a presentation of their system. The system chosen goes through contract negotiations before implementation. This area is covered later in the chapter.

Post-Installation Review

Sometime after the package is installed, a system evaluation is made to determine how closely the new system conforms to plan. System specifications and user requirements are audited to pinpoint and correct any differences.

Software Selection

Software selection is a critical aspect of system development. As mentioned earlier, the search starts with the software, followed by the hardware. There are two ways of acquiring software: custom-made or "off-the-shelf" packages. Today's trend is toward purchasing packages, which represent roughly 10 percent of what is costs to develop the same in house. In addition to reduced cost, there are other advantages:

1. A good package can get the system running in a matter of days rather than the weeks or months required for "home-grown" packages.

2. MIS personnel are released for other projects.
3. Packages are generally reliable and perform according to stated documentation.
4. Minimum risks are usually associated with large-scale systems and programming efforts.
5. Delays in completing software projects in house often occur because programmers quit in midstream.
6. It is difficult to predict the cost of "home-grown" software.
7. The user has a chance of seeing how well the package performs before purchasing it.

There are drawbacks, however, to software packages:

1. The package may not meet user requirements adequately.
2. Extensive modification of a package usually results in loss of the vendor's support.
3. The methodology for package evaluation and selection is often poorly defined. The result is a haphazard review based on a faulty process or questionable selection criteria.
4. For first-time software package users, the overall expectation from a package is often unclear and ill defined.

It can be seen, then, that the quality of a software package cannot be determined by price alone. A systematic review is crucial.

Criteria for Software Selection

Prior to selecting the software, the project team must set up criteria for selection. Selection criteria fall into the categories described here.

Reliability. It is the probability that the software will execute for a specified time period without a failure, weighted by the cost to the user of each failure encountered. It relates to the ease of recovery and ability to give consistent results. Reliability is particularly important to the professional user. For example, a pharmacist relies on past files on patients when filling prescriptions. Information accuracy is crucial.

Hardware may become inoperative because of design errors, manufacturing errors, or deterioration caused by heat, humidity, friction, and the like. In contrast, software does not fail or wear out. Any reliability problems are attributable to errors introduced during the production process. Furthermore, whereas hardware failure is based largely on random failures, software reliability is based on predestined errors.

Although reliable software is a desirable goal, limited progress has been made toward improving it in the last decade. The fact of unreliable software had led to the practice of securing maintenance agreements after the package is in operation. In a sense, unreliability is rewarded.

Software reliability brings up the concept of *modularity*, or the ease with which a package can be modified. This depends on whether the package was originally designed as a package or was retrofitted after its original development for single installation use. A package with a high degree of modularity has the capacity to operate in many machine configurations and perhaps across manufacturers' product lines.[4]

With modularity comes *expandability*, which emphasizes the sensitivity of a software package to handle an increased volume of transactions or to integrate with other programs. The following questions should be considered:

1. Is there room for expanding the master file?
2. How easily can additional fields, records, and files be added?
3. How much of the system becomes unusable when a part of it fails?
4. Are there errors a user can make that will bring down the system?
5. What are the recovery capabilities?[5]

Functionality. It is a definition of the facilities, performance, and other factors that the user requires in the finished product. All such information comes from the user. The following are key questions to consider:

1. Do the input transactions, files, and reports contain the necessary data elements?
2. Are all the necessary computations and processing performed according to specifications?

Capacity. Capacity refers to the capability of the software package to handle the user's requirements for size of files, number of data elements, volume of transactions and reports, and number of occurrences of data elements. All limitations should be checked.

Flexibility. It is a measure of the effort required to modify an operational program. One feature of flexibility is adaptability, which is a measure of the ease of extending the product.

Usability. This criterion refers to the effort required to operate, prepare the input, and interpret the output of a program. Additional points to be considered are portability and understandability. *Portability* refers to the ability of the software to be used on different hardware and operating systems. *Understandability* means that the purpose of the product is clear to the evaluator and that the package is clearly and simply written, is free of

[4] Jan Snyders, "How to Buy Packages," *Computer Decisions*, July 1978, p. 52.

[5] Craig Johannsen, "Software Selection," *Computerworld*, January 1979, p. 33

jargon, and contains sufficient references to readily available documents so that the reader can comprehend advanced contents.

Security. It is a measure of the likelihood that a system's user can accidentally or intentionally access or destroy unauthorized data. A key question is: How well can one control access of software or data files? Control provides system integrity. This topic is covered in detail in Chapter 16.

Performance. It is a measure of the capacity of the software package to do what it is expected to do. This criterion focuses on throughput, or how effectively a package performs under peak loads. Each package should be evaluated for acceptance on the user's system.

The language in which a package is written and the operating system are additional performance considerations. If we plan to modify or extend a package, it is easier if it is written in a language that is commonly known to programmers. Likewise, if the package runs only under a disk operating system and the installation is under a full operating system, then either the package will have to be upgraded to the larger operating system or the system downgraded to handle the package as is. In either case, the change could be costly and counterproductive.

Serviceability. This criterion focuses on documentation and vendor support. Complete documentation is critical for software enhancement. It includes a narrative description of the system, system logic and logic flowcharts, input-output and file descriptions and layouts, and operator instructions. Vendor support assures the user adequate technical support for software installation, enhancements, and maintenance. The user should determine how much on-site technical assistance is provided by the vendor, especially during the first few weeks after the installation.

The user expects on-site training and support as part of most commercial packages. It is vital to inquire about the amount of training provided. The user may require training at several levels—clerical, operations, programming, and management.

Ownership. Who owns the software once it is "sold" to the user? Most of the standard license agreement forms essentially lease the software to the user for an indefinite time. The user does not "own" it, which means that the source code is inaccessible for modification, except by the vendor. Many users enter into an escrow arrangement whereby the vendor deposits the source code with a third-party escrow agent who agrees to release the code to the user if the vendor goes out of business or is unable to perform the services specified in the license.

In acquiring software, several questions should be asked:

1. What rights to the software is the user buying?

FIGURE 14-3 Software Criteria—A Summary

Criterion	Meaning
1. Reliability	Gives consistent results
2. Functionality	Functions to standards
3. Capacity	Satisfies volume requirements
4. Flexibility	Adapts to changing needs
5. Usability	Easy to operate and understand—user-friendly
6. Security	Maintains integrity and prevents unauthorized access
7. Performance	Capacity to deliver as expected
8. Serviceability	Good documentation and vendor support
9. Ownership	Right to modify and share use of package
10. Minimal costs	Affordable for intended application

2. Can the user sell or modify the software?
3. If the vendor is modifying the package especially for the user, can the vendor sell it to others within the same industry the user is in?
4. What restrictions are there to copying the software or documentation?[6]

Minimal Costs. Cost is a major consideration in deciding between in-house and vendor software. Cost-conscious users consider the following points:

1. Development and conversion costs.
2. Delivery schedule.
3. Cost and frequency of software modifications.
4. Usable life span of the package.

A summary of software criteria is presented in Figure 14-3.

The Evaluation Process

In evaluating packages, it is important to consider the knowledge of the user or how much expertise is available for the evaluation phase. The first step is to look at the criteria listed earlier and rank them in order of importance. Mandatory and desirable features should be differentiated. Since it is unlikely that all user requirements can be met, concentration should be on satisfying the primary requirements. The selection of appropriate software becomes a matter of making intelligent compromises.

Sources for Evaluation

Three sources of information are used in evaluating hardware and software: (1) benchmark programs, (2) experience of other users, and (3) product reference manuals.

[6] Ibid., p. 33.

A *benchmark* is a sample program used for evaluating different computers and their software.[7] This is necessary because computers often will not use the same instructions, words of memory, or machine cycle to solve a problem. In the context of software selection, benchmarking may include the following:

1. A determination of the minimum hardware configuration needed to operate a package. A package may be extremely sensitive to its hardware or have considerable flexibility.
2. An acceptance test as specified in the contract.
3. Testing in an ideal environment to determine comparative timings and in a normal environment to determine its influence on other programs.

The more elaborate the benchmarking, the more costly is the evaluation. The user's goals must be kept in perspective. Time constraints also limit how thorough the testing process can be. There must be a compromise on how much to test while still ensuring that the software (or hardware) meets its functional criteria.

Since benchmarks only validate the vendor's claims, other sources of information are necessary. The *experience of other users* with the same system, software, or service is important. Vendors are generally willing to provide a list of "reference accounts" or people with whom to check. Experience shows that the users on such a list happen to be satisfied customers and, therefore, are not typical users. Seeking objective users on one's own can be frustrating, however. Even if the "ideal" user is located, the information is useful only if the user had the same hardware configuration, applications, and software.

An important step in the evaluation process is to read *product reference manuals* that evaluate system capabilities. Since such a search is often laborious, one alternative is to contact organizations that publish reports based on ongoing research and system testing in various sites. For example, Auerbach, Inc. (Philadelphia), publishes looseleaf references on information processing, telecommunications, and computer graphics. The reports elaborate on computer products, services, and prices. Considering the benefits offered, the cost is relatively low.

Evaluation of Proposals

Vendor bids should be reviewed not only to ensure that the information is adequate, but also to determine whether it meets the requirements of the user's RFP. Proposals that fail the test are rejected. After all proposals have been verified, the final vendor is selected by various approaches: (1) ad hoc, (2) scoring or (3) cost-value approach.

---

[7] J. Seaman, "Smart Benchmarking—the Key to Successful Procurement," *Computer Decisions*, March 1981, pp 74–94.

The *ad hoc* approach refers to the user's inclination to favor one vendor over others. This "halo" effect discourages other vendors from applying when the user is known to favor a particular vendor from the outset.

In the *scoring* approach, the characteristics of each system are listed and given a score in relation to a maximum point rating. Then each proposal is rated according to the characteristics. Figure 14-4 rates three proposals according to uniform computer and vendor evaluation factors. Proposal B has the most points for the total performance score and would be the user's first choice.

With the *cost value* approach, a dollar credit amount is applied to the proposal that best meets the user's desirable characteristics. This credit is subtracted from the vendor's quoted price. The proposal with the lowest price is the one selected. To illustrate, assume that the vendor's response for repair is a key criterion. The cost value of a quick response is determined by estimating the cost of hiring additional help to do the required work manually while the system is waiting for service. This amount depends on how quickly the vendor responds to service. Suppose that when the computer is down for a whole day, there is an additional expense of $1,000. Three breakdowns in a year would total $3,000. Eliminating downtime through quick response to service is a credit for the vendor. After appropriate credits (or penalties) are assigned, the total cost of each system is computed. The system with the lowest cost is selected.

FIGURE 14-4 Sample Scoring Approach to Selection

	Proposals		
Elements	A	B	C
CPU			
Arithmetic	3	4	1
Communication capability	20	15	10
Environmental requirements	1	5	4
Input/output capability	4	9	2
Main memory capacity	14	10	12
Multiprogramming capability	3	9	2
Secondary storage capacity	16	21	11
Word size	19	20	14
Subtotal	80	93	56
Vendor			
Delivery time	10	5	12
Maintenance charges	18	20	21
Number installed to date	2	11	6
Performance record	6	9	8
Quality of service	9	9	5
Training	5	10	7
Subtotal	50	64	59
Total (performance score)	130	157	115

Performance Evaluation

Evaluating a system includes the hardware and software as a unit. Hardware selection requires an analysis of several performance categories:

1. *System availability.* When will the system be available?
2. *Compatibility.* How compatible is the system with existing programs?
3. *Cost.* What is the lease or purchase price of the system? What about maintenance and operation costs?
4. *Performance.* What are the capacity and throughput of the system?
5. *Uptime.* What is the "uptime" record of the system? What maintenance schedule is required?
6. *Support.* How competent and available are the vendor's staff to support the system?
7. *Usability.* How easy is it to program, modify, and operate the system?

For the software evaluation, the following factors are considered:

1. The programming language and its suitability to the application(s).
2. Ease of installation and training.
3. Extent of enhancements to be made prior to installation.

In addition to hardware/software evaluation, the quality of the vendor's services should be examined. Vendor support services include the following:

1. *Backup.* Emergency computer backup available from vendor.
2. *Conversion.* Programming and installation service provided during conversion.
3. *Maintenance.* Adequacy and cost of hardware maintenance.
4. *System development.* Availability of competent analysts and programmers for system development.

FINANCIAL CONSIDERATIONS IN SELECTION

When the decision to go ahead with the acquisition has been made, the next question is whether to purchase or lease. There are three methods of acquisition: (1) rental directly from the manufacturer, (2) leasing through a third party or from the vendor, and (3) outright purchase.

The Rental Option

Rent is a form of lease directly by the manufacturer. The user agrees to a monthly payment, usually for one year or less. The contract can be terminated without penalty by a 90-day advance notice. Rental charges are based on 176 usage hours (8 hours per day × 22 working days) per month. Additional usage means higher total charges per month. Computer users favor renting a system for three reasons:

1. Insurance, maintenance, and other expenses are included in the rental charge.
2. There is financial leverage for the user. With no investment in equipment, user capital is freed for other projects. Futhermore, rental charges are tax deductible.
3. Rental makes it easier to change to other systems, thereby reducing the risk of technological obsolescence.

The primary drawback of a rental contract is its high cost because of the uncertainty of rental revenues to the vendor.

The Lease Option

A leased system is acquired through a third party or from the vendor. A third-party purchase ranges from six months with month-to-month renewals to seven years. Longer-running leases have more favorable terms but entail a higher risk as the user is "strapped" with the system. With a short-term lease, the user's risk is low, but lease charges are high.

From the user's view, leasing has several advantages:

1. No financing is required. The risk of system obsolescence is shifted to the lessor (vendor).
2. Lease charges are lower than rental charges for the same period and are also tax deductible.
3. Leases may be written to show higher payments in early years to reflect the decline in value of the system.
4. Leases may or may not include maintenance or installation costs or providing a replacement system in an emergency.

The drawbacks of leasing are:

1. Unless there is a purchase option, the lessee (user) loses residual rights to the system when the lease expires.
2. The lease period cannot be terminated without a heavy penalty.
3. In the absence of an upgrade clause, the user may not be able to exchange the leased system for another system. Also, if interest rates decrease, the user is committed to lease payments at the higher rate.
4. Unlike a purchase system, a leased system does not provide tax benefits from accelerated depreciation and interest deductions in the early years of use. There are no cash savings in a lease arrangement.

The Purchase Option

Purchasing a computer has benefits and drawbacks. Purchasing means assuming all the risks of ownership including taxes, insurance, and technological obsolescence. However, the owner obtains all the services and support that are available under the lease or rental agreement. Compared with renting or leasing, the key advantages of purchasing are:

1. The flexibility of modifying the system at will.
2. Lower continuing cash outlays than those for a leased system due to cash savings from depreciation and investment tax credit.[8] If the equipment is held for five years or more, a credit of 10 percent of the purchase price is deducted from the organization's income tax.
3. A lower total cash outflow if the user keeps the system longer than five years.

The major drawbacks are:

1. Initial high cost in relation to leasing.
2. Insurance expense and various taxes, which are carried by the user. The maintenance agreement is also paid for by the user when the warranty expires.
3. High overall risk. A poorly selected system means adapting to a "problem child." Selling a used computer with flaws could be a real problem.

Each acquisition method has characteristics that are both common and unique. A choice based on these facts meets the *qualitative* test only. *Quantitatively,* an effective method is the net present value (NPV) approach. As discussed in Chapter 8, it allows users to evaluate alternatives while recognizing that a dollar received today is worth more than tomorrow's dollar. A problem at the end of the chapter illustrates the usefulness of this method.

THE USED COMPUTER

Under what circumstances should one consider a used computer? Computers last between five and eight years. Most organizations outgrow their computers long before they become obsolete, however. This means that users are forced to unload equipment at a loss in order to acquire new systems. Savings of 15 to 70 percent can be realized by buying used systems, depending on the model and condition of the system.[9]

Availability is a major advantage to buying used computers. The demand for some systems is so high that promised six-week deliveries can stretch up to six months. For certain highly sought microcomputers, delivery may take as long as four months. Used computer dealers have been known to deliver the same day.

---

[8] There is a limit on the amount of investment tax credit claimed that is tied to the tax liability for the year. The maximum is 100 percent of the first $25,000 of tax liability plus 85 percent of the tax liability that is greater than $25,000. See Wm. Hoffman and Eugene Willis, *1984 Annual Edition West's Federal Taxation* (St. Paul, Minn.: West Publishing).

[9] Tom Wolpert, "Why It Pays to Buy Used," *Venture*, August 1982, p. 12.

Sales in the used computer market are well over $5 billion per year. Independent vendors have been successful in training operators and programmers to use the equipment. They generally rebuild used systems after they have been acquired from the second user.

For stand-alone systems, used computers are ideal for users with inhouse expertise who are located in an area where technical support is adequate, or who are assured of vendor support. Although the biggest drawback to used computers is maintenance, this is readily available from the vendor or independent service firms.

Used computers are acquired through dealers or end users. Most dealers are knowledgeable about the system they sell. The best bargain, however, is buying directly from the end user, provided there is a log that verifies the maintenance record of the system. Checking the maintenance log will reveal how reliable the system has been. The buyer must be sure that the seller has clear title to the system. A qualified consultant can help.

In conclusion, there are savings from acquiring used systems, and more and more organizations are going that route. Furthermore, it is an excellent way to extend the useful life of the computer.

THE COMPUTER CONTRACT

After a decision has been made about the equipment or software, the final step in system acquisition is to negotiate a contract. Unfortunately, the typical user does not negotiate. The assumption is that a contract drafted by a reputable firm is a standard instrument and is not subject to change. To the contrary, every contract is negotiable to some extent. Large users often spend weeks negotiating amenities and terms, using legal counsel or consultants.

The primary law governing contracts is the law of contracts, although contracts can be influenced by other laws, such as the Uniform Commercial Code (UCC). Under the law of contracts, the formation of a contract requires mutual assent (meeting of the minds) and consideration. Performance of a contract is the fulfilling of the duties created by it.

The Art of Negotiation

Many users enter into contract negotiations at the mercy of the vendor, with little preparation. Negotiating is an art.[10] Timing is critical. Strategies must be planned and rehearsed. The leverage enjoyed by either party can change during the course of the negotiations. Figure 14–5 illustrates the negotiation procedure. Part A represents the poorly prepared user, outmaneuvered completely throughout the negotiations. Part B shows a relatively informed user, but one who has a sense of urgency. The user's negotiating leverage

[10] Lynn Haber, "Negotiations Can Spell Out Successful Systems," *Computerworld*, November 21, 1983, p. 22ff.

FIGURE 14-5 Negotiation Procedures

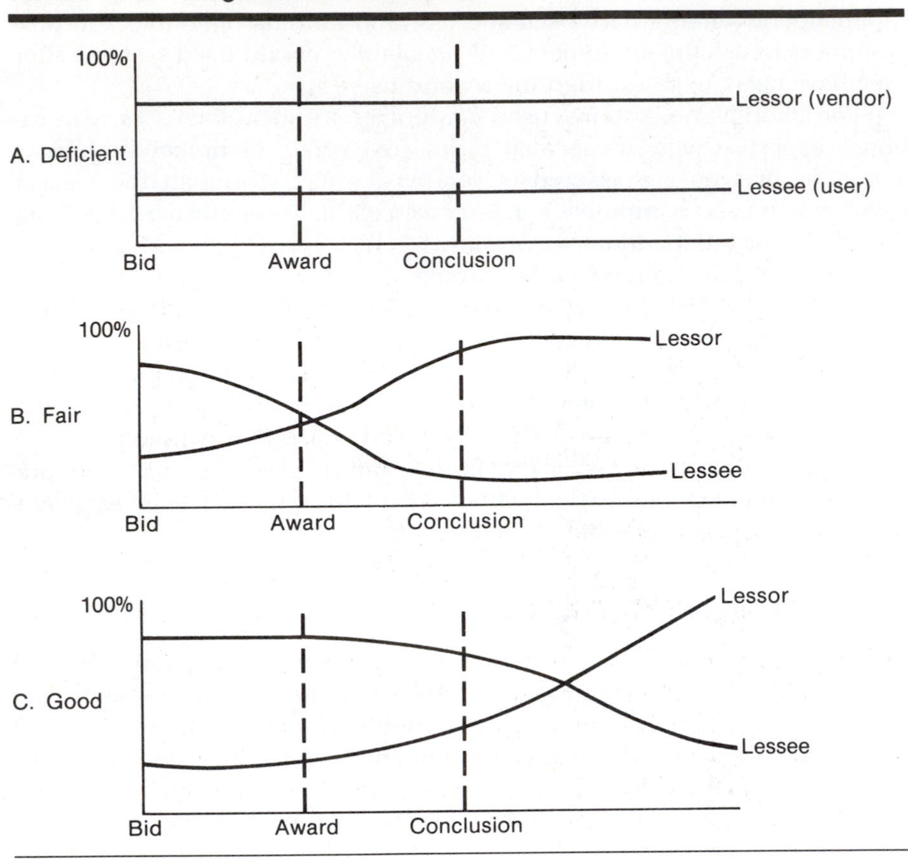

drops to nearly zero as he/she enters the contract-negotiating phase. At this point, the vendor recognizes the user's state of mind and becomes less willing to negotiate in earnest. In part C, the user is following good negotiating procedures and retains fair leverage into the negotiations.

Strategies and Tactics

Various strategies and tactics are used to control the negotiation process. A key strategy is to control the environment. The user's "home field advantage" allows the user's representative to concentrate on the negotiation process in a familiar setting. Other strategies are the following:

1. *Use the "good guy" and "bad guy" approach.* The consultant is often perceived as the bad guy, the user as the good guy. The consultant is the "shrewd" negotiator, whereas the user is the compromiser.
2. *Be prepared with alternatives at all times.* It is a give-and-take approach.
3. *Use trade-offs.* Rank less important objectives high early in the negotiations.

4. *Be prepared to drop some issues.* Certain issues may be better discussed in later sessions.

Contract Checklist

Responsibilities and Remedies

A computer contract should specify the remedies to the parties in the event of default or nonperformance. Remedies should begin with a list of responsibilities that both parties have agreed to assume. Next to each vendor obligation are listed the remedies desired by the user in the event of nonperformance. With such provisions, the contract obligates a vendor to deliver only equipment that operates according to specifications.

There are three major categories of remedies: special remedies, damages, and specific performance. *Special remedies* are the user's first line of defense. For example, if the vendor fails to meet the delivery date, the user has the right to cancel the deal after a stated notice to the vendor. Or, if the system fails the acceptance test due to deficient capacity, the user may expect the vendor to provide additional power or memory at a specific cost (or no cost) within a stated time.

These special remedies provide immediate relief to the user, without litigation. When the vendor hesitates to negotiate special remedies, the user's representative(s) may demand strict *damage remedies*. Such damages may be actual, consequential, or liquidated. *Actual* damages compensate a party for what it should have received in a contractual bargain. *Consequential* damages include all other foreseeable losses that result from the breach of a contract. Because they cover amounts greater than actual damages, vendors make every effort to limit consequential damages in the contract.

Liquidated damages establish a fixed amount in advance payable in the event of default. If the vendor fails to meet the delivery date or the installation or acceptance deadline, liquidated damages are a popular remedy. They actually provide the vendor with a negative financial incentive if they do not perform.

Hardware. A good hardware contract is goal-oriented. It stipulates the results to be achieved with the system. Of greater importance is the definition of system performance in terms of expected functions. The first step in contracting equipment is to identify each component and spell out the performance criteria that the processor must meet under normal operating conditions. Where possible, it should be written in terms understandable to anyone who reviews the contract at a later date.

Software. A software package is a license to use a proprietary process. It is not truly "sold," since a title is not transferred, so it is more or less a "license for use." The user has the right to use the package but does not have title of ownership.

There are several risks inherent in software packages:

1. *Nonperformance or failure to meet specifications.* One remedy is to provide for termination in the contract.
2. *Costs of modification or integration.* The remedy is to specify assistance with modifications for a fee.
3. *Bankruptcy of the vendor.* The remedy is provision for the user to modify the package without penalties, charges, or obligations.

Delivery and Acceptance. A major problem with contracting is failure of the vendor to deliver on schedule.[11] A contract should specify the remedies provided for failure to meet the agreed-upon delivery schedule. The section governing acceptance describes the tests that must be met for the equipment to be accepted by the user. A user should also insist on a period of use under normal operating conditions to ensure the system's performance to standards.

Warranties. Article 2 of the UCC provides three implied warranties by the vendor for goods sold and many lease transactions as well. Essentially, the vendor has title to the goods when sold, the goods are mechantable, and the goods are fit for the purposes they are intended. Because warranties are desirable for customers, vendors include provisions relating to them in agreements, thus suggesting that some warranty is made.

Finances. Contract negotiations involving finances can be summarized by "Let the buyer beware." One of the most difficult contract items to negotiate is the time when payments begin. Most standard contracts specify that payment commences upon installation. In rental arrangements, the contract should specify whether rent is based on a flat monthly fee or on the number of hours used per month or shift. In lease arrangements, however, the contract covers an option to buy, when it can be exercised, and if the optin is assignable. Also, it stipulates whether monthly lease payments can be applied toward the purchase price.

Since the contract is normally signed before delivery, the prices defined in the contract should be protected. If the vendor requires an escalation clause, the user should be able to cancel the contract in exchange for a penalty payment from the vendor as compensation for costs incurred for preparation.

A clause called *force majeure* deals with the suspension of the contract in an event beyond the vendor's control. This includes civil disorders, hurricanes, tornadoes, earthquakes, and acts of war. It is interesting to note that nuclear devastation is normally exempted from a *force majeure* clause.

Guarantee of Reliability. This is a statement by the vendor specifying the following:

---

[11] David Myers, "More Seen to Contracts Than Price, Delivery," *Computerworld*, November 14, 1983, p. 23.

1. *Minimum hours of usable time per day*—that is, the amount of time of computer operation before a shutdown.
2. *Mean time between failures (MTBF)*—the length of time the system will run without breaking down.
3. *Maximum time to repair*—response time to repairing the system.

Negotiations have come a long way during the last decade. One reason is the growing competition in the industry. If one vendor will not negotiate, the chances are that five other vendors will. Another reason is that users are becoming more knowledgeable about computers and are unwilling to be taken in by a standard contract. Standard contracts are designed to protect the vendor. They have been written by knowledgeable legal staff. When a contract is not negotiated, the user is usually on the losing end. Various items can and should be negotiated. If negotiations are done properly, with foresight, planning, and control, the user can not only get a fair contract but also secure a fruitful relationship that will benefit both the user and the vendor.

Summary

1. Computer vendors are classified as hardware, software, and service suppliers. They provide mainframe, operating systems/application programs, and service, respectively.
2. Software is classified as system software for controlling computer operations and application software for solving user-oriented problems. In general, software allows concurrence of operation, resource and information sharing, modularity, and multiplexed operation. It has grown by leaps and bounds, especially since the birth of the microcomputer. Programmer shortages, improved software performance, and economies of scale are other reasons for the growth of software.
3. There are several things to do before selection:
   a. Define system capabilities that make sense for the business.
   b. Specify the magnitude of the problem.
   c. Assess the competence of the in-house staff.
   d. Consider hardware and software as a package.
   e. Develop a time frame for selection.
   f. Provide user indoctrination.
4. The selection process consists of several steps:
   a. Prepare a requirements analysis (understanding user requirements).
   b. Specify system specifications.
   c. Prepare a request for proposal.
   d. Rank vendor proposals (if it is a big job, hire a consultant).
   e. Decide on the best proposal or vendor and invite a demonstration.
5. Software may be custom-made or acquired off the shelf. The latter

approach has several advantages; it is quicker and less costly and generally requires less time to install than in-house software. The major drawback is occasional difficulty meeting user requirements.

6. The criteria for software selection are:
   a. Reliability—gives consistent results.
   b. Functionality—functions to standards.
   c. Capacity—satisfies volume requirements.
   d. Flexibility—adapts to changing needs.
   e. Usability—is user-friendly.
   f. Security—maintains integrity and prevents unauthorized access.
   g. Performance—delivers as expected.
   h. Serviceability—has good documentation and vendor support.
   i. Ownership—has right to modify and share use of package.
   j. Minimal costs—is affordable for intended application.
7. In evaluating packages, we look at the software criteria and rank them in order of importance. Then we select the required features and check them against those offered by the software package. Sources for evaluation are benchmark tests, other users, and product reference manuals.
8. Vendor proposals are evaluated and finalized ad hoc, by scoring the characteristics of each system, or by a cost-value method that essentially applies a dollar credit amount to the proposal that best meets the user's characteristics. In the end, both qualitative and quantitative measures are considered for the final acquisition.
9. There are three methods of acquisition: rental, lease, and purchase. Rental is favored for first-time users who are unsure of what they want. Every service is included but at a high cost. The user has financial leverage and is free to change systems. The lease option lasts longer than the rental option but costs less. No financing is required and the risk of system obsolescence is shifted to the vendor. The main drawback is the heavy penalty for early cancellation. The purchase option offers a lower total cash outflow if the system is kept longer than five years. It also offers a cash savings because of depreciation and investment tax credit. The initial high cost, insurance and maintenance expenses, and overall risks are drawbacks.
10. An alternative to the three acquisition methods is to buy a used computer. Prices are attractive and so is performance, provided the user verifies proper maintenance through a log.
11. Negotiating a computer contract is an art that requires strategies, tactics, and much planning. Time is critical. Before the negotiating session, representatives should be prepared with alternatives and trade-offs and plan to discuss some issues in later sessions when the climate is more amenable for further negotiations.
12. A contract should spell out the following:
   a. Vendor responsibilities and remedies in the event of nonperformance; with hardware, the results to be achieved with the system.

b. Remedies for failure to meet the delivery schedule and failure of the system to pass the user acceptance test.
c. Implied warranties regarding system or software performance.
d. Guarantee of reliability in terms of uptime, mean time between failures, and response time to repairs.

Key Words

Benchmark
Facilities Management (FM)
Force Majeure
Functionality
Lease
Modularity
Portability
Request for Proposal (RFP)
Serviceability
Servicer
Uniform Commercial Code (UCC)
Usability

Review Questions

1. How is the computer industry classified? Summarize.
2. Distinguish between the following:
   a. Hardware and service suppliers.
   b. Concurrence of operation and multiplexed operation.
   c. RFP and vendor proposal.
   d. Portability and understandability.
3. What is the FM concept? Search the literature and write a report on the current demand for FM service.
4. List and briefly summarize the key attributes of software.
5. "Software is classified according to whether it performs internal computer functions or allows use of the computer for problem solving." Explain.
6. What factors must be considered prior to system selection? Expound.
7. It has been suggested that software should be considered before hardware. Do you agree? Why?
8. Elaborate on the major phases in selection. Which phase sets the tone for the vendor contact?
9. Draft a plan to show the makeup of a request for proposal. How does it differ with different sizes of systems?
10. In what ways is the consultant helpful in selection? How would you go about deciding on a consultant? Explain.
11. The chapter lists sources of information to check on vendors. Can you think of other sources? Give it a try.
12. What software criteria are considered for selection? Summarize.
13. How important is the reliability factor in software packages? Discuss.
14. What is the difference between the following?
    a. Functionality and flexibility.

   b. Performance and serviceability.
   c. Reliability and security.

15. Discuss the sources of information used in evaluating hardware and software. Which source do you consider the most reliable? Why?
16. How is the scoring approach different from the cost-value approach in evaluating proposals? Illustrate.
17. Hardware selection requires the analysis of several performance categories. For a first-time user of microcomputers, what category or categories would be most important? Why?
18. There are three methods of acquisition. What are they? Elaborate on the pros and cons of each method.
19. Under what circumstances would one consider buying a used computer? What are the benefits and drawbacks to such an acquisition? Discuss.
20. In what way is computer negotiation an art? Explain.
21. There are strategies and tactics in negotiating contracts. If you were representing a client or acquiring a small computer, how would you go about it?
22. The chapter elaborates on a contract checklist that should be considered in a final computer contract. What goes into the checklist? Are all the items important? Why?

Application Problems

1 A county school system has been processing its payroll through a local bank's computer center for over eight years. One morning in early June, the school superintendent received a telephone call from the bank's vice president with the bad news. The bank could no longer process the school's payroll. He gave the school until December 31 to look for an alternative.

The superintendent contacted a computer consultant with a request for a payroll service or an installation to be in operation by the end of the year. The school had just acquired a IBM/XT, complete with a letter-quality printer, color monitor, and operating system. The price was $6,800. None of the superintendent's staff knew how the system operates. No software was available either.

The consultant checked with several service bureaus. The charges were virtually the same—$0.75 per paycheck. This includes accumulating payroll data and preparing end-of-year withholding tax forms. Payroll checks are prepared for all employees biweekly.

With the availability of a microcomputer, the alternative is to acquire the software to do payroll on the premises. A payroll software package will cost $800 plus an annual maintenance fee of 15 percent. Supplies for printing checks, reports, ribbon, etc. are estimated to cost around $300 per year. Based on other installations, it will take three days to train a clerk to use the system and four hours to process the payroll.

Assignment

a. Should hardware be acquired before software? Why? Explain.

b. What are the unique features of this case? Elaborate.

c. Do a cost/benefit analysis for the two alternatives described. Assume that the hardware has a useful life of four years. Because the school is a nonprofit institution, the cost of money and depreciation are not applicable. Which alternative should be chosen? What are the drawbacks to each alternative? The benefits?

d. In addition to cost, what other factors should be considered in making the final choice?

2

A medium-size commercial bank has been using a service bureau for nine years to handle virtually all applications, including checking, savings, installment and commercial loans, and trust activities. Until last month, it has been paying $285,000 a year for the service, which amounts to 10 percent of the net income. The new contract submitted by the service bureau shows an increase in fees to $315,000.

With the steadily decreasing cost of hardware, improvements in computer processing capabilities, and availability of ready-to-use software packages for banking, the senior vice president was curious to know whether the bank should consider having its own system. A computer consultant was hired to explore the matter.

A computer system that could meet the needs of the bank over the next five years will cost $184,000 to purchase and $65,000 for software, training, and conversion. One full-time person is needed to run the system. The salary is $23,000 plus 30 percent benefits. The system has a useful life of five years. Monthly maintenance is $1,850. If purchased, the system will be on a cash basis. The interest rate on loans at this time is 16 percent.

The hardware may also be leased on a three- or five-year basis. The monthly lease fee is $3,900, which includes maintenance. The five-year lease provides that, upon expiration of the lease, the bank can acquire the hardware for $1.

Assignment

a. Which alternative (service bureau versus in house) would be the most beneficial?

b. Suppose the bank decided to go ahead and process its own applications. Should it purchase or lease? What factors are considered in the choice? Elaborate.

c. Assuming that the bank purchased the equipment, prepare a disaster recovery plan for the system.

Selected References

Bigelow, R. P. "Is There Legal Protection for Software?" *ICP Infosystems,* March 1980, p. 104ff.

Couger, Daniel. "Development to Facilitate Managerial Use of the Computer." *Computing Newsletter,* no. 8 (April 1983), pp 2–4.

Dotto, Lydia, and Lawrence Briner. "The Soft Path, a Consumer's Guide." *Canadian Business,* January 1983, pp. 49–80.

Gruenberger, Fred. "Making Friends with User-Friendly." *Datamation,* January 1981, p. 108.

Haber, Lynn. "Negotiations Can Spell Out Successful Systems", *Computerworld,* November 21, 1983, p. 22ff.

Heckel, Paul. "Developing Software for Microprocessor-based Products." *Mini-micro Systems,* February 1980, pp. 111–15.

Hoffberg, Alan M. "Getting Hardnosed about Software." *Infosystems,* July 1981, p. 37ff.

Holmes, Geoffrey. "Choosing the Right Software Supplier." *Accounting,* April 1981, pp. 67–68.

Johannsen, Craig. "Software Selection." *Computerworld,* January 1979, p. 33.

Lord, Ken. "Software Reviews." *Desktop Computing,* September 1983, pp. 72–76.

Myers, David. "More Seen to Contracts Than Price, Delivery." *Computerworld,* November 14, 1983, p. 23.

Sanders, G. Larry; Paul Munter; and Donald O. Reed. "Selecting a Software Package." *Financial Executive,* September 1982, pp. 39–46.

Seaman, J. "Smart Benchmarking—The Key to Successful Procurement." *Computer Decisions,* March 1981, pp. 74–94.

Snyders, Jan. "How to Buy Packages." *Computer Decisions,* July 1978, p. 52.

Vargo, Paul M. "How to Minimize the Risk of Buying Inadequate Software." *The Practical Accountant,* March 1983, pp. 53–56.

Weidler, Gregory. "Purchase or Lease?" *Journal of Systems Management,* June 1976, pp. 28–35.

Wolpert, Tom. "Why It Pays to Buy Used." *Venture,* August 1982, p. 12.

Chapter 15

Project Scheduling and Software

At a Glance

The material covered so far has addressed the concepts, tools, and technology for building systems. Developing a system requires planning and coordinating resources within a given time. More important, effective project management is needed to organize the available resources, schedule the events, establish standards, and meet conversion deadlines.

A project manager is expected to have managerial and technical skills along with management support for system success. The tools available for project planning are Gantt and PERT charts. More recently, project management software that runs on the personal computer is making a contribution to calculating events and determining the critical path of a project.

By the end of this chapter, you should know:

a. Why systems fail.
b. The steps taken to establish a project.
c. How Gantt and PERT charts are used in project planning.
d. The qualifications and requirements for successful project management.

INTRODUCTION

The process of planning, designing, and implementing computer systems is called a *project*. It is directed by a project manager who uses available resources to produce systems for the organization. In large firms, installing a system may take years and involve hundreds of people. Planning and installing smaller projects on schedule also take time and require control and coordination of resources. It takes an effective manager to organize the available resources, schedule the events, establish standards, and complete the project on time, within budget, and with successful results.

We need to take a close look at project management and its role in system development. In this chapter, we focus on the planning and scheduling functions and on project estimating, organization, and control. Project management is inherently a team effort. The chapter also shows how project managers are selected and teams are managed.

WHY DO SYSTEMS FAIL?

No one likes to fail, especially on a major project. At best, failure can result in waste and a delayed return on investment. At worst, it can drive an organization into bankruptcy or threaten the career of senior officials. Despite the high stakes, however, system projects often fail and there are many reasons. Most project managers have experienced some of the problems illustrated in Figure 15–1. Of these problems, the following are especially critical:

1. *Conflicting objectives.* When everyone on the project team has conflicting objectives, the project is bound to suffer. For example, the user demands a workable system, while the designer strives for a high-powered code. The comptroller interprets success from the bottom line, while the project manager's main goal is to produce an efficient system, whatever the cost.

2. *User's lack of involvement or cooperation.* A frequent complaint is that users are not helpful, are reluctant to assume their responsibilities with the new system, and are critical of the project team. Often promised objectives have been achieved only partially, or the original cost estimate was overoptimistic. As a result, the user is not willing to get involved.

3. *Inexperienced project management.* This is exemplified by poor project management techniques, a poorly motivated project team, poor estimating procedures, cost overruns, incorrect design, and underestimated manpower requirements. Any of these can cause problems for the project.

Another problem stems from the conventional way project managers are selected. For example, an analyst is given responsibility for implementing a substantial application while, in practice, he/she has no management tools or business experience to manage the project.

FIGURE 15–1 Sources of System Failure

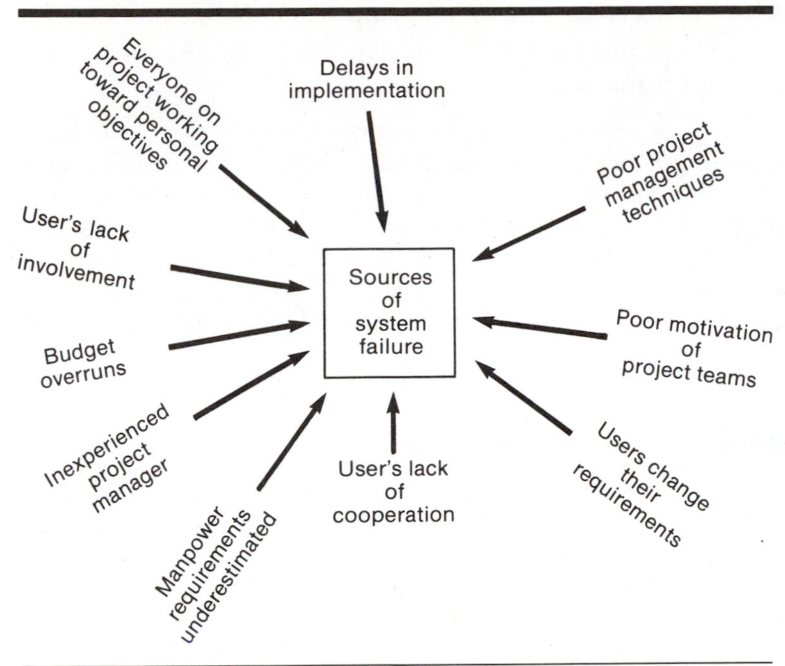

WHAT IS PROJECT MANAGEMENT?

Project management is a technique used by a manager to ensure successful completion of a project. It includes the following functions:

- Estimating resource requirements.
- Scheduling tasks and events.
- Providing for training and site preparation.
- Selecting qualified staff and supervising their work.
- Monitoring the project's program.
- Documenting.
- Periodic evaluating.
- Contingency planning.

From these functions, we can see that project management is a specialized area. It involves the application of planning, organization, and control concepts to critical, one-of-a-kind projects. It also uses tools and software packages for planning and managing each project. Managing projects also requires the following:

1. *Top management commitment* to setting project priorities and allocating resources to approved projects.
2. *Active user participation* to identify information needs, evaluate pro-

posed improvements on a cost/benefit basis, provide committed resources, and be receptive to training when scheduled.

3. *A long-range plan* that includes general project priorities, objectives, schedules, and required resources.

Included in this is pressure from users who require the systems department to accept impractical tasks or deadlines. The result is rushed, compromised projects, contrary to good system development practices. A further difficulty found in many organizations occurs when individual departments acquire microcomputers without knowing about their requirements or consulting with the centralized computer facility. The result is often uncoordinated confusion that makes it difficult to plan or control projects.

A Framework for Project Management

When problems are identified, it is usually toward the end of the project— in system testing or even conversion. The people who are most active during system development are the programmers. This was one reason a great deal of attention has been given to improving the maintainability of programming. Actually, however, the problem typically begins during the initial stages of the project. So there is a need for establishing an overall framework that will lead to a successful implementation.

There are several steps in establishing a project:

1. *Understand the problem*, using experienced staff rather than trainees for the initial investigation. Each project should be studied to evaluate the degree of change, the scope of the project, and the cost of late completion. Projects that involve a high degree of change and require a close working relationship with the user staff normally require a highly managed approach. The smaller, less involved projects that are worked on by one person require only a written objective, a target date, and responsible person. How much management is needed in either project is a decision that involves a delicate balance between risk and cost.

2. *Specify project responsibilities*. The user executive should sponsor the project and be responsible for the right amount of user involvement. A qualified project manager is the leader responsible for project success. He/she reports directly to the user executive.

3. *Select a project manager*. Project managers are responsible for the day-to-day activities and for meeting the objectives within the constraints of time, cost, and quality. The duties include developing a project plan, forming and training the project team, and assigning and coordinating project tasks.

Both technical and leadership skills are required. Technical ability is crucial during the initial (creative) phase of the project, whereas leadership is required throughout the life cycle of the project. Other job qualifications include:

a. Experience in the functional areas encompassed by the project.

b. Ability to recognize problems and communicate ideas and concepts to the project team and the user. The project manager is often a translator, communicating at a level of detail that is understandable.

c. A working knowledge of the system improvement process and how a problem is tranformed into a workable system.

Most project managers are selected from the system staff. This means they have experience in systems analysis and programming and a demonstrated knowledge of the business at hand. An organization that is eager to secure full user commitment to the success of the system might assign a user as a project manager. The advantages of this approach are a commitment to project success through direct responsibility, improved user training, and a better grasp of legal, financial, and other effects of system changes. The main drawback is possible conflict of the assignment with existing line reporting relationships. Regardless of the source from which a project manager is drawn, user commitment is essential for system success.

4. *Establish ground rules* and set performance standards. The standards should be simple to measure, clear, obtainable, and enforceable.

5. *Select the right project.* A first project should be limited in size, have a high probability of success, and show results within one year. Large projects should be segmented into subprojects to demonstrate progress toward completion.

6. *Define the tasks.* The detailed tasks are specified in a plan. In preparing the project plan, the project manager must determine what work is to be done, how it will be done, what resources are available, who will do the work, and how long it will take. This type of planning is extremely important.

The topic that has received considerable attention is estimating the time needed to complete tasks. Too little time is often cited as the reason for project failure. Failure to understand the work to be done, lack of user involvement, overcommitment of manpower, failure to allocate time to special problems such as maintenance, and the "domino effect" caused by unplanned events cause estimating problems. All these emphasize the importance of planning for successful projects.

CRISIS ELIMINATION THROUGH PLANNING

Without planning it is difficult to measure progress. As plans are crystallized, crises should begin to disappear. A project manager must plan the life cycle of the project and delegate authority for its implementation.

Project planning involves plotting project activities against a time frame. One of the first steps in planning is developing a road map structure or a network based on analysis of the tasks that must be performed to complete the project. In the early 1900s, formal planning used a Gantt chart or a milestone chart. By plotting activities on the Y-axis and time on the X-axis, the analyst laid out an overall network specifying interrelationships among actions. Later on, formal planning techniques such as the program evaluation and review technique (PERT) was introduced. Other operations research techniques such as linear programming and queuing theory have also been introduced in allocating resources. In the early 1980s software packages became available for project planning. These tools are discussed next.

Planning Tools

Gantt Charts

Basic planning uses bar charts that show project activities and the amount of time they will take. This activity scheduling method was first introduced in 1914 by Henry L. Gantt as a redimentary aid to plot individual tasks against time. The Gantt chart uses horizontal bars to show the durations of actions or tasks. The left end marks the beginning of the task, the right end its finish. Earlier tasks appear in the upper left and later ones in the lower right.

Figure 15–2 is a Gantt chart for a hypothetical boat-building project. The project is to build a prototype of a new motorboat called *Seacraft Yacht.* The heavy horizontal bars are activities, and the light horizontal bars are tasks. Broken horizontal bars are estimated time delays or slack time. A *task* is a specific job that can be assigned to one person to perform in a specific time. A group of tasks makes up an *activity,* which ends in a *milestone*. In our Gantt chart, designing the hull is a task, whereas signing the boat (requiring three tasks) is an activity. A total of five activities are required for building the prototype. The completion of each activity is a milestone, indicated by an arrow.

In planning this project, several steps are undertaken:

1. Identify the activities and tasks in the stage. Each activity must be identified to plan the completion date and allocate responsibilities among members of the project team. In our example, there are five activities:
   a. Design boat.
   b. Purchase/test engine.
   c. Make boat.
   d. Test boat.
   e. Prepare owner's manual.

2. Determine the tasks for each activity and the estimated completion times. Each activity is broken down into several tasks. In Figure 15–2, designing the prototype involves three tasks: designing the hull, testing the

FIGURE 15-2 Gantt Chart—An Example

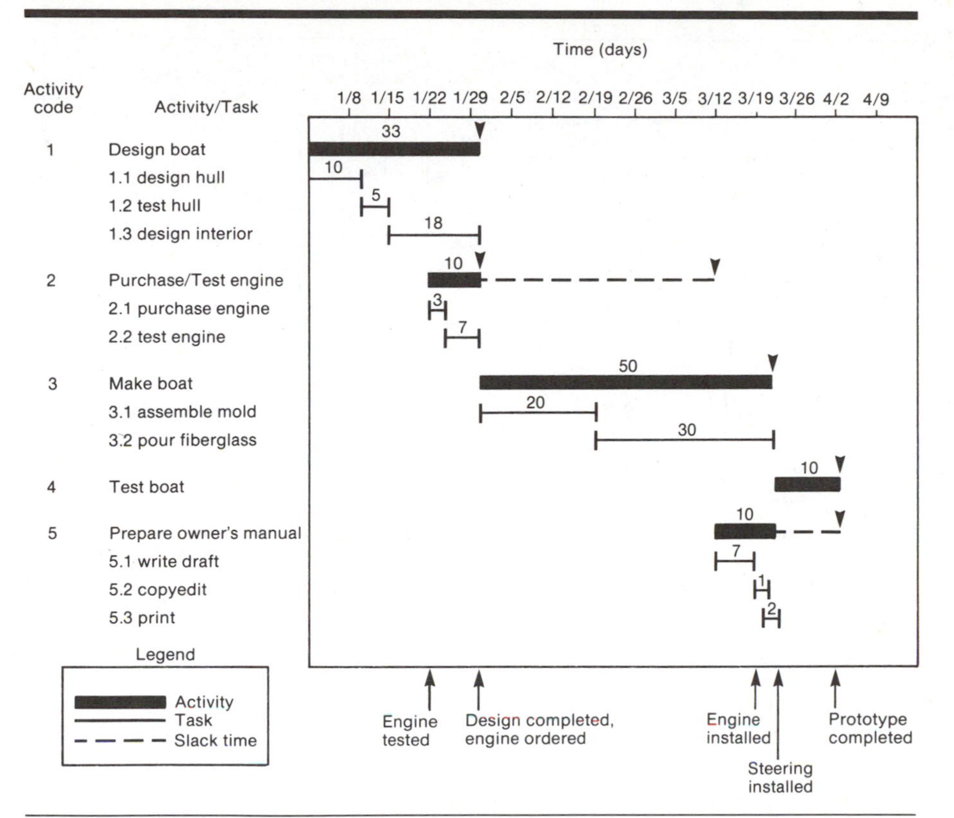

hull, and designing the interior. They are estimated to take 10, 5, and 18 days, respectively. The total (33) is the estimated time for the "design boat" activity. In real-life applications, an allowance for contingencies is provided. This is called *slack time.* Each project allows beween 5 and 25 percent slack time for completion.

3. Determine the total estimated time for each activity and obtain an agreement to proceed. Figure 15-3 shows the number of days budgeted for each activity and a 20 percent activity contingency toward completion.

4. Plot activities on a Gantt chart. All activities, tasks, and milestones are drawn on the Gantt chart, with emphasis on simplicity and accuracy (see Figure 15-2).

5. Review and record progress periodically. The actual amount of time spent on each activity is recorded and compared with the budgeted times. As shown in Figure 15-4, the actual number of days spent on the three tasks in designing the boat is 40 as budgeted. This procedure is applied to the remaining activities of the prototype stage. A summary of progress on the project is sent to management for follow-up.

FIGURE 15-3 Budgeted Activities—An Example

Activity bar chart					Project name: Seacraft yacht Project stage: Build prototype
	Activity	Budgeted days	Actual days	Finish date	
1	Design boat	33			
2	Purchase/test engine	10			
3	Make boat	50			
4	Test boat	10			
5	Prepare manual	10			
	Activity total	113			
	Contingency (20%)	23			
	Grand total	136			

FIGURE 15-4 Activity Time Estimate

Activity time estimate		Date 1/06/85		Project name: Seacraft yacht Project stage: Build prototype Project activity: Design boat
Task/Milestone	Budgeted days	Actual days	Finish date	
Design hull	10	12	1/4	
Test hull	5	4		
Design interior	18	22		
Contingency (20%)	7			
Reviewed by:		2		
Total budgeted time (days)	40	40		

FIGURE 15-5

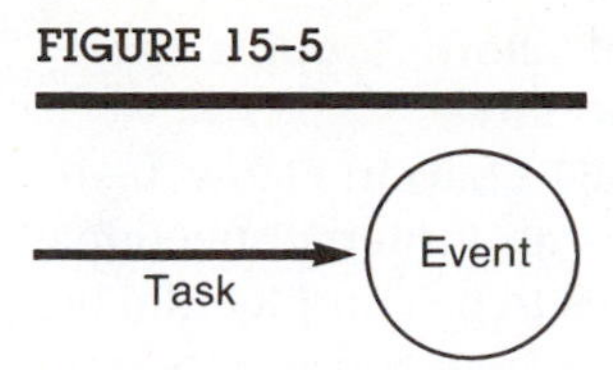

Program Evaluation and Review Technique (PERT)

Gantt charts have one drawback: They do not show precedence relationships among the tasks and milestones of a project; that is, interdependence of tasks is not evident. The critical tasks or those that must be done on time are not evident either. A project manager wants to identify the activities and the amount of time they require, show their interrelationships, specify their sequence, and have a means of monitoring progress on the project. If the Gantt chart is a first-generation project-planning tool, a second-generation tool is the critical path method (CPM). It was introduced in the late 1950s as a tool for scheduling major overhauls of chemical processing plants.

Like the Gantt chart, PERT makes use of tasks. Like milestone charts, it shows achievements. These achievements, however are not task achievements. They are terminal achievements, called *events*. Arrows are used to represent tasks and circles represent the beginning or completion of a task (see Figure 15–5). The PERT chart uses these paths and events to show the interrelationships of project activities.

The events from our boat-building example in Figure 15–2 are shown in the following table:

Activity	Tasks
1	Hull design Hull test Design of interior
2	Engine purchase Engine test
3	Mold assembly Building boat
4	Boat test
5	Draft of manual Copyediting draft Printing of manual

Each task is limited by an identifiable event. An event has no duration; it simply tells you that the activity has ended or begun. Each task must have a beginning and an ending event. A task can start only after the tasks it depends on have been completed. PERT does not allow "looping back" because a routine that goes back to a task does not end.

The list of tasks and events is networked in a PERT chart in Figure 15–6. The arrow length is not significant, but the sequence and interconnections must give a true picture of the precedence of activities to be completed. The numbers on the activity lines are the days required between events (see Figure 15–2).

A PERT chart is valuable when a project is being planned. When the network is finished, the next step is to determine the *critical path*. It is the longest path through the network. No task on the critical path can be held up without delaying the start of the next tasks and, ultimately, the completion of the project. So the critical path determines the project completion date. In our example, it is:

Critical path = Start # − 1 − 2 − 3 − 4 − 6 − 8 − 12
Duration (days) 10 + 5 + 18 + 11 + 50 + 10 = 104 days

If the job started on January 1, then the completion date would be April 14. Any delay in the critical path will affect the schedule. A delay in other paths can be corrected in time for completion of the project on schedule.

In addition to showing the interrelationships among project activities, PERT charts show the following:

1. The activities that must be completed before initiating a specific activity.
2. Interdependencies of the tasks.
3. Other activities that can be completed while a given activity is in progress.
4. The activities that cannot be initiated until after other specific activities are completed. This is called a *precedence relationship*. In Figure 15–6, task 3 (design boat interior) cannot begin until after tasks 1 and 2 (design

FIGURE 15-6 PERT Chart—An Example

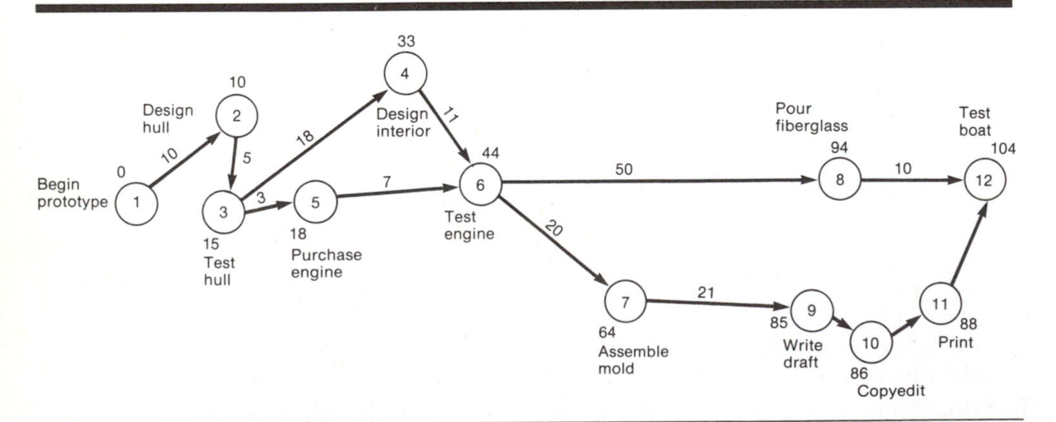

hull and test hull, respectively) have been completed. Similarly, since task 1 takes longer than task 2 to complete, it must be started either first or no later than task 2. It is considered the critical path item (indicated by a heavy line in Figure 15–6), since its delay holds up the project. Conversely, if tasks 1 and 2 could be started at the same time and both are completed on schedule, then task 2 includes five days of slack time.

Event Calculations. With this information, scheduling calculations can be done. We want to analyze the project schedule to find out how early or how late an event can begin without affecting the overall schedule. There are two types of calculations: one for event times and the other for activity times. For the sake of simplicity, let us illustrate event calculations.

To calculate event times, we need the following notations:

t_{ij} = Time it takes to complete task from event i to event j.

E_j = Earliest time at which event j can occur based on the completion of all preceding activities. This is computed by examining E_j and the duration for each preceding event. This algorithm proceeds sequentially, one event at a time, beginning with event 1.

L_j = Latest time at which event j can begin without delaying completion of the project. To determine this time, we work backward through the network.

To calculate the earliest event times (E_j) for each event j, we use the following algorithm:

1. Set $E_1 = 0$ for the starting event.
2. Set $E_j = \text{Max}_i(E_i + t_{ij})$, where maximization occurs over all events i that are immediate predecessors of event j.

Let us use Figure 15–6 to illustrate: In the PERT chart, the values of E_j are shown next to each event. Event 1 is labeled with time 0 as the starting time. Event 2 is labeled with time 0 + 10 = 10 by summing the time of event 1 and the activity time from event 1 to event 2. The values of E_j for the critical path are summarized in Figure 15–7. Note that the value of event 6 (E_6) is the higher value of predecessor events 4–6 and 5–6, which is 44. The same logic applies to event 12. The earliest occurrence time for event 12 is 104. Since it is the last event in the network, the earliest completion time of the project is 104 days.

In network calculations, a backward pass is also made to determine the *latest time* each event can begin without delaying the completion of the project. The algorithm is:

$L_n = E_n$, where n represents the last event in the network

$L_i = \text{Min}_j(L_j - T_{ij})$, where the minimization occurs over all events j that are immediate successors of event i

To calculate, we begin with the last event in the network and work

FIGURE 15-7 The Forward Computational Pass; Early Event Times Calculations

Event Number	Rule and Value
1	$E_1 = 0$
2	$E_2 = E_1 + t_{12} = 0 + 10 = 10$
3	$E_3 = E_2 + t_{23} = 10 + 5 = 15$
4	$E_4 = E_3 + t_{34} = 15 + 18 = 33$
5	$E_5 = E_3 + t_{35} = 15 + 3 = 18$
6	$E_6 = \text{Max } (E_4 + t_{45}), (E_5 + t_{56}) = \text{Max } (33 + 11), 18 + 7)$
	$= (44), (25)$
	$= 104$
8	$E_8 = E_6 + t_{68} = 44 + 50 = 94$
12	$E_{12} = \text{Max } (E_8 + t_{812}), E_{11} + t_{1112}) = \text{Max } (94 + 10), (88 + 10)$
	$= (104, (98)$
	$= 104$

backward through the network. The latest event is event 12. We let $L_{12} = E_{12} = 104$. Then there is one successor for event 8, so:

$$L_8 = L_{12} - t_{812} = 104 - 10 = 94$$

We label event 8 with a 94 for its latest occurrence time and so on backward until all the events are labeled. Note event 3, however, which has two successor events, 4 and 5. Here we have:

$$E_3 = \text{Min}(33 - 18), (18 - 3) = 15$$

We label event 3 with a latest time of 15 days, which is the same as the forward computation time, meaning there is no slack time.

The information calculated from the forward and backward passes helps the project manager identify the critical path, determine the completion date of the project, and compute slack time. When a project is managed, all activities and events on the critical path must be closely monitored, because any slip (extended delays) can mean comparable delays in the completion date. In most projects, no more than 10 percent of all activities are on the critical path.

Project Management Software

For large projects with complex relationships and hundreds of tasks and events, the network cannot be analyzed manually. For over a decade, project management software has been available on in-house computers. Since the early 1980s, several software packages designed to run on the microcomputer have become available. They meet the needs of modern management and facilitate the project management process substantially.

Figure 15-8 is a selected list of project management software. Each software package offers unique features—some with a project network road

FIGURE 15-8 Project Management Software—Selected List

Name	Vendor	Features/Comments
Harvard Project Manager	Harvard Software, Inc. Howard, Mass.	Sophisticated on-screen graphics; support Gantt chart and CPM displays; generates project road map automatically and indicates critical path.
MicroGantt	Earth Data Corp. Richmond, Va.	Interactive, menu-driven command structure; on-screen Gantt chart and calendar
Milestone	Digital Marketing Walnut Creek, Cal.	Gantt-type product; no on-screen network graphics but supports up to 400 activities.
Plantrac	Computerline Ltd. Quincy, Mass.	Draws precedence networks on screen
Project Scheduler	Data Easy Foster City, Cal.	Displays up to seven tasks, task length, and responsibility; project duration is 10 years; no on-screen network diagram
Project Scheduler	Scitor Corp. Sunnyvale, Cal.	Generates hard-copy output of the network diagram.
VisiSchedule	VisiCorp. San Jose, Cal.	Gantt-type product; limited graphic ability

map, others with graphic ability. They run on IBM, Apple, and other personal computers. Some packages provide graphic network diagrams for a clear display of a network's critical path. The software also offers fully detailed, time-phased reports and allows the recalculation of project networks in graphic form on screens.

To illustrate a microcomputer project management package, let us take *VisiSchedule*. This program shows the critical jobs in the project schedule, explains the relationships between jobs, calculates project and manpower costs, and generates reports and schedules as needed. Specifically, the program does the following:

- Determines the jobs that are critical for time and cannot be delayed without delaying the project.
- Determines the jobs with slack time that can be delayed without delaying the project.
- Keeps track of important deadlines and significant milestones throughout the project.
- Maintains schedules in time units of days or weeks, whichever is preferred.
- Maintains requirements and salary information for up to nine different occupations.

- Builds schedules around the holidays, days off, and nonworking weeks that you define.
- Changes any aspects of a job and immediately shows the impact on the overall project.
- Investigates trade-offs between manpower, cost, and time.
- Lets the user track the project's progress by updating the schedule to reflect changes in the plan and the completion of jobs.
- Allows sorting and rearranging jobs in the schedule and reports for the most meaningful presentation.

Communicating with the Program

There are two ways to communicate with the program: by selecting an option and entering data. The VisiSchedule options are listed in menus. A *menu* is a set of options within a status area. Menus list a set of items from which the user may choose. This approach saves the user from having to learn or look up detailed command syntax and being confused by nonsense abbreviations. A typical status area with a menu is shown in Figure 15–9.

An option is selected by moving a cursor to the option or typing the first letter of the option as the program displays it on the screen. To execute the option, the user simply depresses the RETURN or ENTER key. When the program wants the user to enter data, it changes the status area to the appropriate form (Figure 15–10).

When all the information is entered, the program may be instructed to

FIGURE 15–9

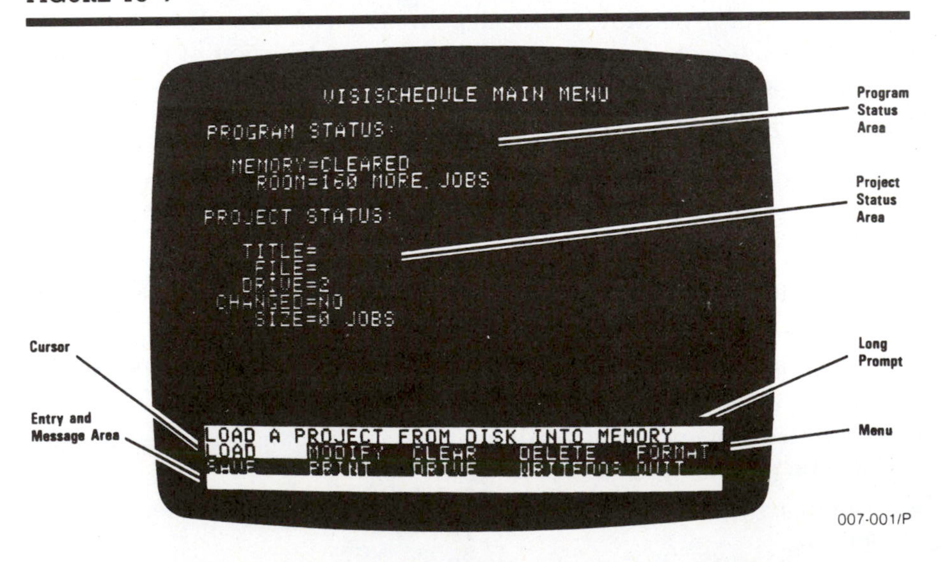

Typical VisiSchedule Menu and the Cursor

FIGURE 15–10 Data Entry Areas

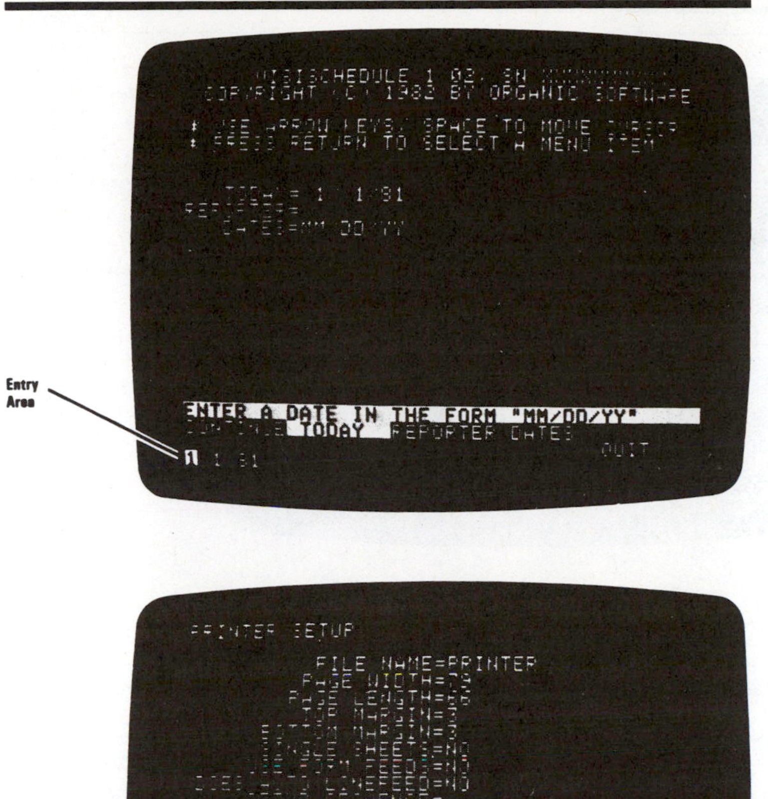

Source: VisiCorp., *VisiSchedule, User Guide,* p. 1–6.

display the project schedule and the schedule menu for modification. As shown in Figure 15–11, the upper part of the screen contains the schedule. The bottom contains the menu for entering and making changes in the schedule. The right side of the screen shows the time line schedule. It starts with a display of the date and week number, followed by the schedule.

PERT is still useful for modeling a project as a network of smaller jobs or activities, which is useful to predict how long it will take to complete that project. A PERT network is only a rough guideline for planning a project,

FIGURE 15-11 Schedule Menu and Symbols

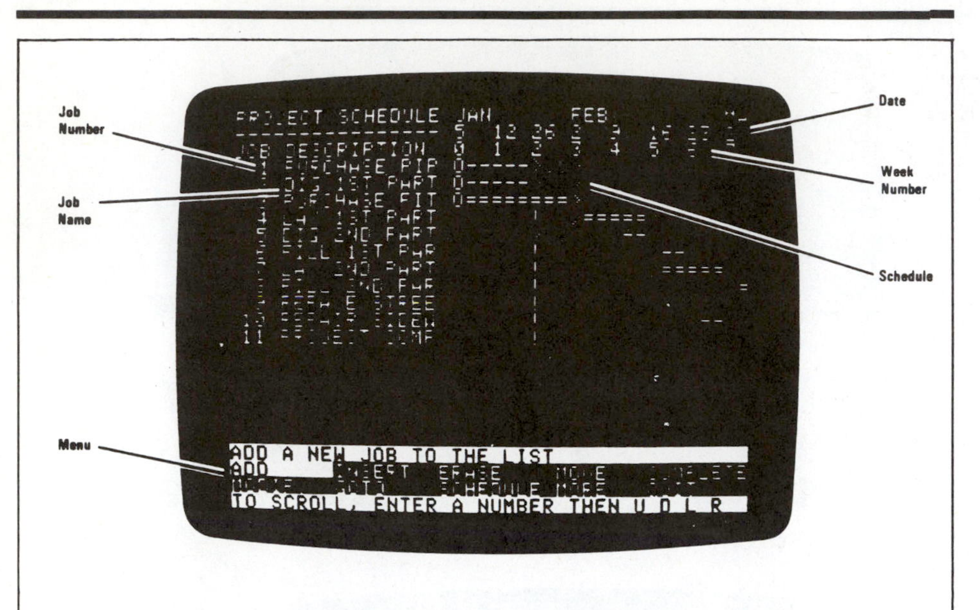

Symbol	Description
> = = = = = = = = = >	A critical job. A critical job cannot be delayed without delaying the entire project.
> - - - - - - - - - >	A non-critical job. There will be slack time associated with a non-critical job. This symbol is used for all jobs when you choose not to show the critical path.
>........>	Slack time for a non-critical job. A job can be delayed up to its total slack time without delaying the project.
>::::::::>	A completed job.
>:::: = = = = >	A partially completed job. The uncompleted portion could be critical or non-critical.
0 = = = = = = = = = > 0 - - - - - - - - - > 0::::::::>	Jobs with no prerequisites. These jobs are scheduled to begin on a specific date, not after the completion of prerequisites.
> = = = = = = = = = X > - - - - - - - - - X >::::::::X	Jobs with no successors. No other jobs name these jobs as prerequisites. The last job in a project normally ends with an X. A project may have multiple finishes.

Source: VisiCorp., VisiSchedule, User Guide, pp. 2–30 and 2–31.

however. It says nothing about who does what or when. Noncritical tasks are given only the earliest and latest permissible start times. These limitations are corrected when an organization installs project management software. The features listed earlier give an indication of flexibility that is available for planning and controlling systems projects. For example, with almost any package, it is possible to create a hierarchy by breaking the main network into subnetworks so that each activity is broken into its own network of activities. Given this flexibility, a software package designed for 200 activities could handle as many as 600,000, though at the expense of many hours of nonstop computing.

PROJECT ORGANIZATION

We have explained the major tools used in project planning. After the tasks have been mapped out and the manpower requirements determined, the next step is to decide on the best way to organize manpower. We shall begin by identifying the staffing and appropriate skills for a project and then suggesting a management style and approach to manage and control the staff.

The Project Team

The term *team* is used here to mean a group of people with similar skills and sharing a common activity. Some of the questions a project manager may ask are: What comprises the project team? Who should be on it? What skills must they have? For large projects, a project team is staffed by systems analysts, programmers, prime user(s), hardware/software suppliers, and even subcontractors. The staff may be retained for the duration of the project.

The skills expected of a project team are the same skills required in computer system development projects:

- Systems analysis.
- Detailed system design.
- Program design.
- System testing.
- Conversion.
- Cost justification.
- Planning and estimating.
- Hardware/software experience.
- Leading teams.

A project team is expected to tap the skills of its members so that one or more members can address the issues that face the project and suggest alternative solutions to keep the project moving the completion. These skills may be secured through a plan that identifies team members and specifies their skills. For larger project, a project team is expected to serve through the maintenance of the new system. Except in unusual situations, team members should not be shared between projects and project managers because of the possibility of priority conflicts.

Reporting Structure

Most people associated with major projects are outside the direct control of the project leader. Ideally, each team member should report directly to the project leader. In practice, however, most team members report to their respective supervisors. This means that the project leader has to use special skills to coordinate a host of persons over whom he/she has no real authority. For this reason, some authors refer to such a position as project *manager* rather than project *leader*.[1]

Management Styles

In discussing project management style, we need to examine two types of work performed in system projects: mechanistic and creative tasks. *Mechanistic* tasks occur primarily during coding, testing, and maintenance, which take up to 80 percent of the system development effort. Managing these tasks is made possible by controlling against predefined standards and formats. *Creative* tasks, on the other hand, are exemplified by program or system design and devising a test strategy, which involve up to 20 percent of the system development effort but could affect more than 80 percent of the outcomes. Compared to mechanistic tasks, a relatively small staff is needed for creative tasks. They report directly to the manager.

With these tasks in mind, a project manager may use democratic, autocratic, or some intermediate management style. The *democratic* approach allows subordinates to think on their own and make suggestions on projects. It is appropriate when the project manager is not sure exactly what is to be done. Yet it could be interpreted as indicative of a weak or inexperienced manager. Other problems with this approach are the likelihood of a slow reaction time and difficulty in expediting work.

In contrast to the democratic approach is the *autocratic* approach. The basic premise states that democracy is the wrong approach to unique projects. The aspects necessary to a system project should be entrusted to an experienced individual rather than a committee. This approach is more rewarding to individuals, although it is difficult to coordinate and control.

The implication of these approaches for project management is that simple mechanistic tasks may be assigned to lower-level staff members with coordination and control by a responsible manager. At the higher staff level,

---

[1] Jeffrey Keen, *Managing System Development* (Reading, Mass.: Addison-Wesley Publishing, 1981), p. 217.

however, innovative people report directly to the project manager, since their work could affect more than 80 percent of the outcome of the project.

These suggestions assume a fairly large project where a hierarchy is important for management control. This scheme does not apply to small system projects. A typical small project could be run by the computer manager, a programmer/analyst, and an operator. This means less staff and fewer levels of management, requiring no formal procedures or project control. Unlike in larger projects, each staff member in a smaller project performs several functions. Consequently, small projects tend to be less thorough and more prone to error. Furthermore, there is pyschological pressure on the project manager to make sure the project will be a success. The excuse of system failure through committee is absent in small projects.

The Project Manager

To plan, coordinate, and control the project team's effort, the project manager is expected to have unique qualities:

1. Flexibility and adaptability to changing situations.
2. Ability to communicate and persuade people affected by the project as well as those working on it about changes or enhancements to incorporate into the plan.
3. Commitment to planning the system development aspect of the project and ability to minimize the risks when assessing different approaches to implementing a project.
4. Understanding technical problems and design details.
5. Understanding the motivations and interests of members of the project staff and the ability to implement ways to improve job satisfaction.

Of all the contributions a project manager can make, maintaining a satisfied staff is the most important for strong project teams. An effective approach is to look into the vocational needs of the staff and try to match them with what the job offers and how it meets their needs. Hundreds of studies have been published on the subject. They all look at the variables that improve job satisfaction and motivate people to continue working in the job. The variables that have been known to be important to the job satisfaction of EDP staff are listed in Figure 15–12.

Looking at these variables, it is clear that success in managing system projects means making the best of project team abilities, getting them interested in their work, assigning them worthwhile jobs, recognizing their efforts in a spontaneous way, and providing opportunities for them to use their abilities, advance, and develop their careers. These factors become goals that a project manager should try to achieve for each member of the project staff and for the group as a whole.

Managing the people who build systems involves more than under-

FIGURE 15–12 Variables Considered Important in Job Satisfaction of the EDP Staff

Variable	Interpretation
1. Ability utilization	I could do something that makes use of my abilities
2. Achievement	The job could give me a feeling of accomplishment
3. Activity	I could be busy all the time
4. Advancement	The job would provide an opportunity for advancement
5. Authority	I could tell people what to do
6. Company policy and practices	The company would administer its policies fairly
7. Compensation	My pay would compare well with that of other workers
8. Co-workers	My co-workers would be easy to make friends with
9. Creativity	I could try out some of my own ideas
10. Independence	I could work alone on the job
11. Moral values	I could do work without feeling that it is morally wrong
12. Recognition	I could get recognition for the work I do
13. Responsibility	I could make decisions on my own
14. Security	The job would provide for steady employment
15. Social service	I could do things for other people
16. Social status	I could be "somebody" in the community
17. Supervision-human relations	My boss would back up his people (with top management)
18. Supervision-technical	My boss would train his people well
19. Variety	I could do something different every day
20. Working conditions	The job would have good working conditions

Source: Rene V. Dawis, L. H. Lofquist, and D. J. Weiss, *A Theory of Work Adjustment* (a revision), Minnesota Studies in Vocational Rehabilitation, 1968.

standing the methodology or estimating algorithms. It is understanding the people themselves. Their behavior can be more important to a project's success than the correct development methodology. The project manager must be the *kingpin in personnel motivation.* He/she must build teams of mature people who are motivated and can be trusted to see a project through to completion. This is the essence of project management.

Summary

1. System projects fail for many reasons: conflicting objectives, user's lack of involvement, inexperienced project management, budget overruns, and changes in user requirements. These problems make it important that projects are properly planned, managed, and implemented.
2. Project management is the application of planning, organizing, and control concepts to critical one-of-a kind projects. Managing projects requires top management commitment, active user participation, and a long-range plan.
3. In establishing a project, several steps are considered:

   a. Study the problem to evaluate the scope, degree of change, and cost of late completion.
   b. Specify project responsibilities through a qualified project team.
   c. Select a project manager with experience in the functional areas, ability to recognize problems and communicate ideas, and working knowledge of the system improvement process.
   d. Establish ground rules and standards for handling projects.
   e. Select the right project, especially if it is the first project for the firm.
   f. Define the tasks to be done and plan accordingly.

4. A project manager plans the life cycle of the project and eliminates crisis through proper planning. Planning means plotting activities against a time frame and developing a network based on an analysis of the tasks that must be performed to complete the project.

5. Two planning tools are used in project planning:
   a. *Gantt chart* uses horizontal bars to show the duration of actions or tasks. Broken bars are estimated time delays or slack time. A task is a specific job to be performed; a group of tasks make up an activity that ends in a milestone.
   b. *Program evaluation and review technique (PERT)* uses tasks and events to represent interrelationships of project activities. Each task is limited by an identifiable event that has no duration. The list of tasks and events is networked in a PERT chart. The numbers of the activity lines are the days required between events. When the network is completed, the next step is to determine the *critical path*—the longest path through the network. It determines the project completion date.

6. Project management software is available for virtually every size of project. It produces a network's critical software packages load on the personal computer; some also have on-screen graphic capabilities.

7. The skills required of a project team are the same as those for computer system development projects. This means that there should be at least one team member qualified to address each issue related to the project.

8. Two types of work are performed in systems projects: mechanistic and creative. Mechanistic tasks take up to 80 percent of the system development effort, whereas creative tasks take up to 20 percent of system development effort but could affect more than 80 percent of the outcomes.

9. The democratic approach to project management allows subordinates to think on their own and make suggestions, although the drawbacks are slow reaction time and difficulty in expediting work. In contrast, autocratic management is more rewarding to individuals than committees, although it is difficult to coordinate and control.

10 Success in project management requires flexibility to changing situa-

tions, the ability to communicate, and a commitment to planning and understanding technical problems and the motivations of staff members. Of all the contributions a project manager can make, maintaining a satisfied staff is the most important for strong project teams. The project manager must be the kingpin of personnel motivation. This is the essence of project management.

Key Words

Activity
Critical Path
Gantt Chart
Menu
Milestone
Precedence Relationship
Program Evaluation and Review Technique (PERT)
Project Manager
Slack Time
Task
VisiSchedule

Review Questions

1. In your own words, why do systems fail? How would one reduce potential failure in system development? Explain.
2. From what we have learned about system development and the analyst-user interface, how important is the user's involvement for successful system implementation? What other factors are important? Be specific.
3. Define the following terms:
   a. Project management.
   b. Task.
   c. Milestone.
   d. Critical path.
4. Distinguish between the following:
   a. Event and milestone.
   b. Gantt and PERT.
   c. Task and activity.
   d. Precedence and successor relationships.
5. Discuss the steps for establishing a system project. Which step do you think is the most critical? Why?
6. What skills and qualifications are required of a project leader? Explain.
7. What is a Gantt chart? How would you develop one? How does it differ from a PERT chart? Explain.
8. Explain how a task leads to an activity and an activity to a milestone.
9. Illustrate the steps taken in planning a project. What charts or forms are used? What information do they contain?
10. Think of a problem area where a Gantt chart may be used. List the steps and the procedure used in developing the chart.
11. What information does a PERT chart show? Explain the two methods of calculating events.

12. Review the computer journals and report to class two applications for project management. Explain briefly what each application does and how they differ.
13. What is the main function of a project team? What skills should team members provide? Explain.
14. If you were a project manager developing a mailing list for a large retail store, what management style would you adopt? Why? Justify your preference.
15. "A project manager must be the kingpin of personnel motivation." Do you agree? Discuss in detail.

Application Problems

1

The First National Bank of Kendall contracted a computer service to install an automated teller machine (ATM) in a new shopping plaza five miles away. The computer service that processes checking and savings transactions presented the vice president of operations with a critical date calendar as shown in Exhibit 15–1.

Assignment

Prepare a Gantt chart based on the information provided.

2

A systems project includes designing a stock status routine, followed by writing two programs (A and B). A reorder routine is also designed, and two programs (R1 and R2) are to be written at the same time. Each program is tested after being written. Program R1 is tested only after the stock file is generated, which, in turn, must follow the stock status design. The system as a whole is tested after all programs have been tested.

Assignment

a. Draw a PERT chart and schedule the required activities around the following conditions:

Each program takes two weeks to write and one week to test.
Stock status design takes three weeks and reorder design takes two weeks.
The stock file takes four weeks to generate.
There are two full-time programmers who test their own programs and are not involved in stock file or stock status design.

[cut off]15–1 ATM Critical Date Calendar

	Week 3/8	3/15	3/22	3/29
	1	2	3	4
[cut off]k	Live date set Ad slicks obtained (card and forms production)	Bank personnel assigned	Short name and name/address printouts received; maintenance work begun Work started on validation and disclosure forms	Building contractor and Diebold meet on building specs Alarm system planned
Servicer	OLDS rep assigned Contract received	Survey completed Equipment, form-proofs, sample plastic ordered		Deliver manuals
Phone Co.				
Diebold				Building specs.
Other Burroughs				
General Data Comm.				
Ad agency				

EXHIBIT 15–1 *(Concluded)*

	Week 5/17	5/24	5/31	6/7
	11	12	13	14
Bank	Credit criteria due BCF form due	Second card edit due	Pin and pan contacts assigned Autodialer questionnaire Review final card issue Data entry training	Building completed
Servicer		Second card edit due	Card tape to vendor Data entry training Order demo. and special cards	First 10 cards produced and tested
Phone Co.				
Diebold				Cards tested
Other Burroughs				
General Data Comm.				
Ad agency				

4/5	4/12	4/19	4/26	5/3	5/10
5	6	7	8	9	10
Marketing campaign meeting	Plan ad campaign	CIF training begun Artwork proofs approved	CIF clean-up (emphasis on DDA)	Continue CIF DDA cleanup	First card edit maintenance begun
Marketing campaign meeting	Supplies ordered CIF conversion/training				First card edit produced—instructions given
		Circuit due			
		CRT's installed			
		Data set due			
Marketing campaign meeting	Ad campaign				

6/14	6/21	6/28	7/5	7/12	7/24
15	16	17	18	19	20
Interchange	Supplies due Ad campaign begins	Cards due demo and G/L accts. opened Employee kickoff	Cards mailed ATM live to employees Uniforms and activity	Pins distributed to customers CIF clean-up completed	* * * * * * * * * * * * * * * * * *
			Demonstrators response team balancing training		* * * * * * *
	ATM circuit due	Atm installed			* * * * * * * * * * * * *
	ATM date set due				

The total system takes 1.5 weeks to test, which involves the two programmers.

Remember that there are two design activities, four program writing and testing activities, stock file creation, and a total system test.

b. What is the critical path duration? Can the project be completed within the time allotted by the critical path? Explain.

Selected References

Diamond, Daniel S. "Project Management Via PC." *Business Computing*, December 1983, p. 30ff.

Harrell, Clayton Jr. "Sure-Handed Project Management, Part I. Reducing the Risks." *Computer Decisions*, November 1983, p. 260ff.

Harrison, William D. "For Stronger Project Team: Working the Human Side." *Computerworld*, May 21, 1984, pp. ID15–16ff.

Justice, Karen. "Systems to Keep You on Schedule." *ICP Interface Administrative and Accounting*, Winter 1983, pp. 25–27ff.

Newkirk, Claire. "Project Estimating—What's So Tough about It? *ICP Software Business Review*, December/January 1984, p. 26ff.

Potts, Paul. "Project Management: Getting Started." *Journal of Systems Management*, February 1983, pp. 18–19.

————. "Project Manager: Technician or Administrator." *Journal of Systems Management*, January 1982, pp. 36–37.

Kerzner, Harold. "Project Management in the Year 2000." *Journal of Systems Management*, October 1981, pp. 26–31.

Rolefson, Jerome F. "Project Management—Six Critical Steps." *Journal of Systems Management*, April 1978, pp. 10–17.

Strehlo, Kevin. "When the Objective Is Efficient Project Management." *Personal Computing*, January 1984, p. 132ff.

Yasin, Rutrell. "Project Manager's Role Is Changing." *MIS News*, February 1, 1984, pp. 34–35.

Chapter 16

Security, Disaster/Recovery, and Ethics in System Development

At a Glance

Every candidate system must provide built-in features for security and integrity of data. Without safeguards against unauthorized access, fraud, embezzlement, fire, and natural disasters, a system could be so vulnerable as to threaten the survival of the organization.

To do an adequate job on security, a systems analyst must analyze the risks, exposure, and costs and specify measures such as passwords and encryption to provide protection. In addition, backup copies of software and recovery restart procedures must be available when needed. A disaster/recovery plan that has management support should also be prepared. Then no matter what the disaster, the firm can recover.

The strength behind system integrity and success is ethics and professional standards of behavior. When ethics are compromised, regardless of technology, the very fabric of a system is in question.

By the end of this chapter, you should know:

a. The various threats to system security and their defenses.
b. How to do risk analysis and specify measures.
c. The importance of disaster recovery planning and how such a plan is initiated.
d. The meaning and importance of ethics in system development.

Disaster/Recovery Planning

THE PLAN
- The Team
- Planning Tasks
- The Manual

Ethics in System Development

ETHICS CODES AND STANDARDS OF BEHAVIOR

INTRODUCTION

Just when the computer age will arrive is no longer a question; it is here. Its impact is everywhere, but not without a price. The end user is concerned about security along with increased dependence on the computer. In system development, the project manager and the analyst must consider measures for maintaining data integrity and controlling security at all times. This involves built-in hardware features, programs, and procedures to protect candidate systems from unauthorized access.

In this chapter, we address the issues of data and system security and suggest some control measures. We also look at ways of planning for and recovering from disasters so that the organization can continue to operate. Underlying the entire system development process is the issue of ethics and ethical standards that govern the behavior of analysts, designers, and project managers. Ethics is becoming an important topic in systems analysis and should be addressed at this point.

SYSTEM SECURITY

Newspapers, journals, and television are rife with stories about computer criminals embezzling millions of dollars, "hackers," and kids electronically breaking into computers across the nation. Here are two examples:

1. A Wells Fargo bank employee embazzled $21 million. The employee was performing the entire reconciliation function of the branch and knew exactly the operating procedures of the system.[1]

2. In July 1983, the FBI caught a group of teenager computer enthusiasts armed with a personal computer, a modem, and "home-grown" computer knowledge. They had broken into more than 60 business and government computers, including those of the Defense Department, the Los Alamos National Laboratory, and the Sloan-Kettering Cancer Center in New York.

Points of vulnerability in today's computer world are numerous. As shown in Figure 16–1, they relate to hardware, communication links, and terminals. This explains the many break-ins in system security, costing companies over $300 million a year. Stealing secrets from computers is more profitable than robbing banks. The average computer theft is $430,000 compared to $10,000 for a bank holdup or $19,000 for fraud and embezzlement.[2] Episodes like the ones cited have made security one of the fastest-growing areas of concern in the computer industry. An estimated $300 million and $425 million were spent on safeguards in 1982 and 1983, respectively. A Dallas-based oil firm invested $500,000 to overhaul its com-

---

[1] Arnold M. Cohn, "Total Information System Security," *Journal of System Management*, April 1983, p. 17.

[2] Ben Harrison, "Planning for the Worst," *Infosystem*, June 1982, p. 54.

FIGURE 16–1 Points of Vulnerability in Computer-Based Systems

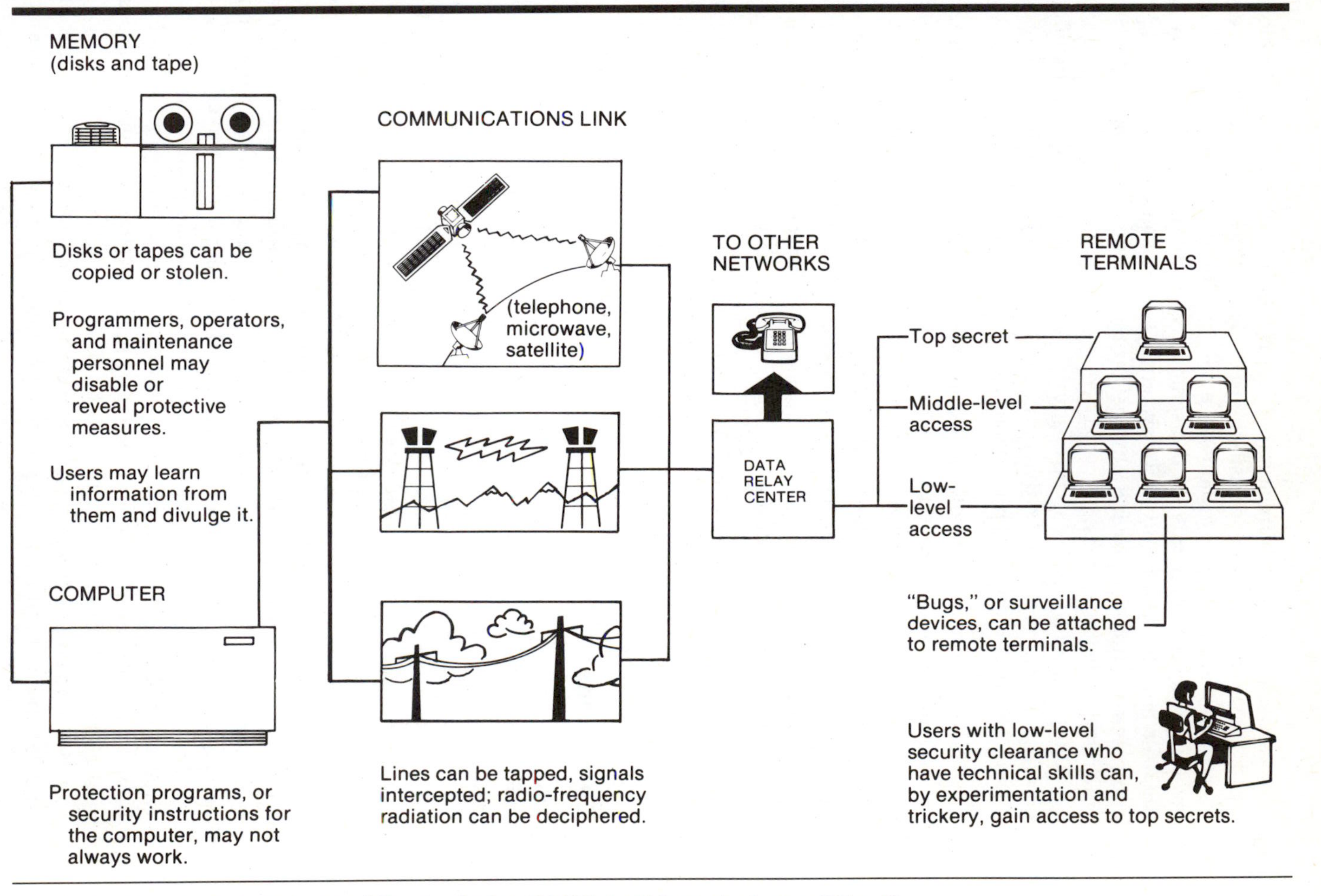

MEMORY
(disks and tape)
Disks or tapes can be copied or stolen.
Programmers, operators, and maintenance personnel may disable or reveal protective measures.
Users may learn information from them and divulge it.
COMPUTER
Protection programs, or security instructions for the computer, may not always work.
COMMUNICATIONS LINK
(telephone, microwave, satellite)
Lines can be tapped, signals intercepted; radio-frequency radiation can be deciphered.
TO OTHER NETWORKS
DATA RELAY CENTER
REMOTE TERMINALS
Top secret
Middle-level access
Low-level access
"Bugs," or surveillance devices, can be attached to remote terminals.
Users with low-level security clearance who have technical skills can, by experimentation and trickery, gain access to top secrets.

Source: Adapted from Arielle Emmett, "Thwarting the Data Thief," *Personal Computing*, January 1984, p. 97.

puter security plan.[3] This means that security is critical in system development. The analyst has a responsibility to design a workable security system to protect the system from damage, error, and unauthorized access. The level of protection depends on the sensitivity of the data, the reliability of the user, and the complexity of the system. A well-designed system includes control procedures to provide physical protection, maintain data integrity, and restrict system access.

Tight system security can be costly, but appropriate security is justified compared to the catastrophe that could result from no protective measures. There are three motives behind security:

1. The near-total dependence of organizations on computer-based information makes it imperative that a system be protected on a regular basis.
2. Data are a major asset and should be protected. In a data base environment where computer files are centralized, security becomes critical.
3. Demonstrating effective security measures reinforces management support for designing and implementing candidate systems.

In discussing system security, we need to explore the major threats to systems, the types of failures and physical protection of the data base, and appropriate control measures. First we define key terms.

Definitions

The system security problem can be divided into four related issues: security, integrity, privacy, and confidentiality. They determine file structure, data structure, and access procedures.

System security refers to the technical innovations and procedures applied to the hardware and operating systems to protect against deliberate or accidental damage from a defined threat. In contrast, *data security* is the protection of data from loss, disclosure, modification, and destruction.

System integrity refers to the proper functioning of hardware and programs, appropriate physical security, and safety against external threats such as eavesdropping and wiretapping. In comparison, *data integrity* makes sure that data do not differ from their original form and have not been accidentally or intentionally disclosed, altered, or destroyed.

Privacy defines the rights of the users or organizations to determine what information they are willing to share with or accept from others and how the organization can be protected against unwelcome, unfair, or excessive dissemination of information about it.

The term *confidentiality* is a special status given to sensitive information in a data base to minimize the possible invasion of privacy. It is an attribute

---

[3] Charles Alexander, "Crackdown on Computer Crime," *Time*, February 8, 1982, p. 60. See also J. R. Cook, J. D. Eure, M. A. Johnston, and H. J. Matford, "DPMA Chapters Speak Out on DP Securtiy, *Data Management*, May 1982, p. 42.

of information that characterizes its need for protection. System security is the technical means of providing such protection. In contrast, privacy is largely a procedural matter of how information is used.

Data privacy and security are issues that go beyond the scope of system development. They are actually a societal concern. An organization that depends heavily on the use of data bases requires special controls to maintain viable information. These controls are classified into three general categories:

1. Physical security or protection from fire, flood, and other physical damage.
2. Data base integrity through data validation techniques.
3. Control measures through passwords, encryption, and monitoring users on a regular basis.

Each of these categories is discussed next.

Threats to System Security

A procedure for protecting systems makes sure that the facility is physically secure, provides a recovery/restart capability, and has access to backup files. If we list in order of probability (most probable first) the threats to system security or data integrity, research shows that the most damage comes from errors and omissions—people making mistakes. The threat of external attack on a computer system is virtually last. This means that in establishing a priority sequence, one would probably want to *start from within the firm and work out*.[4] The list of potential threats is:

1. Errors and omissions.
2. Disgruntled and dishonest employees.
3. Fire.
4. Natural disasters.
5. External attack.

Threats and their usual defenses are illustrated in Figure 16–2.

Errors and omissions cover a broad range of miscues. Some result in incredible but short-lived windfalls such as the Sears Roebuck customer who had $1 billion credited to her account, transactions routed to the wrong place, or those that cause double ordering.

When huge quantities of information are stored in one data base, sensitive data can easily be copied and stolen. Information can also be entered directly into a computer without any written record or proper authorization and can be changed without a trace. A dishonest programmer can bypass control and surreptitiously authorize his/her own transactions.

---

[4] A. Richard Immel, "Data Security," *Popular Computing*, May 1984, pp. 65–68.

FIGURE 16-2 Threats and Their Usual Defenses

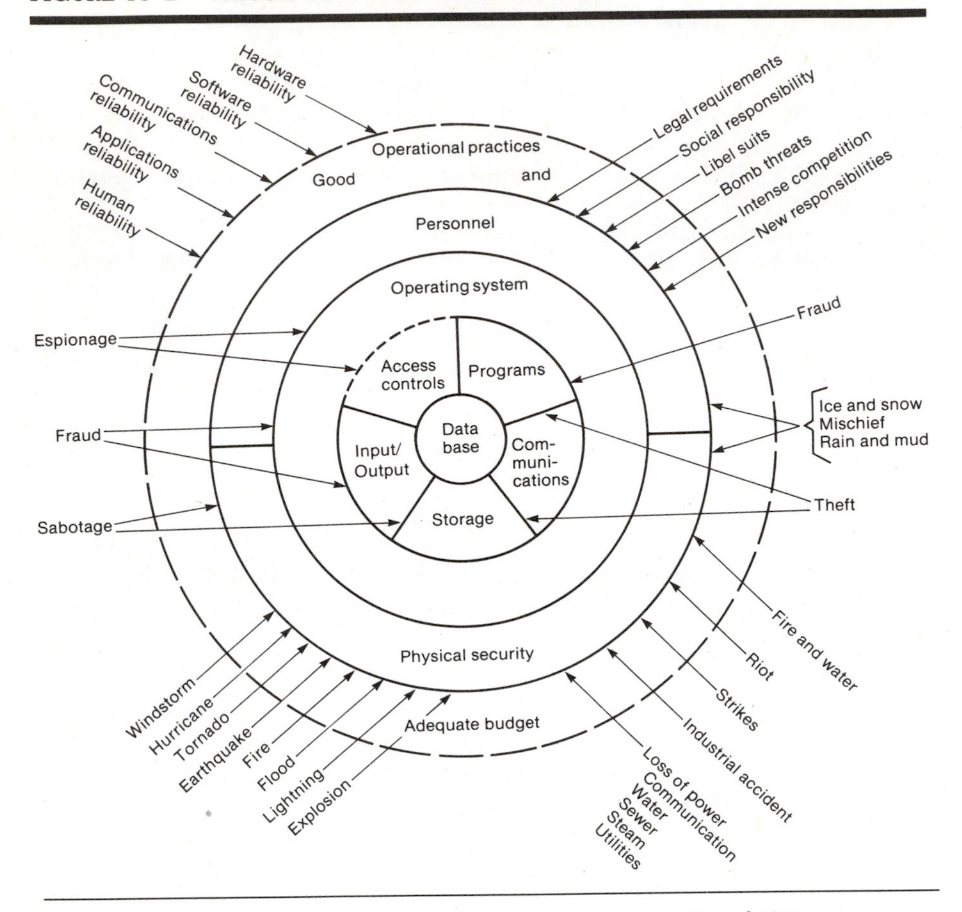

Source: American Federation of Information Processing, *Systems Review Manual*, 1974, p. 9.

Dishonest employees have an easier time identifying the vulnerabilities of a software system than outside hackers because they have access to the system for a much longer time and can capitalize on its weakness. For example, the employee first discovers a weakness that is inherent in the system. If he/she is in inventory control, merchandise may be shipped to a nonexistent warehouse.

Such computer crimes tend to be insider crimes by people who abuse positions of trust and responsibility. According to a recent survey, an estimated $70 billion are lost each year to computer-related crime, fraud, and embezzlement; 75 percent of this is attributed to insiders.[5] This lack of ethics is widespread in the profession and society in general.

Fire and other man-made disasters that deny the system power, air

[5] Robert Batt, "White Collar Crime," *Computerworld*, December 26, 1983, p. 49ff.

conditioning, or needed supplies can have a crippling effect. In the design of a system facility, there is a tendency to place fire-fighting equipment where the most dollars are rather than where the most flammable material is. Proper planning for safeguards against such disasters is critical, especially in organizations that depend on centralized data base systems.

Natural disasters are floods, hurricanes, snowstorms, lightning, and other calamities. Although there is no way to prevent them from occurring, there are measures to protect computer-based systems from being wiped out. For example, it is disastrous to put computer installations on the ground floor in seacoast towns or in buildings without adequate fire protection.

System reliability is also important in system security design. For example, a facility plagued by hardware outages, bug-ridden software, or a deficient communication network can cause chaos for the end user.

The Personal Computer and System Integrity

Personal computers have been viewed as a step backwards in accounting controls. The casual operating environment of a microcomputer makes it relatively easy to make changes in accounting systems that require rigid controls. There is also a tendency to put everything on the microcomputer with hardly a backup. A third problem is the lack of audit trails in most off-the-shelf software packages. It is difficult to reconstruct transactions for audit purposes. Finally, as more personal computers are linked to company mainframes so remote users can access data, the potential increases for altering the data deliberately or by mistake. The consequences of all these things might not be fatal, but it is wise to set up access controls in the form of software that limit access to the mainframe. The measures taken depend largely on where and how the microcomputer is being used.

It is becoming obvious that the personal computer is adding security problems to system installations. With the use of microcomputers in the corporate environment, the potential for misuse of information becomes enormous. Many of today's operating systems contain no password; a would-be thief can copy at will. A person with a microcomputer at a remote location who knows how to bypass the codes and passwords can use a phone line and illegally retrieve information without leaving any clues to his/her identity.

Risk Analysis

Given the threats to system security, system designers should assess each of the system's data aggregations in light of possible threats against them. The purpose of risk analysis is to determine the probability of problems occurring, the cost of each possible disaster, the areas of vulnerability, and the preventive measures to adopt as part of a security plan.[6]

---

[6] J. P. Curry "Analyzing Computer Security Risks," *Canadian Datasystems*, July 1981, pp. 69–70.

FIGURE 16-3 Risk Analysis—An Illustration

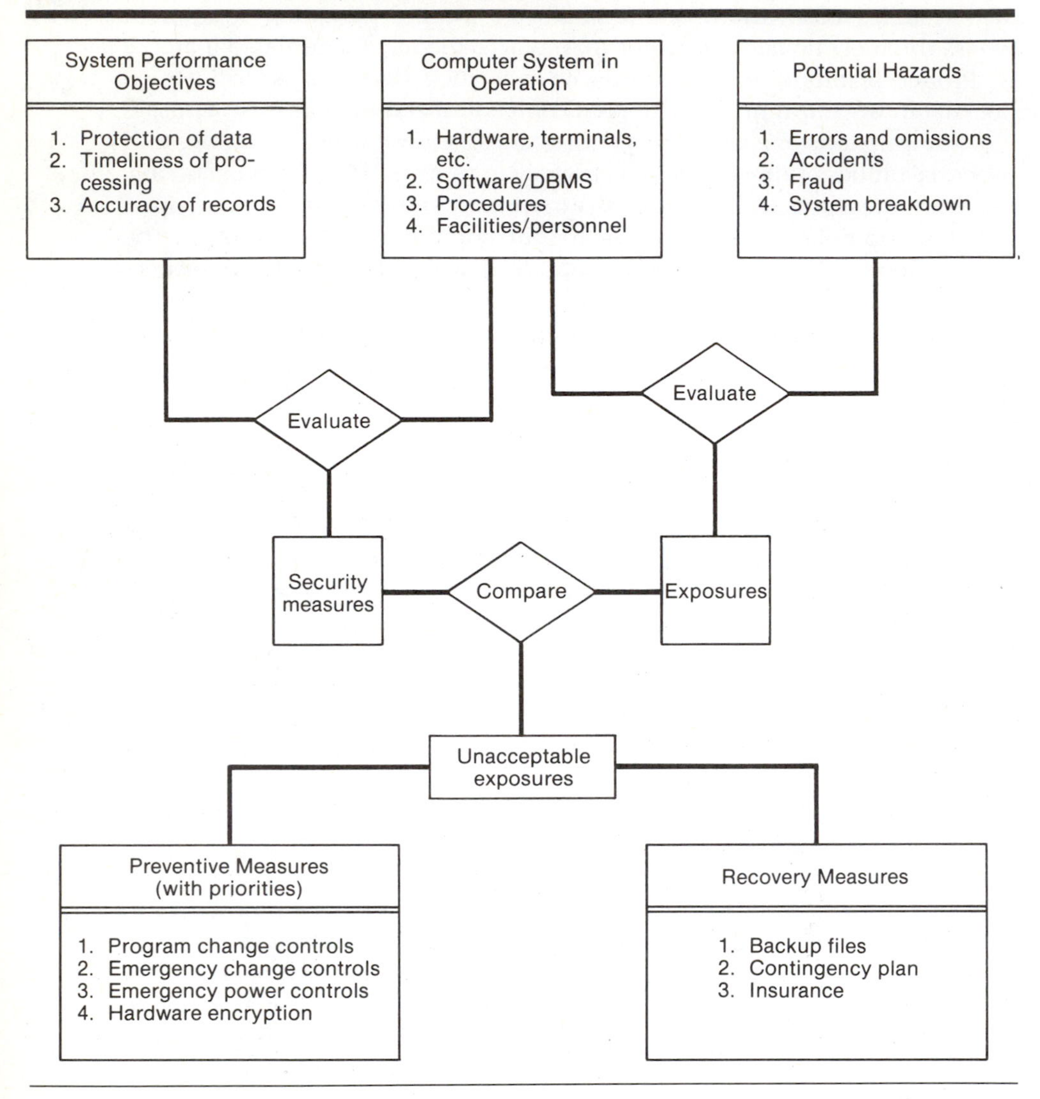

Figure 16-3 is an illustration of risk analysis. First, the designer lists the objectives of the system and evaluates them against the existing computer facility to determine the security requirements. The facility, in turn, is evaluated against the potential hazards to determine specific exposures. Security measures are then compared with specific exposures to pinpoint unacceptable exposures. The outcome is a draft specifying the preventive and recovery measures to be adopted for effective system security.

Risk analysis is not a foolproof plan for security. It merely makes the user aware of the exposures and their respective costs and control measures. A special risk analysis matrix that specifies the risks, costs and effects, and probability of exposure helps the designer determine the actions to be taken and how quickly they must be taken. Figure 16–4 shows the matrix. Note that the first risk has the highest cost-probability product and, therefore, is the highest priority for security. Also, the higher the probability for a given occurrence, the more appropriate are preventive measures. For low-risk factors, ability to recover is usually an adequate measure. If ability to recover is difficult and consequential losses are excessive, however, insurance is a must.

The two key elements in risk analysis are the value or impact of a potential loss and the probability of loss. As shown in Figure 16–4, an attempt is made to match a potential loss with the probability of its occurrence to decide on the action to be taken. The goal is to identify the threat that results in the greatest monetary loss and provide protection to the appropriate degree.

Control Measures

After system security risks have been evaluated, the next step is to select the measures (layers of protection) that are internal and external to the facility. These measures are generally classified under the following: (1) identification, (2) access control, (3) audit controls, and (4) system integrity.

Identification

There are three schemes for identifying persons to the computer:

1. *Something you know,* such as a password. A password is the most commonly used means for authenticating the identity of people. Passwords should be hard to guess and easy to remember. They should not be recoverable, except from the mind of the password holder. The process of accepting a password should not permit the recovery of passwords.[7]

Security is often lax in system installations. Many users copy down a difficult password or give passwords to associates, making them subject to potential unauthorized access. Experience has shown that many illicit entries to systems are due to written passwords. If a password is written, it is no longer a password. It becomes something possessed (rather than known), and the knowledge of the hiding place becomes the password.

Another scheme under the "something you know" category is the picture badge, which identifies people who bring work to the center. Although it positively identifies the carrier of the information, the badge does not verify that the person is authorized to submit a job or receive reports from the system.

2. *Something you are,* such as fingerprints or voice prints. Although fin-

[7] Sigmund Porter, "A Password Extension for Improved Human Factors," *Computers and Security,* (North-Holland Publishing, Amsterdam, 1982), pp. 54–56.

FIGURE 16–4 System Controls and Security

Nature of Risk	Potential Effect(s)	Estimated Cost of Risk	Probability (high = 0.75 average = 0.50 low = 0.25 nil = 0.10)	Cost probability (× 1,000)*	Preventive/ Remedial Action	Costs of Safeguards/ Comments
1. Unauthorized disclosure of employee data	*a.* Lawsuits against firm *b.* Loss of goodwill	*a.* $10,000,000 *b.* $4,500,000	High	10,800	*a.* Reevaluate the integrity of present staff *b.* Establish penalty structure for unauthorized access to employee data	Cost to customers is critical but need not be considered right away
2. Theft of information (of use to competitors)	*a.* Erosion of market position *b.* Estimated savings to competitor	*a.* $12,000,000 *b.* $10,000,000	Average	11,000	*a.* Strict control of access to vital files *b.* Personnel bonding	*a.* Implement procedure for signing out files *b.* Tighten recruitment practices and procedures
3. Illegal sale of computer time	Increased computer costs	$135,000	Low	33.7	*a.* Spot check computer use	No specific action necessary; risk/small loss outweighed by staff morale considerations
4. Complete loss of data base	*a.* Unable to bill customers *b.* Production stoppage within four days	*a.* $1,600,000 *b.* $700,000	Low	5,075	*a.* Ensure backup copy is kept of complete data base *b.* Ensure against loss during recovery	$2,500 per year
5. Computer facility destroyed	*a.* Replacement of computer *b.* Site reconstruction	*a.* $3,470,000 *b.* $1,400,000	Nil	487	*a.* Ensure backup *b.* Maintain a fallback system *c.* Insure site and equipment *d.* Impose fire precautions, etc.	Nil, $4,900 per year

* Highest product indicates highest priority.

FIGURE 16–5 Guidelines to Control Access

1. Have a single entrance to the operating area monitored around the clock.
2. Install intruder alarms.
3. Install a key lock, a cipher lock, or a badge-operated lock on the door of the operating area.
4. Employ a guard during operating hours of the computer center.
5. Issue badges with new encoding; change locks periodically.
6. Control use of files through a librarian.
7. Identify keys to operations area with a registration number logged in a control book when issued and marked DO NOT DUPLICATE.

gerprinting is commonly used in law enforcement, it is ill suited for the MIS environment. Voice prints, on the other hand, are making headway as a reliable method for verifying authorized users. The technique essentially analyzes a person's voice against prerecorded voice patterns of the same person. An exact match allows access to the system.

3. *Something you have,* such as the credit card, key, or special terminal. Magnetic stripe credit card readers on terminals identify the operator to the system. The card along with a password gives added assurance of the identification of the user.

Access Control

Various steps are taken to control access to a computer facility (see Figure 16–5). One way is to use an encoded card system with a log-keeping capability. The card serves as a key to unlock doors, including tape storage and other classified areas. The card is essentially a magnetic key and a "keyport" is a lock. Inserting the card into the lockport unlocks the door. A card that includes a photograph of the bearer may double as an employee ID badge.

Encryption. An effective and practical way to safeguard data transmitted over telephone lines is by *encryption.* Data are scrambled during transmission from one computer or terminal to another. The process transforms data in such a way that they become utterly useless to unauthorized users. It eliminates the authentication problem by preventing a determined "intruder" from injecting false data into the channel or modifying messages.

The encryption concept is simple. A plaintext (unenciphered) message is transmitted over an unprotected communications channel. To prevent unauthorized acquisition of the message, it is enciphered with a reversible transformation to produce a cryptogram or ciphertext. When it arrives at an authorized receiver, it is decrypted back into the plaintext data form (see Figure 16–6).[8]

[8] J. Michael Nye, "A Primer on Security," *Mini-Micro Systems,* July 1981, p. 166ff.

FIGURE 16-6 Encryption/Decryption

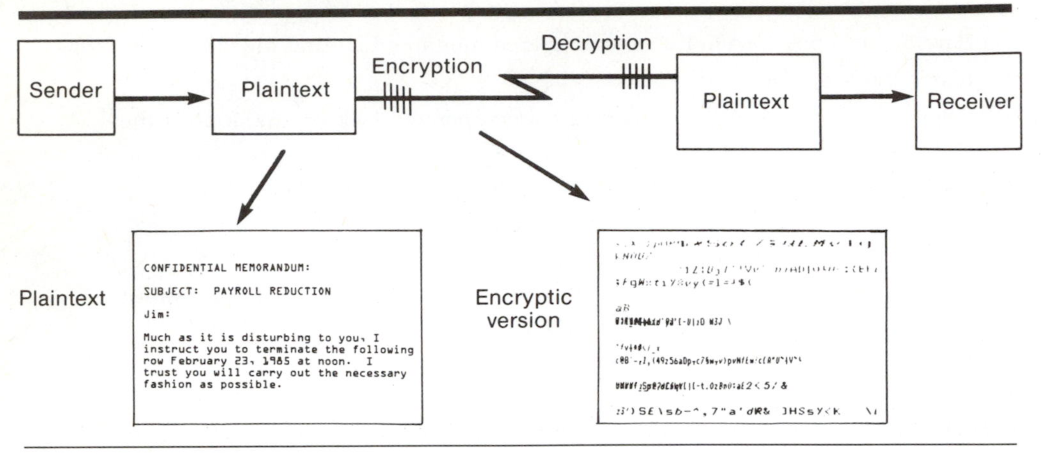

Most of today's encryption is based on the National Bureau of Standards encryption algorithm, known as the Data Encryption Standard (DES). It is a general technique developed in 1977 and used in many commercial network security systems. A system that assures that encryption and decryption are done without human intervention is virtually secure from unauthorized access.[9] Encryption devices for personal computers are available at the chip level and in the software. They are designed specifically to transmit encryptic data; others protect software by making it difficult, if not impossible, to copy.

For computer personnel fraud and embezzlement, several controls may be instituted. For example, the use of programs and changes must be authorized and documented at all times. Other programs and data base files should be stored in a library and accessed only when needed. Other guidelines for protection against embezzlement and fraud are listed in Figure 16-7.

Audit Controls

Audit controls protect a system from external security breaches and internal fraud or embezzlement. The resources invested in audit controls, however, should balance with the sensitivity of the data being manipulated. One problem with audit controls is that it is difficult to prove their worth until the system has been violated or a company officer imprisoned. For this reason, auditability must be supported at all management levels and planned into every system.

One of the most vulnerable places for a system is the MIS department. Programmers can pirate, modify, and even sell software for personal gain.

[9] Matthew A. Kenny, "Cryptography Comes out of the War Room to Aid in Securing Computer Systems," *Data Management*, July 1981, pp. 19–21.

FIGURE 16–7 Security Guidelines against Fraud

1. Classified programs should be run only with the correct password.
2. There should be at least two persons in the computer room at all times.
3. Critical forms such as checks should be locked in a safe location and subject to an inventory system.
4. Control final program assemblies so that only the appropriate program is installed.
5. Periodically compare disk programs to control copies on another medium.
6. Review the software library periodically to ensure that a complete set of source and object programs and operating documentation exists for all applications.

To audit the maintenance process properly, there must be an audit trail from the change requests to the production programs. Various audit software is available to do the job properly. Generalized audit software helps the auditor examine files and data bases for consistency, correctness, and completeness. There are also programs to trace the flow of data through a program and the activity that they generate. Specialized audit software, on the other hand, probes into specifics. For example, financial analysis programs reduce the volumes of data in the data base to a manageable amount for analysis. They can perform further statistical analysis to determine out-of-range transactions or nonrandom runs.

In summary, the complexity of systems makes automatic auditing necessary. Neither the auditor nor the user can verify the system activities adequately, so the system must check itself. The internal controls required mean that programmers and analysts build controls into every system. Developing a corporate auditing policy will ensure that future systems meet the minimum requirements for security and control against fraud and embezzlement.

System Integrity

System integrity is a third line of defense that concentrates on the functioning of hardware, data base and supportive software, physical security, and operating procedures. The most costly software loss is program error. It is possible to eliminate such error through proper testing routines. Parallel runs should be implemented whenever possible. Physical security provides safeguards against the destruction of hardware, data bases, and documentation; fire, flood, theft, sabotage, and eavesdropping; and loss of power through proper backup (see Figure 16–8).

The proper use of the file library is another imporant security feature. This involves adequate file backup and reliable personnel to handle file documentation when needed. File backup means keeping duplicate copies of the master and other key files and storing them in suitable environmental conditions. For tape files, a common procedure is to save the old master file

FIGURE 16-8 Physical Security Measures

FIRE

1. Install sensors for early detection of heat, smoke, or fire.
2. Install wall-mounted extinguishing systems in storage areas and facilities.
3. Maintain direct communication links to central protection agency and/or fire station.
4. Provide fireproof vaults for critical data, tapes, and other materials.
5. Explicit fire instructions must be posted in obvious locations.
6. Copies of computer files, software, documentation, and hardware configuration should be stored off the premises.

FLOOD

1. Water pipes should be located away from computer facilities and storage.
2. Building should be located on high ground to avoid rising water tables or heavy rain damage.
3. Seal ceiling and walls against water seepage.

after each update. The most recently created file is called the *son*, the previous one the *father*, and the one previous to the latter the *grandfather*. Since it is a costly procedure, the decision to proceed along these lines has to be balanced against potential loss if the files are destroyed.

Recovery/Restart Requirements. Restoring a damaged data base is generally done by a rollforward or rollback procedure. The *rollforward* approach involves updating a prior valid copy of the data base with the necessary changes to produce a current version of the data base. The *rollback* approach starts with the current invalid state and removes the record(s) of the activity (roll back) to produce the prior valid state of the data base. Either approach depends largely on the software to bring the backup copy up to date and determine the cause of the failure.

Backup can be extremely important in a recovery/restart procedure. If the data base is physically damaged, one could not roll back because of the damaged data base—only roll forward. To do so, a prior valid backup copy is required. The procedure differs for sequential and data base environments.

For a sequential file, the grandfather-father-son approach to a backup is followed. As shown in Figure 16–9, each newly created file is backed up by a prior valid master and transaction files.

In a data base environment, the master file is not generally copied as it is updated. Instead, transactions are posted directly to the file, which

FIGURE 16–9 Grandfather-Father-Son Approach to Backup

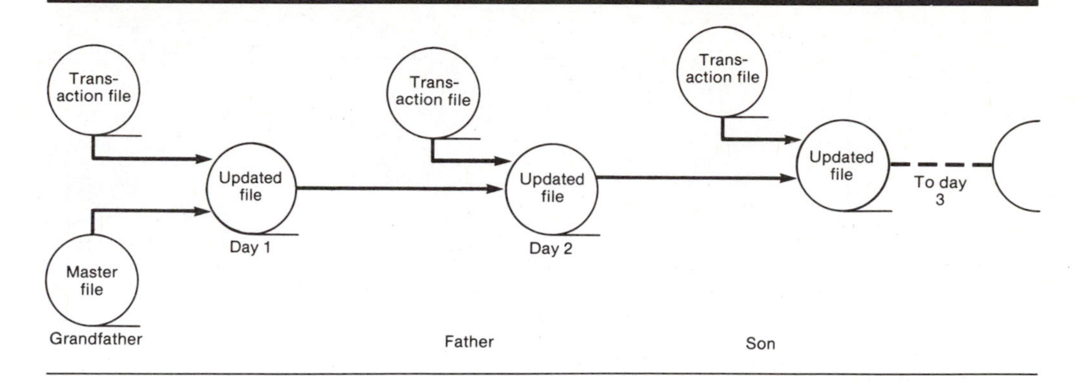

replaces the original data. In such a system, a backup procedure creates files in case they are lost, just as in sequential file systems. Figure 16–10 illustrates this approach. Prior to processing, a backup copy of the data base is generated. Over the next three days, all transactions are processed, copied on a transaction log, and combined. On the fourth day, the data base "crashes," using a prior valid data base and copies of the transaction log. Reprocessing is initiated to restore the data base.

System Failures and Recovery

In a data base environment, there are three types of failures: catastrophic, logical, and structural. A *catastrophic* failure is one where part of a data base is unreadable. It is restored using the rollforward method of recovery.

A *logical* failure occurs when activity to the data base is interrupted (e.g., a power failure) with no chance of completing the currently executing transactions. When the system is up and running again, it is not known whether or not modifications are still in memory or were made to the actual data. Though still readable, the data base may be inaccurate. In this case, rolling back the file to the prvious updated point of the data base (prior valid copy) and adding all changes are required. Many of today's interactive systems provide automatic recovery (rollback) when the system is restarted, which removes the human function from the process.

The third type of failure of a data base is *structural* damage. A typical example is a pointer incorrectly stored in a record that points to unrelated or nonexistent data. If the problem cannot be corrected by software utility, the data base must be recovered to the most recent up-to-date point before the damage occurred.

In summary, a system must be durable. It must survive all threats, including machine crashes, program bugs, and user errors. Recovery measures can undo data base changes that must be removed and redo changes

FIGURE 16-10 Backup Procedure in a Data Base Environment

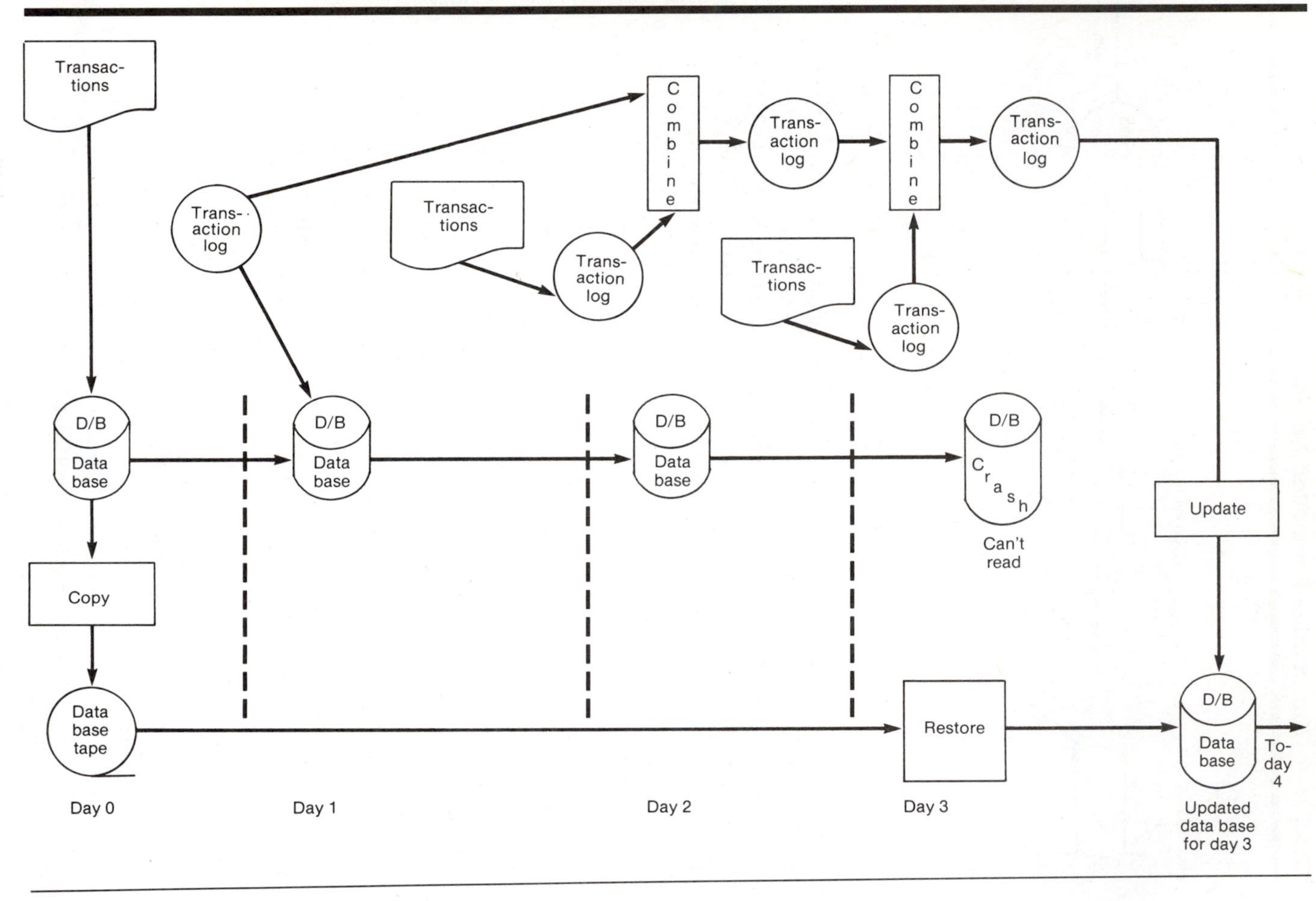

that are lost. The responsibility for recovery may lie with the designer, the project manager, or both, depending on the requirements of the system.

DISASTER/RECOVERY PLANNING

What happens if your system burns out or if a disgruntled employee inflicts serious damage to it? How do you recover? Where do you get your information? Studies have shown that a company's very survival is threatened within 48 hours after the loss of computer operations. Even a limited disruption or disaster can result in substantial financial losses and even the threat of litigation.[10] Direct financial losses result from the loss of sales and production. Indirect financial losses include long-term loss of customers, uncollected receivables, and undetected fraud. Loss of control over vital data compromises data integrity and business decisions.

Disaster/recovery planning is a means of addressing the concern for system availability by identifying potential exposure, prioritizing applications, and designing safeguards that minimize loss if a disaster occurs. It means that no matter what the disaster, you can recover. The business will survive because a disaster/recovery plan allows quick recovery under the circumstances.

How does one guard against such losses? There are several alternatives. As summarized in Figure 16–11, they range from having an entire facility in one location with a complete redundancy of hardware to leasing a site with no computer but adequate electricity and air conditioning to support a computer facility on a temporary basis. After an alternative has been determined, a decision must be made about the applications to be processed, the hardware to process the applications, and what would be relocated after a disaster.

In disaster/recovery planning, management's primary roll is to accept the need for contingency planning, select an alternative measure, and recognize the benefits that can be derived from establishing a disaster/recovery plan. Top management should establish the disaster/recovery policy and commit corporate support staff for its implementation. The user's role is also important. The user's responsibilities include the following:

1. Identifying critical applications, why they are critical, and how computer unavailability would affect the department.
2. Approving data protection procedures and determining how long and how well operations will continue without the data.
3. Funding the costs of backup.

---

[10] Richard A. Bernstein, "A Contingency Plan for Recovery from a Disaster," *Mid-Continental Banker*, December 1983, p. 16.

FIGURE 16–11 Alternative Plans against Disaster

Alternative	Features	Advantages	Limitations
"Fortress" approach	Entire facility in one location	Full redundancy of hardware, good environmental control, tight physical security	Expensive to install and maintain; may be swept away by major flood or terrorist assault
"Cold" backup service	Lease terms on a site with no computer but with adequate electric power, air conditioning, and telecommunication to support a computer facility	Low cost, high security	Difficulty replacing hardware, especially custom-made equipment; limited use, usually up to six weeks; shared membership of site complicates security and privacy; easy to outgrow site due to undercapacity
"Warm" backup service	Contract with a computer service bureau to use its facility in case of disaster	Time saving, good environment, security	High cost, restrictiveness of use, compatibility problems
Mutual backup approach	Two companies from different industries but similar systems agree to back up each other	Low cost, flexibility, reliability, security	Problem finding a partner, possible contractual problems
Private cold site	Own an empty computer facility in a remote area with features similar to "cold" backup service	Guaranteed access, no compatibility problems, relatively low cost	Problem obtaining and installing necessary equipment after a disaster
Private warm backup site	Ranges from having a complete facility left idle to a duplicate computer installation that can be switched to service immediately	Reliability, quick availability	High cost, possible loss of management and control

Source: Adapted from Marguerite Zientara, "How to Recover from a Disaster," *Computerworld*, September 13, 1982, p. 1ff.

The Plan

When a disaster/recovery procedure is planned, several questions have to be answered:

1. How long would it take to rebuild the computer center or aspects of it?
2. What type of accommodation should we look for in a backup installation? How quickly is it available?
3. What equipment is needed to keep the corporation functioning?
4. How would reports be transmitted to the user? That is, is there going to be a telecommunications network or simply a courier service?
5. What utilities (electric power, air conditioning, etc.) are required when a disaster occurs?
6. Would there be sufficient experienced staff available for proper recovery?

When these questions are answered and management gives its support for a disaster/recovery plan, the next step is to initiate a plan that involves four phases:

1. Appoint a disaster/recovery team and a team coordinator to develop the plan or procedure.
2. Prepare planning tasks.
3. Compile a disaster/recovery manual.
4. Dummy run to test the procedure.

The Team

A disaster/recovery team should include a cross section of system designers, users, and computer operators. Under the leadership of a coordinator, the team's main functions are to organize the project, monitor progress on the plan, and oversee its completion. The team meets periodically to ensure that the plan is kept up to date, considers new vulnerabilities or exposures to loss, and implements new technology or procedures as needed. More specifically, the objectives of a disaster/recovery team include the following:

1. Secure backup sites for occupation and use.
2. Contract for hardware to meet minimum processing needs.
3. Supply working copies of all operating systems and application programs to meet minimum processing requirements.
4. Supply communication facilities to make reports promptly available to the users.
5. Supply consumables and administrative support.

Planning Tasks

Disaster/recovery planning tasks are prepared in a cycle similar to that of system development. Briefly, the cycle entails the following:

1. *Definition* phase sets the objectives of the disaster/recovery project.
2. *Requirements* phase evaluates applications against disaster/recovery objectives, determines what is to be included in the plan, and specifies priorities. The team takes inventory of the hardware, software, telecommunications, backup and clerical procedures, utilities, and personnel assignments.
3. *Design* phase evaluates design alternatives, potential vendors, and prices and chooses the final design.
4. *Testing and implementation* phase runs backup systems, compares results, and corrects errors. During implementation, procedures are written, sites are prepared, and maintenance plans are developed.[11]

The Manual

Once the team has completed the assignment, a disaster/recovery manual is prepared and copies are made available to team members and management. All copies of the manual should be updated as needed. The number of manuals should be kept to a minimum with a log of holders and the number of each manual assigned. The more manuals that are printed, the more likely it is that some will be overlooked during the maintenance and update procedure. This is especially true in a large corporation.

ETHICS IN SYSTEM DEVELOPMENT

Increased dependence on computers has brought with it problems of vulnerability to dishonest and unethical practices. Internally, the ease with which employees can manipulate data has tempted many to steal funds and sensitive information at the touch of a button. Networking has further complicated protection. The issues of computer crime, system security, and ethics have a significant impact on the role of the systems analyst. The analyst has an obligation to maintain ethical standards in system development.

Consider the following examples of how a systems professional might act unethically:

- The analyst knows that a user's requirement can be adequately met with a simple, inexpensive system yet installs a state-of-the-art system that takes twice the time (and consulting fees) to implement.
- A systems analyst accepts a microcomputer from a vendor in return for recommending a system sold or made by the vendor.

---

[11] For detailed coverage of disaster/recovery planning, see R. P. R. Gaade, "Picking up the Pieces," *Datamation*, January 1980, pp. 113–18.

- A new client competes with a past client for whom the analyst has already installed a system. The analyst knows what system will give the new client the competitive edge and offers to install it in return for a gift of company stock.

Apprehension regarding unauthorized access and invasion of privacy have forced a recognition in the computer industry of the need for an operational code of ethics to emphasize good sense and the law. This code stresses the right to privacy and recommends against the collection of sensitive details such as religious or political affiliations. It also recognizes the need for security safeguards to prevent breach of confidentiality of the file.

Programs are now available to fill "holes" in computer defenses. For example, computer audit programs allow auditors to probe the record of transactions on a computer for irregularities and to uncover unusually large or frequent transfers of funds. Encryption devices also provide defense by scrambling data across the line until they reach their destination.

Although unethical acts have been focused on mainframes, a "small-time" level of thievery is also growing. For example, copying unprotected software packages such as dBASE II is no problem for anyone with a personal computer and a blank diskette. It is estimated that there are 1.5 illegal copies, which cost as much as a blank diskette ($4), for every authentic version of Visicalc, which retails for $250.[12]

Ethics Codes and Standards of Behavior

Concern over the ethical behavior of analysts and computer professionals has led to the development of standards and codes of behavior by a number of professional associations. Three associations are worth mentioning: the Association for Computing Machinery (ACM), the Data Processing Management Association (DPMA), and the Institute for Certification of Computer Professionals (ICCP). The codes of ethics deal with issues such as competency, honesty, and confidentiality. Excerpts from these codes are shown in Figure 16–12. Unfortunately, professional associations are limited in how they can "punish" code violators. The most serious action taken is revocation of membership. Of course, loss of status could prompt an employer to take more drastic steps toward suspension.[13]

The basic belief among analysts and MIS professionals is that the problems of unethical behavior result not from a lack of honesty or integrity, but from ignorance of what constitutes an unethical act. In many respects, there are ethical considerations without rules.[14] So the purpose of professional standards of conduct is to make the analyst aware of how to act to

[12] Editorial, "Roaming High Tech Pirates," *Time*, February 8, 1982, p. 61.

[13] David H. Freedman, "Ethics," *Infosystems* , August, 1983, pp. 34–36.

[14] Ibid, p. 36.

FIGURE 16-12 Codes of Ethics—Examples

Association for Computing Machinery (ACM)

An ACM member should maintain and increase his competence through a program of continuing education encompassing the techniques, technical standards, and practices in his fields of professional activity.

An ACM member shall disclose any interest of which he is aware which does or may conflict with his duty to a present or prospective employer or client.

An ACM member shall not use any confidential information from any employer or client, past or present, without prior permission.

Data Processing Management Association (DPMA)

In recognition of my obligation to my employer, I shall: Make every effort to ensure that I have the most current knowledge and that the proper expertise is available when needed; Avoid conflicts of interest and ensure that my employer is aware of any potential conflicts; Protect the privacy and confidentiality of all information entrusted to me.

Institute for Certification of Computer Professionals (ICCP)

One has a special responsibility to keep oneself fully aware of developments in information processing technology relevant to one's current professional occupation.

One shall not hold, assume, or consciously accept a position in which one's interests conflict or are likely to conflict with one's current duties unless that interest has been disclosed in advance to all parties involved.

One shall exercise maximum discretion in disclosing, or permitting to be disclosed, or using to one's own advantage, any information relating to the affairs of one's present or previous employers or clients.

Source: David H. Freedman, "Ethics," *Infosystems*, August 1983, p. 35.

promote professionalism. Codes of ethics do not substantially change a dishonest person, but they create a body of peer pressure that tells the individual that unethical behavior is not condoned.

Perhaps the largest contributing factor to the unethical behavior of analysts and information system professionals is lack of education about ethical issues and standards. The computer revolution has come so rapidly that educators have barely kept abreast of it, let alone developing norms on how analysts should behave. Furthermore, there seems to be a professional "disincentive" for educators to address the issue. Ethics courses are new to business schools and few instructors know how to incorporate them into the data processing curriculum.

In reviewing the state of computer technology and system development, we find an informational "elite" exercising more and more power because of its control over vast information resources. Every day one hears horror stories about computer break-ins, security leaks, fraud, and the like. We are witnessing a rapidly growing market in stolen information. Virtually anyone can buy information about anybody. The likelihood of being prosecuted is remote because the courts do not see such offenses as serious crimes. How could one be accused of stealing information when it is still in the corporate data base?[15]

An ethical society has little to worry about. Twenty-three states already have computer crime laws in force. Some of these laws view stealing or abusing data bases as a felony or crime. Technology is merely a vehicle; it is a means. The objective is the betterment of society through ethical standards and professionalism that supports these standards.

Summary

1. Security is critical in system development. The amount of protection depends on the sensitivity of the data, the reliability of the user, and the complexity of the system. The motives behind security are to keep the organization running, protect data as an asset, and seek management support for more installations.
2. There are three categories of controls in data security: physical security (protection from fire, flood, etc.), data base integrity, and control measures (passwords, encryption). Potential threats to system security include errors and omissions, disgruntled and dishonest employees, fire, and natural disasters. Errors and omissions cause the most damage.
3. Personal computers have been adding security problems to system installations. Many of today's operating systems have no passwords. There is also a tendency to put everything on the microcomputer with hardly a backup.
4. Risk analysis helps assess the probability and cost of possible disasters, pinpoint unacceptable exposures, and adopt preventive measures as part of a security plan. The goal is to identify the threat that results in the greatest monetary losses and provide protection to the appropriate degree.
5. After system security risks have been evaluated, the next step is to select security measures. These measures are classified as follows:
   a. *Identification.* it is a scheme for identifying persons to the system based on "something you know" such as a password or a picture

[15] Jake Kirchner, "August Bequai, Fighter for Ethics," *Computerworld*, May 21, 1984, pp. 1–4ff.

badge, "something you are" such as a fingerprint or voice print, or "something you have" such as a credit card, key, or special terminal.

b. *Access control.* Controlling access to the computer facility is secured through encoded cards or similar devices. Encryption prevents intruders from accessing data by scrambling messages across telephones to their destination.

c. *Audit controls.* Auditability must be supported at all levels of management. Audit controls protect a system from external security breaches and internal fraud or embezzlement. Various software programs are available to help in the audit function.

d. *System integrity.* This line of defense safeguards the functioning of hardware, data base, software, physical security, and operating procedures. Proper backup of software and hardware is extremely important.

6. A damaged data base is restored by a rollforward or rolback procedure. In a rollforward procedure, a prior copy of the data base is updated with the changes to generate a current set. Rollback starts with the current invalid copy and removes the records of activity to produce the prior valid data base. This is another example of the importance of backup.

7. Disaster/recovery planning is a means of addressing the concern for system availability by identifying potential exposures, prioritizing applications, and designing safeguards that minimize loss if a disaster occurs. The safeguards range from having an entire facility in one location with a full redundancy of hardware to leasing a site with no computer but adequate electricity and air conditioning to support a computer facility on a temporary basis. Regardless of the alternative, management and user support is critical. The master plan involves four phases:

   a. Appoint a team and a coordinator to develop the plan.
   b. Prepare planning tasks.
   c. Compile a disaster/recovery manual.
   d. Dummy run to test the procedure.

8. The analyst has an obligation to maintain ethical standards in system development. Programs are now available to fill "holes" in computer defenses. Encryption devices also provide defense by scrambling data across the line to their destination.

9. Various associations have made an effort to formalize ethics codes and standards of behavior for programmers and system developers. The goal of these codes is to make the professional aware of the do's and don'ts rather than to monitor and punish violators. Schools and training programs can do a lot to improve consciousness of ethics in developing systems and dealing with users.

Key Words

Confidentiality
Data Integrity
Data Security
Decryption
Encryption
Ethics
Hacker
Password
Plaintext
Rollback
Rollforward
System Security

Review Questions

1. In your own words, what is the purpose of this chapter? How is it related to system development? Expound.
2. "Points of vulnerability in systems today relate to hardware, communication lines, and terminals." Do you agree? Explain.
3. Review the journals on system development or security and suggest two motives behind security.
4. Distinguish between the following:
   a. Data security and data integrity.
   b. Privacy and confidentiality.
   c. Rollforward and rollback.
   d. Logical and structural failure.
5. What are the major threats to system security? Which one(s) is the most serious? Why?
6. In what way is the personal computer a step backward in system security? Illustrate.
7. Why do we need to conduct risk analysis? When would this type of analysis be cost justified? Explain the makeup and procedure behind risk analysis.
8. List and briefly explain the control measures in system security.
9. Visit a local computer center and report to class the security features incorporated in the facility.
10. Why is a password inadequate for security control? In your opinion, does it provide better or worse security than encryption? Explain.
11. What is encryption? How does it work? What type or level of system would incorporate this technology? Illustrate.
12. In a data base environment, how is a backup procedure used? Illustrate.
13. What types of failures are encountered in a data base environment? Explain briefly.
14. Much has been written about disaster/recovery planning. What is it? Why is it important? Who initiates the planning? What procedure is involved? Explain.

15. Disaster/recovery planning tasks are prepared in a cycle similar to that of system development. Explain.
16. Ethics is a relatively new topic in system development. Research the literature during the past two years and present a summary of progress made in this area.

Application Problems

1

An air-conditioning firm is planning to install a terminal in each of its stores for online stock control. A batch of parts is tagged to a job number and only to an authorized repair person to draw against that job number.

Assignment

Develop a procedure and security check that can be integrated into the candidate system for the control of stock issuing.

2

Uptown Savings & Loan Bank has problems with address maintenance. Customer complaints prompted an investigation of the nature and source of errors of customers' addresses in the savings application. A preliminary check discovered the following:

a. Almost 10 percent of the savings accounts contained misspelled names or wrong addresses, zip codes, or account numbers.

b. 28 percent of the savings accounts had no zip codes, and 6 percent of those with zip codes were incorrect.

c. Common last names such as Jones or Smith caused the most problems with address maintenance. For example, a notification to the bank by Mary Jones of a change of address resulted in a change in Martha Jones's address.

d. The address maintenance problem was so acute that of the 19,000 savings accounts, 617 accounts were in the "address unknown" status.

The bank president is concerned abou the inaccuracy of the savings file because it gives a poor image and could result in losing deposits. More and more customers have been closing their accounts, adding tension to an already tense situation.

Assignment

a. What can be done to correct the address maintenance problem?

b. What safeguards can be introduced to prevent a recurrence of these problems?

c. How can the zip code problem be rectified on the records?

d. As a systems analyst, what measures would you take to "clean up" the file?

---

3 A credit union is worried about the potential loss of its master files from accidental or intentional destruction. The computer system and the tape library are located on the second floor of a 50-year-old building adjacent to the employee cafeteria. The union owns and operates an in-house system with no backup facility in case of emergency. The systems department is asked to review and report in the current security safeguards of the data files. The vice president in charge recently sent his officers a letter that read, in part, "What protection do we have if the grease in the frying pan in the cafeteria should catch on fire and spread to our data files? What good is it to carry heavy insurance against fire and theft when there is no way we can recover our data for continuing operations?"

Certainly, much of what he said is true. If the tape library is accidentally destroyed, most of the data on retirement, health, and other benefits would be impossible to recover. Thus, an effective security plan must be developed to include the following:

a. Safeguard vital data records against accidental or intentional destruction.

b. Provide duplicate records of vital data as a backup.

c. Provide a way of reconstructing the master files in the event of a national or natural disaster.

Assignment

a. Outline a master plan to provide security and backup procedures at the credit union headquarters.

b. Devise a built-in plan to minimize the possibility of loss of vital records by sabotage or wiretapping.

c. What tape files should be duplicated? Retained?

d. For how long and where should sensitive data files be stored? Why?

Selected References

Alexander, Charles. "Crackdown on Computer Crime." *Time,* February 8, 1982, pp. 60–62.

Batt, Robert. "White Collar Crime." *Computerworld,* December 26, 1983, p. 49ff.

Bernstein, Richard A. "A Contingency Plan for Recovery from a Disaster." *Mid-Continental Banker,* December 1983, p. 16.

Campbell, Robert P. "Locking up the Mainframe." *Computerworld,* October 10, 1983, pp. ID1–3ff; October 17, 1983, pp. ID1–3ff.

Cohen, Arnold M. "Total Information System Security." *Journal of Systems Management,* April 1983, pp. 14–17.

Cook, J. R.; J. D. Eure; M. A. Johnston; and H. J. Matford. "DPMA Chapters Speak Out on DP Security." *Data Management,* May 1, 1982, pp. 42–46.

Cullen, Kathryn M. "Systems for Authorized Access to Information." *Administrative Management,* May 1982, pp. 35–38.

Curry, J. P. "Analyzing Computer Security Risks." *Canadian Datasystems,* July 1981, pp. 69–70.

Editorial, "DPMA Code of Ethics and Standards of Conduct for Information Processing Professionals." *Data Management,* October 1981, pp. 58–61.

Editorial, "Roaming High Tech Pirates." *Time,* February 8, 1982, pp. 60–63.

Emmett, Arielle. "Thwarting the Data Thief." *Personal Computing,* January 1984, p. 97.

Freedman, David H. "Ethics." *Infosystems,* August 1983, pp. 34–36.

Friedman, Stanley. "Contingency and Disaster Planning." *Computers & Security.* North-Holland Publishing, Amsterdam, 1982, pp. 34–40.

Gaade, R. P. R. "Picking up the Pieces." *Datamation,* January 1980, pp. 113–18.

Gilliam, Les. "Implementing a Disaster Recovery Plan." *Computerworld,* December 5, 1983, p. 97ff.

Harrison, Ben. "Planning for the Worst." *Infosystems,* June 1982, pp. 52–62.

Heide, Dorothy, and James K. Hightower. "Organizations, Ethics, and the Computing Professional." *Journal of Systems Management,* November 1983, pp. 38–42.

Henkel, Tom, and Peter Bartolite, eds. "Protecting the Corporate Data Resources." *Computerworld,* November 28, 1983 (special report), pp. 1–29.

Immel, A. Richard. "Data Security." *Popular Computing,* May 19, 1984, pp. 65–68.

Johnson, Deborah G. "Privacy, Power, and Property: Ethical Dilemmas for Computer Professionals." *Small Systems World,* June 1983, pp. 17–22.

Kenny, Matthew A. "Cryptography Comes out of the War Room to Aid in Securing Computer Systems." *Data Management,* July 1981, pp. 19–21.

Kirchner, Jake. "August Bequai, Fighter for Ethics." *Computerworld,* May 21, 1984, pp. ID1–4ff.

Morris, Robert S. "Do-It-Yourself Disaster Recovery Planning." *Small Systems World,* October 1980, pp. 23–26ff.

Nye, J. Michael. "A Primer on Security." *Mini-Micro Systems,* July 1981, p. 166ff.

Patterson, Wm. "Where Is Technology Taking Us?" *Industry Week,* May 30, 1983, pp. 34–36.

Porter, Sigmund. "A Password Extension for Improved Human Factors." *Computers & Security.* North-Holland Publishing, Amsterdam, 1982, pp. 54–56.

Rames, David. "Recovering from Disasters." *Computer Decisions,* September 1981, p. 109ff.

Sherwin, Douglas S. "The Ethical Roots of the Business System." *Harvard Business Review,* November/December 1983, pp. 183–92.

Spiro, Bruce E. "Ethics—The Next Step Is Crucial." *Data Management*, November 1983, pp. 32–33.

Weiss, Eric A. "Self-Assessment Procedure IX." *Communications of the ACM*, March 1982, pp. 181–94.

Westin, Alan. "New Eyes on Privacy." *Computerworld*, November 28, 1983, pp. ID11–13ff.

Zientara, Marguerite. "How to Recover from a Disaster." *Computerworld*, September 13, 1982, p. 1ff.

Glossary of Terms

Abstract system Conceptual or nonphysical entity.

Action entry The lower right quadrant of a decision table; indicates the response to the question entered in the condition entry.

Action stub Lower left quadrant of a decision table; outlines in narrative form the conditions that may exist.

Activity In system development life cycle—a group of logically related tasks that make it possible to accomplish a specific objective; a group of related tasks.

Aggregate Two or more data items handled as a unit.

Alias An alternative name used to stand for an identified data structure within a data dictionary notation.

Alpha testing Verifying and studying software errors and failures based on simulated user requirements.

Analysis Breaking a problem into successively manageable parts for individual study.

Attribute A data item that characterizes an object.

Audit trail A feature of data processing systems that allows for the study of data as processed from step to step; an auditor can then trace all transactions that affect an account.

Ballot box design A type of form designed so that all the user has to do is check the applicable box.

Benchmark In system testing—a test run on a candidate system to measure how long it takes to run a selected application.

Beta testing Subjecting modified software to the actual user site (live) environment.

Bounded rationality Coined by Herbert Simon—the notion that humans have a limited capacity for rational thinking; rationality for determining information requirements is bounded by limited training, prejudice, and the attitude of the user.

Brainstorming A technique for generating new ideas; a participant is asked to define ideal solutions and then select the most feasible one.

Break-even analysis The point at which the cost of the candidate system and the present one are equal.

Bubble chart In data base design—a diagram that uses circles to represent the nature and direction of relationships between data items.

Candidate system The newly developed system designed to replace the current system.

Caption A word on a form that specifies what information to write in the space provided.

Cash-flow analysis A procedure designed to keep track of accumulated costs and revenues on a regular basis.

Chaining In a data base—linking records; establishing relationships among data items.

Closed question A question in which the response(s) is presented as a set of alternatives.

Closed system A system that is isolated from environmental influences; see *Open system* for contrast.

Cohesion Strength within a module; degree of relationship between elements within a module.

Computer output microfilm (COM) Recording system output on microfilm or microfiche, usually for archival storage.

Concatenated key Two or more keys linked to identify or access a record.

Condition entry Upper right quadrant of a decision table; provides answers to questions asked in the condition stub quadrant.

Condition stub Upper left quadrant of a decision table; sets forth in question form the condition that may exist.

Confidentiality Special status given to sensitive information in a computer system to minimize the possible invasion of privacy.

Contrived observation An observation set up by the observer in a place like a lab.

Control In a system—the element or component that governs the pattern of activities of the system.

Conversion Process of changing from an existing system to a new one.

Cost/benefit analysis The process of comparing projected savings and benefits to projected costs to decide whether a system change is justified.

Couple A connection between modules; a symbol representing data items moved from one module to another.

CPM See *Critical path method.*

Critical path Events in a PERT network that, if behind schedule, will cause the final event in the network to be late.

Critical path method (CPM) A planning and scheduling method that determines trade-offs between relative costs and alternative completion dates for a project.

Cryptography A system of secret communications to improve the security of confidential computerized files.

Data aggregate See *Entity.*

Data base A store of integrated data capable of being directly addressed for multiple uses; it is organized so that various files can be accessed through a single reference based on the relationship among records in the file rather than the physical location.

Data base administrator (DBA) A specialist whose main tasks are to protect and manage the data base, resolve user conflict, and maintain and update the system.

Data base management system (DBMS) The software that determines how data must be structured to produce the user's view; manages, stores, and retrieves data and enforces procedures.

Data definition language (DDL) Describes how data are structured in the data base.

Data dictionary A structured repository of data about data; a list of terms and their definitions for all data items and data stores of a system.

Data element The smallest unit of a record; roughly equivalent to a field.

Data flow Movement of data in a system from a point of origin to a specific destination—indicated by a line and arrow.

Data flow diagram (DFD) Graphic representation of data movement, processes, and files (data stores) used in support of an information system.

Data independence Changing hardware and storage procedures or adding new data without having to rewrite application programs.

Data integrity The extent to which the data used for processing are reliable, accurate, and free from error.

Data item Represents one or more bytes; describes some attribute of an object (for example, social security number, sex, or age).

Data manipulation language (DML) In data base—it specifies for the DBMS what is required; the techniques used to process data.

Data model In data base—a framework or a mental image of how the user's view looks.

Data security Protection of data from loss, disclosure, modification, or destruction.

Data set A file.

Data store In a data flow diagram—a storage area for collecting data input during processing; the symbol used in an open rectangle.

Data structure A logically related set of data that can be decomposed into lower-level data elements; a group of data elements handled as a unit.

Decision support systems (DSS) A "what if" approach that uses an information system to assist management in formulating policies and projecting the likely consequence of decisions.

Decision table A table of contingencies for defining a problem and the actions to be taken; a method of presenting the logic of a computer program that tells what action must be taken when a given condition is met or not met.

Decision tree Graphic representation of conditions and outcomes resembling the branches of a tree.

Decomposition Partitioning a system into detailed functions to be studied in relative isolation.

Decryption Conversion of an enciphered message into plaintext.

Delphi method A debate by questionnaire in which participants fill out the forms and the results are given to them with a follow-up questionnaire; results are again summarized and fed back to them until their responses have converged enough.

Design Process of developing the technical and operational specifications of a candidate system for implementation.

Desk checking Manually going over a program's source code to determine whether errors or inconsistencies exist.

DFD See *Data flow diagram*.

Dichotomous question It is a yes or no question; a question answerable by yes or no; a question offering two answer choices.

Differentiation In open systems—a tendency toward increasing the specialization of functions of system components.

Direct-access organization Records in a file are accessible and can be replaced anywhere in the existing file with no particular regard to the sequence in which the file is arranged.

Direct cost Cost that is normally applied directly to the operation in question; for example, the purchase of a new tape for $40 is a direct cost.

Direct observation A situation where the analyst actually observes the subject or the system at work.

Documentation A means of communication; a written record of a phase of a specific project; it establishes design and performance criteria for phases of the project.

Dynamic system model A model that depicts an ongoing, constantly changing system.

Eavesdropping Unauthorized access to information through wiretapping, learning a security code by watching displays, taking pictures with a camera, or using bugging devices.

Encryption Scrambling signals that represent data during transmission or over networks.

Enhancement Adding, modifying, or redeveloping the code to support changes in specifications.

Entity Also called a *data aggregate*; something of interest to the user about which to collect or store data; represents a number of data elements.

Entropy Loss of energy; a system running down due to loss of energy, often leading to a temporary state of disorganization.

Entry In a decision table—the answers to questions or the actions resulting from the answers to conditions entered in the stub.

Equifinality Achieving goals through differing courses of actions and by a variety of paths.

Equilibrium Balance; see also *Steady state.*

Ergonomics The science of modifying machine specifications to match human comfort.

Expectancy theory Stresses important relationships between (1) effort and performance and (2) performance and rewards; in systems work—the value the user places on perceived rewards from a candidate system determines the motivation to accept and use the system.

Facilities management (FM) A company's computer facility managed by an outside agency under contract.

Fanfold form A multiple-unit form joined together in a continuous strip with perforations between pairs of forms.

Feasibility study A procedure that identifies, describes, and evaluates candidate systems and selects the best system for the job.

Feedback The part of a closed-loop system that automatically brings back information about the condition being controlled.

Field A specified area of a record used for a particular category of data—for example, a group of card columns used to represent a wage rate or a set of bit locations in a computer word used to express the address of the operand; see also *Data item.*

File Collection of related records organized for a particular purpose; also called a *data set.*

File volatility Refers to the proportion of records changed during a time period.

Fill-in-the-blanks question A question or an item that seeks a specific, factual response not restricted to a set of choices.

Fixed cost Sunk cost that does not vary with the volume of processing; examples are rent of a building or a computer system and a supervisor's salary.

Flat form A single-copy form; a form that has no carbon copy.

Flow system model A model that shows the flows of material, energy, and information that hold the system together.

Flowchart A graphic picture of the logical steps and sequence involved in a procedure or a program.

Force majeure Deals with the suspension of a contract due to events beyond the vendor's control such as hurricanes, earthquakes, and acts of war.

Form A physical carrier of data, of information.

Forms control Coordination of forms design and use among users of forms in the organization.

Forms design Evaluating present documents and creating new or improved forms that offer useful information for action.

Functional structure An approach that assigns a group of analysts to serve a specific system such as personnel, production, or marketing.

Functionality A definition of the facilities, performance, and other factors that the user requires in the finished product.

Future value Time value of money in the form of interest on the funds invested over a specific time period.

Gantt chart A static system model used for scheduling; portrays output performance against time.

Hacker A devout user of the computer; usually for "kicks" or the sheer pleasure of trying to find out how far to push the computer before it reaches its limits.

Hierarchical structure Breaking down a large project into a series of successively smaller, manageable parts through iteration and according to a logical sequence.

Hierarchy plus input/processing/output (HIPO) A documentation tool that graphically portrays functions in a chart from a general level down to detailed levels.

Homeostasis A balanced state; a system that maintains a dynamic equilibrium among its inputs, processing, and outputs.

Identifier A key that uniquely identifies a record.

Implementation In system development—a phase that focuses on user training, site preparation, and file conversion for installing a candidate system.

Indexed-sequential A method of organizing a direct-access storage device that combines some of the features of sequential and direct processing.

Indirect cost Overhead; results of operations not directly associated with a given system or activity; examples are insurance, maintenance, heat, and air conditioning.

Indirect observation An observation that relies on mechanical devices such as cameras and videotapes to capture information.

Informant A person in a department or division (usually an employee) who offers information to an observer because of his/her position, seniority, or experience.

Information A meaningful set of data that tells something about the data relationships.

Information system The tools, procedures, and technology that generate user-initiated information.

Initial investigation An exploratory activity that determines whether a user request is valid and feasible before a course of action is recommended.

Input The data to be processed; the processes of transferring data from external storage to internal storage.

Input/Output control system (IOCS) Machine-based software that responds to DBMS requests to retrieve information from physical storage as specified by the application program.

Instance of an entity The value of an attribute for a specific occurrence.

Intangible cost Readily identified but not easily quantified.

Interdependence Dependence of a system's components on one another for effective functioning.

Interview A data-gathering or data-verification approach; talking with people in an organized manner and with a purpose.

Inverted list organization A structure in which there are separate lists for each type of data element in the data base.

Investment period In a candidate system—when initial costs exceed the costs of running the current system.

IPO chart HIPO-associated chart.

Key See *Identifier*.

Kitchen sink strategy A plan where the user overstates (throws in the sink) his/her needs from a candidate system.

Logic error Deals with problems such as incorrect data fields, division by zero, and invalid combinations.

Logical record A record that maintains a logical relationship among all data items in the record.

Magnetic ink character recognition (MICR) Input method that codes and identifies checks, deposit slips, and documents preencoded with special magnetic ink.

Maintenance Restoring something to its original condition.

Management information system (MIS) An integrated approach to the design and use of a computer-based information system that provides summary information and highlights exception conditions for corrective decision making.

Mean time between failure (MTBF) The average time the system is available to users before breaking down.

Menu A selected list of options that the user chooses from and then types an option for a computer operation.

Milestone Steppingstones; sets of activities make up a project.

Model A logical or mathematical representation of a system that encompasses features of interest to the user.

Modularity In systems maintenance—a system is constructed in modular units of a limited size to simplify maintenance when necessary; in software reliability—the ease with which a package can be modified.

Multiple-choice question An item that requests the respondent to choose one out of a series of specific choices.

Natural observation An observation in the interviewee's place of work or surrounding.

NCR paper No carbon required; chemically treated paper that does not require carbon for carrying impressions from the top form to the copies underneath.

Net benefit analysis Total benefits minus total costs.

Net present value Discounted benefits minus discounted costs.

Network structure Events and activities in PERT/CPM; in data base—a data structure that allows 1:1, 1:M, or M:M relationships between entities.

Normalization A process of replacing a given file with its logical equivalent; the object is to derive simple files with no redundant elements.

On-site observation See *Natural observation.*

Open-ended interview An interview that poses questions that do not require a specific or a brief response.

Open system A system that performs interactions across its boundary, receives inputs from and delivers outputs to the outside.

Operating system In data base—machine-based software that facilitates the availability of information or reports through the DBMS.

Opportunity cost Benefits forgone as a result of choosing an alternative.

Organization Implies structure and order; a procedure that determines how components must be arranged to achieve objectives.

Organization chart A chart that shows the official structure of the organization in terms of functional units or in terms of superior-subordinate relationships.

Output Data that have been processed; the end result (product) of the system under study.

Overhead Allocated costs that include insurance expense, maintenance expense, heat, light, and power; costs that are neither direct nor indirect; costs that are tagged to the general administration of the business.

Parallel run Putting the new system into operation in conjunction with the continued operation of the old system.

Parameter A variable that is given a constant value for a specific purpose or process.

Paraprofessional A trained aide who assists professionals such as analysts or lawyers.

Participant observation The process of collecting data through natural observation.

Partitioning Dividing a problem into smaller, separate elements for easier understanding or solution; see also *Hierarchical structure.*

Password Identity authenticator; a key that allows access to a program, system, or procedure.

Payback analysis A method of determining how long it will take a system to generate enough savings to cover developmental costs.

PERT See *Program evaluation and review techniques.*

Phase A set of tasks or activities that, when competed, brings a project to a critical milestone.

Physical design A design that produces the working system by defining design specifications that tell programmers exactly what the candidate system must do.

Physical record The way data are physically recorded on a storage medium.

Physical system A tangible entity that may be static or dynamic in operation—for example, an office or a bulldozer.

Pilot testing Preliminary testing of a program or a system using simulated or representative data.

Plaintext In encryption—an unenciphered message.

Planning Studying a project course of action and determining what is to be done to meet stated goals.

Pointer See *Identifier.*

Policy A generalization that prescribes what an organization ought to do.

Pool-oriented structure An arrangement that allows analysts to work on any system assignment in the firm; once completed, they return to the pool for another assignment.

Portability Ability of the software to be used on different hardware and operating systems.

Post-implementation review Evaluation of a new system after it has been in operation to determine its actual performance against expectations.

Precedence relationship In PERT—activities that cannot be initiated until after a specific activity is completed.

Present value Current value of money, determined by discounting future economic values backward in time to the present.

Privacy The right of users, individuals, and organizations to decide what information they are willing to share with or accept from others.

Process A procedure that transforms input into useful output; in a data flow diagram—indicated by a bubble or a circle.

Program directive An authorization document to proceed with the design and implementation of a candidate system.

Processing See *Process.*

Project evaluation and review technique (PERT) A flow system model used to manipulate various values as a basis for determining the critical path, to interpret these relationships and to relate them back to the real world as a control technique.

Project manager A person who uses a combination of techniques to facilitate planning, scheduling, and control of system projects.

Project-oriented structure An arrangement that forms a team of analysts to work on one project at a time; it is descriptive of small installations with limited projects.

Project proposal See *Project directive.*

Prompt A symbol on a computer screen asking the user for a command or response.

Prototyping A working system to explore implementation or processing alternatives and evaluate results.

Quality assurance Developing controls to ensure a quality product; defining factors that determine system quality and the criteria that the software must meet to contain these factors.

Questionnaire A data-gathering instrument that requests specific information that can be quantitatively tabulated, usually from a large sample.

Ranking scales A questionnaire item that asks the respondent to determine preference for the importance of a set of items.

Rating scale In a questionnaire—a multiple-choice item that offers a range of responses along a given dimension.

Recency effect A behavior that is altered due to information that was discovered recently.

Record A collection of aggregates or related items of data treated as a unit.

Redundancy A situation in which two or more pieces of information in a file are the same.

Relation Two-dimensional table.

Relational structure Data and relationships represented in a flat, two-dimensional table.

Reliability Dependability or level of confidence; in systems work—the need to gather dependable information for use in making decisions about the system being studied.

Request for proposal (RFP) A report by the user requesting selected vendors to bid on a proposed system.

Resident expert A natural teacher, an employee who takes time to learn a system and become "expert" in its operation.

Response time The time required by a system to react to an input stimulus.

Return period A point when a candidate system provides a greater benefit (profit) than the old system.

Rollback In recovery procedure—updating a prior valid copy of the data base with the necessary changes to produce a current version.

Rollforward In recovery procedure—records of activity are removed from a current invalid state to produce the prior valid state of the data base.

Root In data base—a parent with no owners.

Rule In forms design—a rule (line) guides the human eye in reading and writing data groups and separates them on the form.

Sabotage Physical destruction of computerized files, hardware, or the computer facility itself.

Savings Reduction or elimination of expenditures.

Schema A map of the overall structure of a data base.

Schematic model A two-dimensional chart depicting system elements and their linkages.

SDLC See *System development life cycle.*

Security The protection of data or hardware against accidental or intentional damage from a defined threat.

Sequential organization Sorting in physical, contiguous blocks within files on tape or disk.

Sequential Testing Checking the logic of one or more programs in a candidate system, where the output of one program will affect the processing done by another program.

Serial access Accessing a record only after the record(s) preceding it has been scanned.

Serviceability A criterion for software selection focusing on documentation and vendor support.

Servicer A computer service bureau.

Set The relationship between records.

Slack time Time spent on subsidiary tasks that do not affect the duration of a whole project.

Snapout form See *Unit set/snapout form.*

Source code A procedure or format that allows enhancements on a software package.

Source language The language in which a program or a software package is written.

Static system model A model that exhibits one pair of relationships like activity-time or cost-quantity, such as the Gantt chart.

Steady state In dynamic systems—a self-adjusting and self-regulating function that helps the system take corrective steps to maintain balance.

Strategic planning In system planning—establishing relationships between the organization plan and the plan for a candidate system.

Stress testing Subjecting the new system to a high volume of data over a short time; the purpose is to ensure that the system does not malfunction under peak loads.

String testing Testing one program to determine whether it conforms to related programs in the system; testing each portion of a system against the entire module before the system as a whole is ready to be tested.

Structure chart Graphic representation of the control logic of processing functions or modules representing a system.

Structured analysis A set of techniques and graphic tools that allow the analyst to develop a new kind of system specification that is easily understandable to the user.

Structured English Strongly worded formal English statements used for communicating processing rules or describing the structure of a system.

Structured interview Also called a *directive interview*; an approach in which the questions and the alternative responses are fixed.

Structured observation An observation where the observer looks for and records a specific action such as the number of soup cans a shopper picks up before choosing one.

Structured questionnaire A questionnaire that requests specific information and a guided response.

Structured walkthrough Interchange of ideas between peers who review a product presented by its author and agree on the validity of a proposed solution to a problem.

Subschema A map of the programmer's view of the data he or she uses; derived from the schema.

Subsystem A series or group of components that perform one or more operations of a more complex system.

Sunk cost See *Fixed cost.*

Syntax error A program statement that violates one or more rules of the language in which it is written.

System A regular or orderly arrangement of components or parts in a connected and interrelated series or whole; a series or group of components necessary to some operation.

System design Detailed concentration on the technical and other specifications that will make the new system operational.

System development The process of identifying the user's needs and designing a system that meets those needs through implementation.

System development life cycle A structured sequence of phases for implementing an information system.

System flowchart A graphic representation of a system showing the overall flow of control in processing at the job level; specifies what activities must be done to convert from a physical to a logical model.

System integrity The proper functioning of hardware, programs, and physical security, and the required degree of safety against eavesdropping and wiretapping.

System privacy See *Privacy.*

System security See *Security.*

System specifications Key information for programming, testing, and implementing the project.

System testing Testing the whole system by the user after major programs and subsystems have been tested.

Systems analysis Reduction of an entire system by studying the various operations performed and their relationships within the system; an examination of a business activity with a view to identifying problem areas and recommending alternative solutions.

Systems analyst A methods person who starts with a complex problem, breaks it down for analysis, and designs a better system based on specifications set in advance.

Tangible cost Costs that are known to exist and their financial value is easily quantified.

Task The smallest unit of work that is assigned to one person and controlled through a project management routine.

Team-oriented structure An approach that assigns a programmer to a team with responsibility for a specific project; the team is headed by a lead programmer who reports to a project leader.

Technical writer A person who develops procedures manuals, describing technical specifications, and user manuals.

3/5 space In forms spacing, 3 applies to the number of lines per vertical inch; 5 applies to the number of characters that fit in one horizontal inch.

Top-down design A design that consists of a hierarchy of modules; each module has a single entry and a single exit subroutine.

Top-down programming An approach in which the top module is first tested, then program modules are added from the highest level to the lowest level.

Tuple A group of related fields.

Turnaround The elapsed time between the receipt of the input and the availability of the output (results).

Unit set/snapout form A form with an original copy and successive copies with carbon sheets interleaved between each copy; the set is glued together and handled as a unit.

Unit testing Testing changes made in an existing or new program.

Unobtrusive observation An observation that takes place in a contrived way, such as behind a one-way mirror.

Unstructured interview An approach in which the questions and the alternative responses are open-ended.

Unstructured questionnaire An approach that allows respondents to freely answer questions in their own words.

Usability A criterion in software selection—easy to operate and user-friendly.

User acceptance test A test that verifies for the user that the system's procedures operate to system specifications and the integrity of vital data is maintained.

Validation Checking the quality of software in both simulated and live environments.

Validity In interviewing—the extent to which the questions asked are worded so as to elicit the information the interviewer is after.

Variable A measurable quantity that has a definite numerical value at every instant.

Variable cost Cost that varies with the volume of processing or number of shifts per day; examples are employee wages and costs of supplies and raw materials.

VisiSchedule A software package used for showing the critical jobs in the project schedule, explaining the relationships between jobs, calculating project and manpower costs, and generating reports and schedules as needed.

Index

This book has been set Linotron 202, in 10 and 9 point Zapf Book Light, leaded 2 points. Part numbers are 36 point Lubalin Graph Medium and part titles are 36 point Lubalin Graph Bold. Chapter numbers are 27 point Lubalin Graphic Medium and chapter titles are 27 point Lubalin Graph Bold. The size of the type page is 31 by 49 picas.